cis-trans **Isomerization in**
Biochemistry

Edited by
Christophe Dugave

Related Titles

Michael Haley, Rik R. Tykwinski (eds.)

Carbon-Rich Compounds

From Molecules to Materials

2006
ISBN 3-527-31224-2

Andrei K. Yudin (ed.)

Aziridines and Epoxides in Organic Synthesis

2006
ISBN 3-527-31213-7

Vincenzo Balzani, Alberto Credi, Margherita Venturi

Molecular Devices and Machines

A Journey into the Nanoworld

2003
ISBN 3-527-30506-8

Norbert Krause, A. Stephen K. Hashmi (eds.)

Modern Allene Chemistry

2002
ISBN 3-527-29773-1

Tanetoshi Koyama, Alexander Steinbüchel (eds.)

Biopolymers

Vol. 2: Polyisoprenoids

2001
ISBN 3-527-30221-2

Alexander Steinbüchel, Stephen R. Fahnestock (eds.)

Biopolymers

Vol. 7: Polyamides and Complex Proteinaceous Materials I

2002
ISBN 3-527-30222-0

Stephen R. Fahnestock, Alexander Steinbüchel (eds.)

Biopolymers

Vol. 8: Polyamides and Complex Proteinaceous Materials II

2003
ISBN 3-527-30223-9

cis-trans Isomerization in Biochemistry

Edited by
Christophe Dugave

WILEY-
VCH

WILEY-VCH Verlag GmbH & Co. KGaA

Chemistry Library

The Editor

Christophe Dugave
CEA/Saclay
Department of Protein Engineering and Research
Bâtiment 152
91121 Gif-sur-Yvette
France

■ All books published by Wiley-VCH are
carefully produced. Nevertheless, authors,
editors, and publisher do not warrant the
information contained in these books,
including this book, to be free of errors.
Readers are advised to keep in mind that
statements, data, illustrations, procedural
details or other items may inadvertently
be inaccurate.

Library of Congress Card No.: applied for

British Library Cataloguing-in-Publication Data
A catalogue record for this book is available
from the British Library.

**Bibliographic information published by
Die Deutsche Bibliothek**
Die Deutsche Bibliothek lists this publication
in the Deutsche Nationalbibliografie; detailed
bibliographic data is available in the Internet at
<http://dnb.ddb.de>.

© 2006 WILEY-VCH Verlag GmbH & Co. KGaA,
Weinheim

Typesetting   Kühn & Weyh, Satz und Medien,
Freiburg
Printing   betz-druck GmbH, Darmstadt
Bookbinding   Litges & Dopf Buchbinderei GmbH,
Heppenheim
Cover   Adam Design, Weinheim

Printed in the Federal Republic of Germany.
Printed on acid-free paper.

ISBN-13:   978-3-527-31304-4
ISBN-10:   3-527-31304-4

Contents

cis-trans Isomerization in Biochemistry. Edited by Christophe Dugave
Copyright © 2006 WILEY-VCH Verlag GmbH & Co. KGaA, Weinheim
ISBN: 3-527-31304-4

Preface

Life is governed by a relatively small number of chemical reactions that exploit a limited variety of simple concepts. However, their combination has led to an amazing chemical diversity which is still beyond the reach of the organic chemist, even in the most complex supramolecular systems, despite a huge set of synthetic methods. Among these basic processes and reactions, *cis-trans* isomerization (CTI) is undoubtedly one of the finest ways to tune the physical and chemical properties of biomolecules and hence to control their biological activities. Moreover, CTI of simple molecules generates molecular diversity in the form of geometric isomers with particular structures and properties.

Surprisingly, chemists were rather slow to study CTI, and it is only fairly recently that it has been taken into account in attempts to understand biological processes at the molecular level. The Swiss chemist Alfred Berthoud first investigated the light-driven CTI of alkenes and proposed a radical mechanism that accounted for the reversibility of the phenomenon, a theory which is still taught today.

Since the first attempts to understand CTI in simple molecular systems, a huge amount of work has been done to investigate CTI processes in biology, such as chromophore isomerization in chromoproteins. The finding that CTI concerns not only double bonds but also pseudo double bonds and restrained single bonds has led to extensive study of protein folding, modulation of the activity of peptides and proteins, and the construction of sophisticated supramolecular structures. CTI was also found to be implicated in the organization of metal complexes and supramolecular systems, though the mechanisms are basically different from those proposed for CTI of organic molecules.

The study of CTI has given rise to a large number of publications, which are the fruit of active collaborations between scientific teams whose expertise ranges from in silico quantum molecular mechanics to medicine. For these reasons, CTI processes concern not only chemists and biochemists, but also physicists and physicians.

Over the past 20 years there have been considerable changes in the way we consider CTI. In view of the results obtained in chemistry and biochemistry, CTI appears to be much more than a simple tuning of the properties of the molecule itself, and important remote effects have been highlighted. It is now obvious that the inversion about either a double bond or a restrained single bond generates extensive changes in molecular size and shape, and in stereoelectronic properties, all of which function cooperatively. This has considerable consequences for the behavior of larger molecular systems such as membranes and proteins, and is suspected to be a particular way of storing potential energy usable not only for chemical reactions but also for macroscopic movement.

Light-driven CTI also plays a central role in the transduction of light into a chemical signal and so is the starting point of light perception in primitive organisms and in the vision of more complex organisms. CTI has also emerged as the basic concept underpinning holographic information storage of extraordinary capacity and resolution. However, we should never forget that billions of years before chemists utilized CTI to tune gel–sol phase transitions, nature used this simple reaction to modulate membrane permeability to enable adaptive responses to stress and environmental change.

Beyond protein folding, the discovery of peptidyl prolyl isomerases (PPIases) and related proteins has opened the way to novel concepts in biology: the notion of chaperone-assisted receptor binding is an emerging field of research which sheds light on receptor function and protein–protein interactions. The recent discovery of a secondary amide peptide bond *cis-trans* isomerase (APIase) heralds new advances in this field.

Recently, the French Nobel prizewinner Jean-Marie Lehn proposed the use of CTI as a source of molecular diversity in dynamic combinatorial chemistry. The prospect of using a dynamic fully reversible process such as CTI for the evolutionary selection of ligands is extremely attractive and should lead to fundamental advances in this field of research.

Althought this book will not tackle the technological uses of CTI nor its application in supramolecular chemistry, the impressive advances in the development of molecular devices that produce a microscopic motion as well as a macroscopic movement must be cited herein.

Progress in the study of CTI should lead not only to better understanding of one of the main molecular bases of life, but also to the development of fascinating tools for studying biomolecules. These main lines of research are not incompatible, since one talks of supramolecular systems able to release bioactive molecules at the right place through a controlled CTI process. I am confident that work on CTI will yield important applications beneficial to humankind, and I sincerely hope that this book, which collates most of the recent data on CTI in biology, organic and inorganic chemistry, will help scientists to work with this aim in mind.

Saclay, January 6th 2006   *Christophe Dugave*

List of Contributors

Alzir Azevedo Batista
Universidade Federal de São Carlos
Departamento de Química
Campus São Carlos, Rodovia
Washington Luís
São Carlos – SP 13565-905
Brazil

Chryssostomos Chatgilialoglu
I.S.O.F.
Consiglio Nazionale delle Ricerche
Via P. Gobetti 101
40129 Bologna
Italy

Christophe Dugave
CEA/Saclay
Department of Protein Engineering
and Studies
Building 152
91191 Gif-sur-Yvette
France

Carla Ferreri
I.S.O.F.
Consiglio Nazionale delle Ricerche
Via P. Gobetti 101
40129 Bologna
Italy

Gunter Fischer
Max Planck Research Unit for
Enzymology of Protein Folding
Weinbergweg 22
06120 Halle/Saale
Germany

Muriel Gondry
CEA/Saclay
Department of Protein Engineering
and Studies
Building 152
91191 Gif sur Yvette
France

Hideki Kandori
Nagoya Institute of Technology
Department of Materials Science
and Engineering
Showa-Ku
Nagoya 466-8555
Japan

Yoshinori Kakitani
Kwansei Gakuin University
Faculty of Science and Technology
2-1 Gakuen
Sanda 669-1337
Japan

cis-trans Isomerization in Biochemistry. Edited by Christophe Dugave
Copyright © 2006 WILEY-VCH Verlag GmbH & Co. KGaA, Weinheim
ISBN: 3-527-31304-4

Yasushi Koyama
Kwansei Gakuin University
Faculty of Science and Technology
2-1 Gakuen
Sanda 669-1337
Japan

John J. Lopez
Ecole Polytechnique Federale de
Lausanne (EPFL)
Institute of Chemical Sciences and
Engineering
1015 Lausanne
Switzerland

Luis Moroder
Max-Planck-Institut für Biochemie
Am Klopferspitz 18
82152 Martinsried
Germany

Manfred Mutter
Ecole Polytechnique Federale de
Lausanne (EPFL)
Institute of Chemical Sciences and
Engineering
1015 Lausanne
Switzerland

Hiroyoshi Nagae
Kobe City University of
Foreign Studies
9-1 Gakuen-Higashimachi, Nishi-ku
Kobe 651-2187
Japan

Salete Linhares Queiroz
Universidade de São Paulo
Instituto de Química de São Carlos
Av. Dr. Carlos Botelho 1465
São Carlos – SP 13560-970
Brazil

Christian Renner
Nottingham Trent University
Nottingham NG11 8NS
UK

Ute F. Röhrig
Ecole polytechnique fédérale de
Lausanne
Institut des sciences et ingénierie
chimiques
BCH – LCBC
1015 Lausanne
Swizerland

Ursula Rothlisberger
Ecole polytechnique fédérale de
Lausanne
Institut des sciences et ingénierie
chimiques
BCH – LCBC
1015 Lausanne
Swizerland

Ivano Tavernelli
Ecole polytechnique fédérale de
Lausanne
Institut des sciences et ingénierie
chimiques
BCH – LCBC
1015 Lausanne
Swizerland

Gabriele Tuchscherer
Ecole Polytechnique Federale de
Lausanne (EPFL)
Institute of Chemical Sciences and
Engineering
1015 Lausanne
Switzerland

Stephan Wawra
Max Planck Research Unit for
Enzymology of Protein Folding
Weinbergweg 22
06120 Halle/Saale
Germany

Marc Zimmer
Connecticut College
Department of Chemistry
New London
Connecticut 06320
USA

1
Nomenclature

Christophe Dugave

Molecules in which free rotation around one or more bonds is restricted may exist as distinct stable rotamers in proportions that depend on the free enthalpy difference $\Delta G°$ of each rotamer. They can interconvert provided the intrinsic rotational barrier $\Delta G^{\ddagger}$ is not too high (Fig. 1.1). In the simplest case, there are two marked energy minima separated by energy barriers to rotation, which often implies either the effective breaking of a chemical bond (i.e. C=C photoisomerization in ethylene derivatives) or a disruption of conjugation (i.e. isomerization of a Y–C=X system, X and Y being heteroatoms). Therefore, there are only two geometrical isomers for one given system and theoretically 2^n possible isomers for a molecule that contains n isomerizable systems (e.g. retinal).

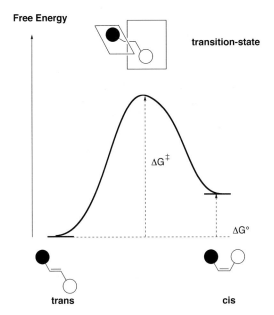

Fig. 1.1 Schematic representation of general *cis-trans* isomerism of a double bond.

cis-trans Isomerization in Biochemistry. Edited by Christophe Dugave
Copyright © 2006 WILEY-VCH Verlag GmbH & Co. KGaA, Weinheim
ISBN: 3-527-31304-4

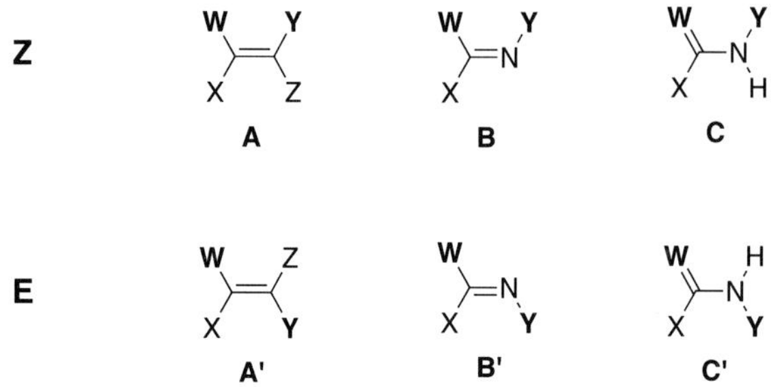

Fig. 1.2 General Z/E nomenclature for the description of geometrical isomers in π-systems.

The first proposed nomenclature suggested that isomers should be called *cis* when W and Y are on the same side of the double bond and *trans* when they are on the opposite side (Fig. 1.2), provided W ≠ X and Y ≠ Z. However, this nomenclature was limited to the particular case where W and Y are identical. The more recent nomenclature of Cahn–Ingold–Prelog, based on the German *Zusammen* (Z) and *Entgegen* (E) notation, was extended to systems where W and Y are different substituents (Fig. 1.2). There is no direct relation between the two nomenclatures since they depend on the nature of substituents; and so the Z isomer is not necessarily *cis*. Moreover, the order of priority is determined by the atomic number of each atom connected to the C=C double bond [1]. Although the E/Z nomenclature may also be applied to compounds B/B' and C/C', these are considered as conformational isomers, whereas compounds A/A' are configurational isomers.

Z/E isomerism is not limited to true double bonds and may be used when sp² electrons of a heteroatom are conjugated with a π-system to form a planar pseudo double bond. In particular, in the case of amides, the *cis* isomer is called E. Although the general tendency now is to use the E/Z nomenclature in chemistry, despite their inaccuracy *cis* and *trans* are still utilized by biochemists because they give a more readily understandable description of molecular shape, in particular for amides in peptides and proteins. When the chains are connected through a motif containing more than three dihedral angles (i.e. carbamates), the *syn–anti*

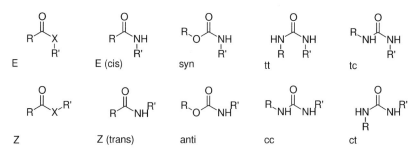

Fig. 1.3 Usual nomenclature for geometric isomers of esters, amides, carbamates, and ureas (*t*: *trans*, *c*: *cis*).

and *cis–trans* nomenclatures are usually applied since they refer to the relative position of substituents (Fig. 1.3).

Cis-trans isomerism may also occur with true single bonds. In fact, preferred conformational minima for $\omega = 180°$ (anti) and $\pm 60°$ (gauche) are usually found in alkanes (Fig. 1.4A). However, in severely crowded compounds, backbone valence angles are smaller than tetrahedral and therefore the Prelog–Klyne nomenclature is the standard (Fig. 1.4B) [2]. However, this notation is unhelpful for energy minima for ω of about 90 and 150°, a situation which is common for Si_nX_{2n+2} polysilanes and which has resulted in a proliferation of nonstandard symbols and notations. Recently, Michl and West have suggested the use of new labels that account for particular conformations found in polymers, disulfides, etc. (Fig. 1.4C). They also recommend specifying the positive or negative sense (right or left) since these conformers are chiral. This notation also accounts for strongly deformed π-systems with the *syn* ($\omega \approx 0°$) and *anti* ($\omega \approx 180°$) configurations [3].

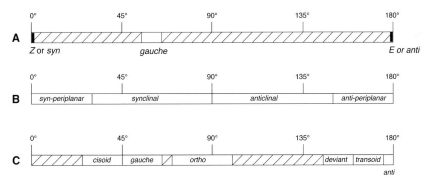

Fig. 1.4 Proposed labels for favored dihedral angles in the Cahn–Ingold–Prelog (A), Prelog–Klyne (B), and Michl–West (C) nomenclatures.

Metal complexes display a wide variety of coordination geometries that permit the existence of several geometric isomers. The situation is rather more complicated thanwith organic molecules since the three-dimensional arrangement of coordinates around the metal core leads to the multiplication of possible diastereomers. The *cis/trans* notation is usually employed but this nomenclature is based on a spatial reference: "*cis*" means "adjacent" while "*trans*" means "opposite" (Fig. 1.5) [4].

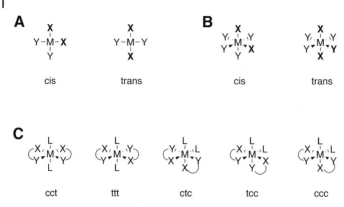

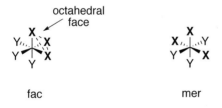

Fig. 1.5 Possible geometric isomerism around a metal core in a square planar (A) and two octahedral complexes (B, C).

The number of possible diastereomers depends on the variety of ligands and sometimes requires use of the one-letter code (*cis/trans* is noted *c/t*). This nomenclature may be applied to square planar complexes and to square planar pyramidal and octahedral complexes, but not to tetrahedral complexes where a given position is equivalent to any other. Moreover, geometric isomerism often implies the existence of optical isomerism.

The new labels *fac* and *mer* were introduced to reflect the relative position of three identical ligands around the octahedral structure. Thus, placing the three groups on one face of the octahedron gives rise to the facial isomer, and placing the three groups around the center gives rise to the meridional isomer (Fig. 1.6). When there are only two different ligands, the *cis/trans* and *fac/mer* nomenclature may be mixed in order to describe the complex geometry unambiguously [4].

octahedral
face

fac   mer

Fig. 1.6 *Fac* and *mer* isomers in octahedral metal complexes.

Syn/anti nomenclature is mainly employed for octahedral complexes when geometric isomerism arises from the presence of a fused ring. Therefore, the *syn* isomer has adjacent fused rings whereas the *anti* isomer has opposite fused rings [5].

In summary, molecular variety leads to a multiplicity of stereoisomerisms, which in turn has given rise to several specific nomenclatures. These will be utilized throughout this handbook with the overriding purpose of clarity, rather than strict accuracy.

References

1 G. Lekishvili, *J. Chem. Inf. Comput. Sci.* **1997**, *37*, 924–928.

2 W. Prelog, V. Klyne, *Experientia* **1960**, *16*, 521–523.

3 J. Michl, R. West, *Acc. Chem. Res.* **2000**, *33*, 821–823.

4 S. L. Queiroz, *Química Nova* **1998**, *21*, 193–201.

5 P. Comba, G. N. DeIuliis, G. A. Lawrance, S. M. Luther, M. Maeder, A. L. Nolan, M. J. Robertson, P. Turner, *Dalton Trans.* **2003**, 2188–2193.

2
General Mechanisms of *Cis-Trans* Isomerization: A Rapid Survey

Christophe Dugave

2.1
Introduction

A database search yields more than 20 000 references that contain "Z–E isomerization," "*cis-trans* isomerization," or "geometric isomerization" as keywords, and
the general tendency is an increase in the number of papers devoted to the kinetic
aspects of *cis-trans* isomerization (CTI) in all fields. The main isomerization pathways have probably been discovered, though many remain the object of intense
theoretical (see Chapter 7) and experimental research (see Chapters 4–6, 8–10, 13,
and 14). In the present chapter, general CTI mechanisms will be divided into
homolytic and heterolytic cleavage of the π-bond which allows isomerization,
though some molecular motifs such as amides are able to switch from *cis* to *trans*
via both processes. An overview of CTI in metal complex (mainly thermal, photochemical, and oxidative isomerizations) will be the purpose of Chapter 14 and will
not be detailed here.

2.2
Homolytic *Cis-Trans* Isomerization

Since the elucidation of the photoisomerization of alkenes in 1928, numerous
CTI pathways have been proposed. In fact, many unsaturated compounds may
isomerize via different pathways depending on the conditions. For example, polyenes may photoisomerize via either the π,π^* singlet or triplet excited states and
also via photosensitization by singlet–singlet and triplet–triplet intersystem crossing [1] via a perpendicular radical transition state that accounts for the formation
of the least stable Z isomer [2]. CTI of olefine and polyene systems has been thoroughly investigated using a wide variety of models including stilbenes and stilbene analogs, retinal derivatives and carotenoids, etc., as well as simple cycloalk-
enes (Fig. 2.1) [3], leading to the parallel emergence of novel theories and powerful techniques to probe the behavior of molecular systems in the 10^{-14} to 10^{-11} s
range, such as femtosecond laser spectroscopy [4] (see Chapter 4).

cis-trans Isomerization in Biochemistry. Edited by Christophe Dugave
Copyright © 2006 WILEY-VCH Verlag GmbH & Co. KGaA, Weinheim
ISBN: 3-527-31304-4

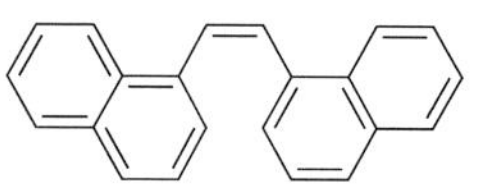

Fig. 2.1 Some model compounds used for studying photo-isomerization processes: Z-1,2-bis-α-naphthylethylene **1**, retinal **2**, β-carotene **3**, cyclooctene **4**.

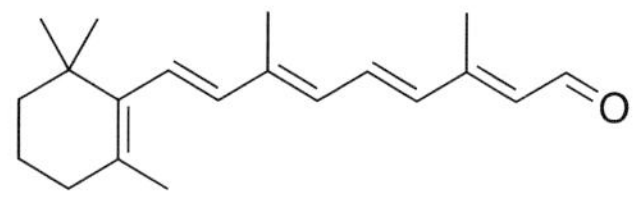

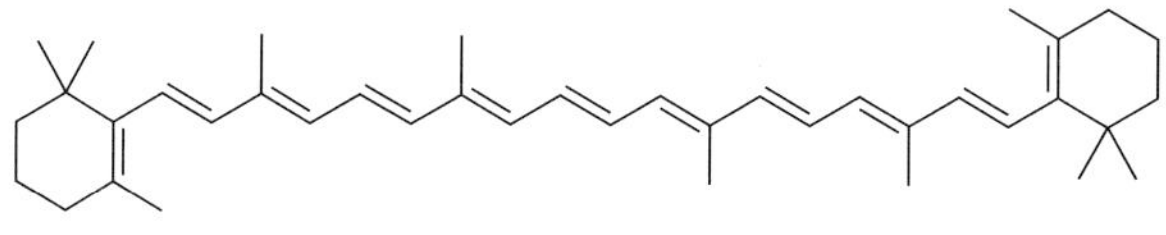

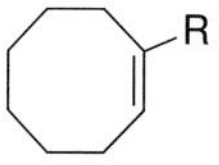

However, there are many other CTI pathways for the simple polyenes: aborted heterogeneous hydrogenation, radical reactions initiated by radical generators including photosensitization by ketones [5] and paramagnetic molecules (e.g. oxygen, atomic bromine and iodine, nitrogen oxide) and heat [6,7]. Recently, a new mechanism for the iodine/light-catalyzed CTI of stilbene has been proposed and seems to imply the formation of a complex between iodine and the alkene that leads to a single radical adduct [8]. A similar process was proposed for the thiyl radical-mediated CTI of polyenes [9]. A relevant example is retinal, which may isomerize experimentally through many of these distinct pathways. Moreover, many pathways play an important part in the CTI of polyenes in vivo where they have a central role in vision, metabolism, and accidental alteration of biomolecules, in particular phospholipids (Fig. 2.2).

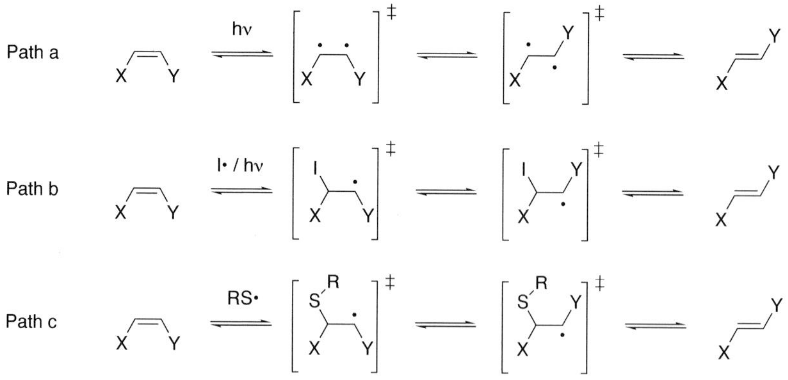

Fig. 2.2 Possible mechanisms of *cis-trans* (*Z-E*) isomerization of olefins and related compounds via heterolytic cleavage of the C=C bond: photoisomerization (path A).

As a general rule, the more the molecule is conjugated, the lower the energy barrier to isomerization. While nonconjugated alkenes require typically 97–164 kcal mol^{-1} to isomerize via the lowest triplet state and the π,π^* singlet state, respectively, the calculated energy barrier to CTI of retinal Schiff bases lies between 23 and 60.6 kcal mol^{-1} depending on the C=C bond and the protonation state of the imine [10]. It is well known that *cis*-polyacetylene isomerizes to the all-*trans* compound upon heating to 150 °C. In the same way, diarylazo compounds require less energy to isomerize from *trans* to *cis* than stilbene derivatives, reflecting the optimal wavelength needed to induce CTI, for example $\lambda_{max} = 319$ nm for azobenzene and $\lambda_{max} = 294$ nm for stilbene.

Z-E isomerization via simple geometric inversion (one-bond flip, OBF, Fig. 2.3A) involves the torsional relaxation of the perpendicular excited state via an adiabatic mechanism which implies a non-volume-conserving process. This is not compatible with the ultrafast CTI in polyenes, in particular retinyl chromophores, and two other possible ways of photo-CTI have been proposed over the past 15 years [11].

The hula twist mechanism (HT, Fig. 2.3B), first validated with carotenoids, is not consistent with the time-scale of photoisomerization of chromoproteins since CTI of the retinal chromophore, which is inserted deep inside the protein, necessitates a major reorganization of the peptide molecular framework. Therefore, a new volume-conserving mechanism, called bicyclic pedal (BP, Fig. 2.3C), was proposed. In fact, all these mechanisms are still a topic of discussion since chromoprotein photo-intermediates highlighted by recent studies do not confirm this hypothesis. In particular, several photo-products of the retinal Schiff base in the

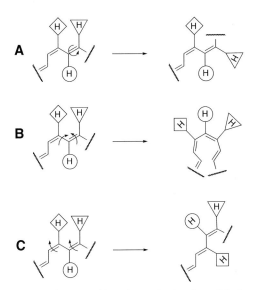

Fig. 2.3 Three possible pathways for the photo-CTI of polyenes: (A) one-bond flip (OBF); (B) hula twist (HT); (C) bicyclic pedal (BP).

rhodopsin protein family display a slightly constrained structure which is not taken into account by such theories.

In many cases, the CTI process competes with rearrangements and cyclizations that occur during the radiationless transition. Z-Stilbene is well known to give dihydrophenanthrene as a photoproduct along with E-stilbene (Fig. 2.4A). Reversible photocyclization is even the dominant reaction in fulgide [12] and merocyanine [13] systems [3] (Fig. 2.4B,C).

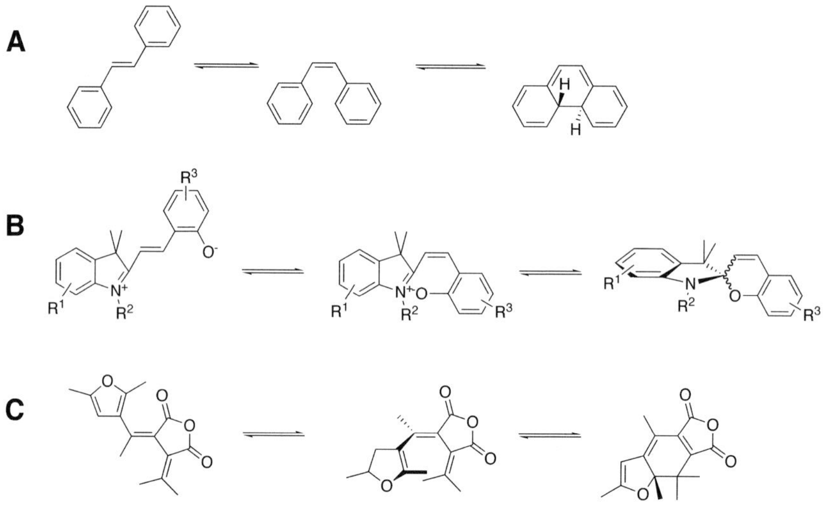

Fig. 2.4 Photoisomerization of E-stilbene gives a mixture of Z-stilbene and dihydrophenantrene (A), whereas merocyanines isomerize directly to the enantiomeric spiropyran forms (B) and E-fulgides transform into the corresponding cyclic adduct.

2.3
Heterolytic *Cis-Trans* Isomerization

Although diazene compounds undergo photoisomerization in a similar way to alkenes [14,15], they also interconvert from Z to E and E to Z via simple doublet inversion (Fig. 2.5 path d), as also observed with other nitrogen-containing compounds such as nitroso derivatives. Moreover, the ultrafast isomerization of azosulfides implies a cleavage/recombination mechanism though radical anion cleavage seems to operate in the Z isomer exclusively, preventing isomerization from Z to E [16].

CTI driven by conjugation transfer (including deconjugation and tautomeric effects) gives rise to a wide range of mechanisms that may explain isomerization of many molecular motifs such as push–pull olefins, acrylates, imines and enamines, amides, and related compounds. Push–pull olefins, which are substituted by electron-donating and electron-withdrawing groups simultaneously, can isomerize spontaneously by simple transfer of conjugation which decreases the double bond character of C=C (Fig. 2.5 path e). The tautomeric effect in the enol/ketone and enamine/imine equilibria plays a similar role, since electron delocalization is disrupted by a change in polarity (Fig. 2.5 path f) [3].

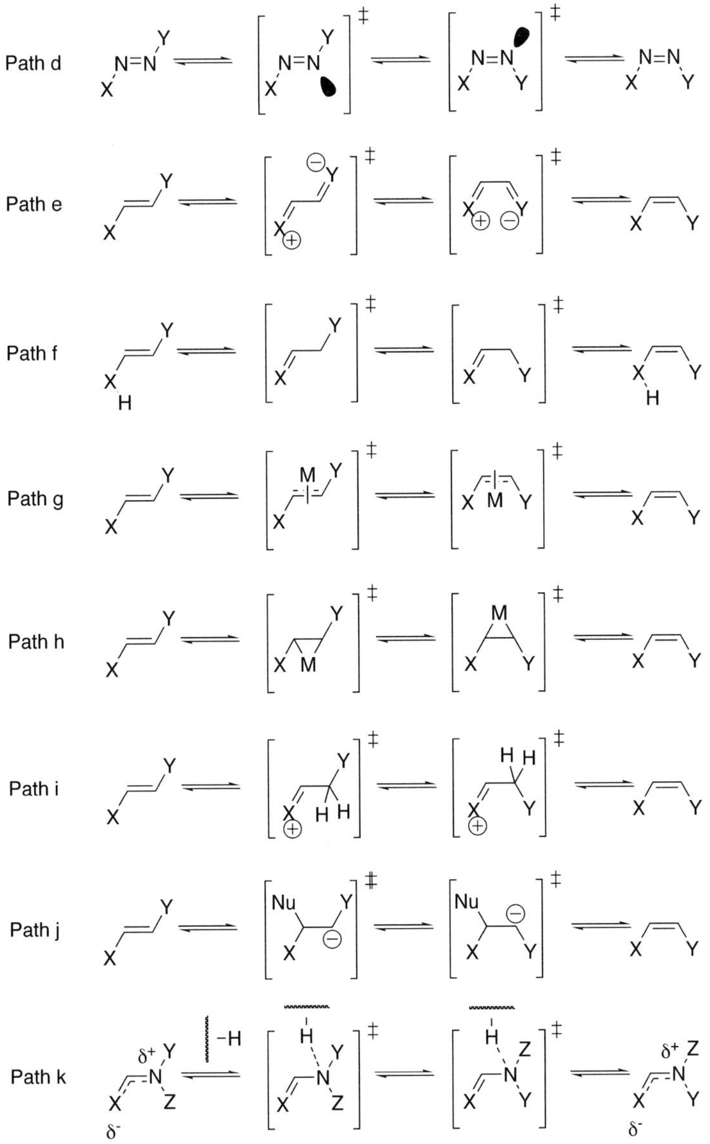

Fig. 2.5 Possible mechanisms for CTI via the heterolytic disruption of the conjugation of double bonds and pseudo double bonds (radical cleavage/recombination pathway was omitted there since it is not a true CTI process).

Several exogenous entities may also facilitate a *cis-trans* interconversion. Lewis acids and transition metals may disrupt conjugation by simply "hijacking" the π-electrons (Fig. 2.3 path g). Transition metal π-bases also operate via an insertion inside a triangular intermediate (path h) [17]. Brønstedt acids (path i) and nucleophiles (path j) can also facilitate CTI via the formation of a tetrahedral intermediate. Amide and analogous compounds are undoubtedly the most versatile compounds in terms of possible mechanisms of CTI since most of the proposed pathways, including radical mechanisms, may account for the experimental results. This probably reflects the very low energy barrier (typically 5–30 kcal mol^{-1}) to CTI. In peptides and proteins, amide CTI seems to occurs via a simple disruption of the double-bond character (path k) putatively through the creation of either intra- or intermolecular H-bonds [3,18]. The geometric deviation from double-bond planarity helps to lower the energy barrier to isomerization and also plays an important role in energy storage, as observed with photointermediates of chromoproteins.

Rotation around hindered or retrained single bonds usually implies that the molecule reaches the energy barrier that restricts interconversion from one conformer to another. In general terms, steric hindrance is the main limitation and heating is then sufficient to cross the barrier, although additional interactions such as H-bonds, stereoelectronic effects and ionic interactions may either hamper or facilitate the rotation.

References

1 J. Saltiel, D. F. Sears Jr., D. H. Ko, K. M. Park, *CRC Handbook of Organic Photochemistry and Photobiology,* W. M. Horspool, and P.-S. Song, Ed. CRC Press, Boca Raton, FL, USA, **1995**, 3.

2 A. Simeonov, M. Matsushita, E. A. Juban, E. A. Z. Thompson, T. Z. Hoffman, A. E. Beuscher IV, M. J. Taylor, P. Wirsching, W. Rettig, J. K. McCusker, R. C. Stevens, D. P. Millar, P. G. Schultz, R. A. Lerner, K. D. Janda, *Science* **2000**, *290*, 307–313.

3 C. Dugave, L. Demange, *Chem. Rev.* **2003**, *103*, 2475–2532.

4 N. Tamai, H. Myasaka, *Chem. Rev.* **2000**, *100*, 1875–1890.

5 A. Gupta, R. Mukhtar, S. Seltzer, *J. Phys. Chem.* **1980**, *84*, 2356–2363.

6 J. Schwinn, H. Sprinz, K. Drössler, S. Leistner, Brede, *Int. J. Radiat. Biol.* **1998**, *74*, 359–365.

7 H. Jiang, N. Kruger, D. R. Lahiri, D. Wang, J.-M. Watele, M. Balazy, *J. Biol. Chem.* **1999**, *274*, 16235–16241.

8 T. S. Zyubina, V. F. Razumov, *High Energy Chem.* **2001**, *35*, 100–106.

9 C. Ferreri, C. Costatino, L. Perrota, L. Landi, Q. G. Mulazzani, C. Chatgilialoglu, *J. Am. Chem. Soc.* **2001**, *123*, 4459–4468.

10 E. Tajkhorshid, B. Paizs, S. Suhai, *J. Phys. Chem. B.* **1999**, *103*, 4518–4527.

11 R. S. H. Liu, *Acc. Chem. Res.* **2001**, *34*, 555–562.

12 Y. Yokoyama, *Chem. Rev.* **2000**, *100*, 1717–1739.

13 G. Berkovic, V. Krongauz, V. Weiss, *Chem. Rev.* **2000**, *100*, 1741–1753.

14 H. Suginome *CRC Handbook of Organic Photochemistry and Photobiology,* W. M. Horspool and P.-S. Song, Ed. CRC Press, Boca Raton, FL, USA, **1995**, 824–840.

15 Y.-C. Lu, E. Wei-Guang Diau, H. Rau, *J. Phys. Chem. A.* **2005**, *109*, 2090–2099.

16 P. Guiriec, P. Hapiot, J. Moiroux, A. neudeck, J. Pinson, C. Tavani, *J. Chem. Phys. A.* **1999**, *103*, 5490–5500.

17 S. H. Meiere, F. Ding, L. A. Friedman, M. Sabat, W. D. Harman, *J. Am. Chem. Soc.* **2002**, *124*, 13506–13512.

18 G. Fischer, *Chem. Soc. Rev.* **2000**, *29*, 119–127.

3
Mechanisms of *Cis-Trans* Isomerization around the Carbon–Carbon Double Bonds via the Triplet State

Yasushi Koyama, Yoshinori Kakitani, and Hiroyoshi Nagae

It has been shown that *cis-trans* isomerization (CTI) around a carbon–carbon double bond takes place not in the singlet state but in the triplet state. Schiff base of retinal is also the case, although protonated Schiff base isomerizes via the singlet state (see Ref. [1] for a review). The lowest-triplet (T_1) state can be generated through intersystem crossing in retinoids, and through singlet fission of a particular singlet ($1B_u^-$) state in carotenoids [2]. Since the quantum yield of triplet generation is generally lower in carotenoids ($\sim 10^{-3}$), a triplet sensitizer is necessary to study CTI via the T_1 state.

The excited-state properties of retinoids and carotenoids have been reviewed, together with chlorophylls, in relation to their biological functions [1]. In the present review, we will briefly summarize the previous results concerning polyenes (Sections 3.2.1 and 3.3.1), and add most recent results along these lines (Sections 3.2.2 and 3.2.3, and 3.3.2–3.3.4), focussing on a retinoid (retinal) and two carotenoids (*β*-carotene and spheroidene).

In accord with the purpose of this book, to provide detailed and critical reviews of most recent highlights, we have tried to incorporate as much data as possible so that they can tell stories by themselves. We hope readers will enjoy reading the figures and tables as well.

3.1
A Concept of a Triplet-Excited Region

The concept of a "triplet-excited region" emerged during studies on Raman spectra and isomerization of retinoids and carotenoids in the T_1 state (see Ref. [3] for a summary). The triplet-excited region was defined as a region where large changes in bond order take place toward its inversion upon the ground (S_0) to T_1 transition; that is, a double bond becomes more single bond-like and a single bond becomes more double bond-like. It has a span of approximately six conjugated double bonds, it is located in the central part of a conjugated chain, and it triggers CTI. This is a basic concept in understanding the isomerization properties of poly-

cis-trans Isomerization in Biochemistry. Edited by Christophe Dugave
Copyright © 2006 WILEY-VCH Verlag GmbH & Co. KGaA, Weinheim
ISBN: 3-527-31304-4

enes with various conjugation lengths. The excited-state properties of polyenes are strongly dependent on the number of conjugated double bonds, n.

The triplet-excited region could have been predicted theoretically before it was actually found spectroscopically [3]. It originates from strong correlation of the parallel spins in the T_1 manifold. Figure 3.1 shows the results of Pariser–Parr–Pople calculations including singly and doubly excited configurational interactions (PPP-SD-CI calculation) for polyenes having $n = 5, 7, 9$, and 11 [4]. Polyenes having $n = 5–7$ and 9–11 may correspond to the cases of the typical retinoids and carotenoids, respectively. The calculations predict the following: large changes in π-bond order should take place in the central part of the conjugated chain. In the very central part, the inversion of π-bond order should take place, and the changes in π-bond order decrease in both peripherals. The triplet-excited region, defined in terms of those changes in π-bond

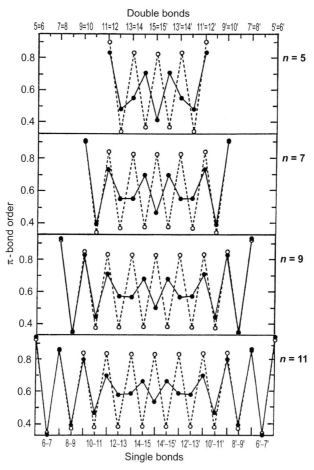

Fig. 3.1 π-Bond orders in the S_0 (open circles) and T_1 (filled circles) states of model polyenes with the number of conjugated double bonds, $n = 5, 7, 9$, and 11 calculated by the PPP-SD-CI method, the details which are described in Ref. [4].

order, must cover the entire molecule in retinal, but only the central part in carotenoids. Its span is about six conjugated double bonds.

Since CTI is expected to be triggered by the elongation of the double bonds and by the steric hindrance of the *cis*-bend structure, in retinal, which has a short conjugated chain, the quantum yield of CTI may be high in the central-*cis* isomers where the *cis*-bend is within the triplet-excited region, but it may be low in peripheral-*cis* isomers where the *cis*-bend is out of the triplet-excited region. In carotenoids with a long conjugated chain, the quantum yields in central-*cis* isomers must be much higher than those in peripheral-*cis* isomers.

The picture of triplet-excited region was slightly modified after it was determined spectroscopically in terms of stretching force constants that include the contributions of both the σ and π bonds (*vide infra*).

3.2
Triplet-State Isomerization in Retinal

3.2.1
Cis-Trans Isomerization Examined by Electronic Absorption and Raman Spectroscopies and by High-Performance Liquid Chromatography Analysis

Scheme 3.1 shows the configurations of the all-*trans* and *cis* isomers of retinal. Concerning the position of the *cis*-bend, the 7-*cis* and 13-*cis* isomers can be classi-

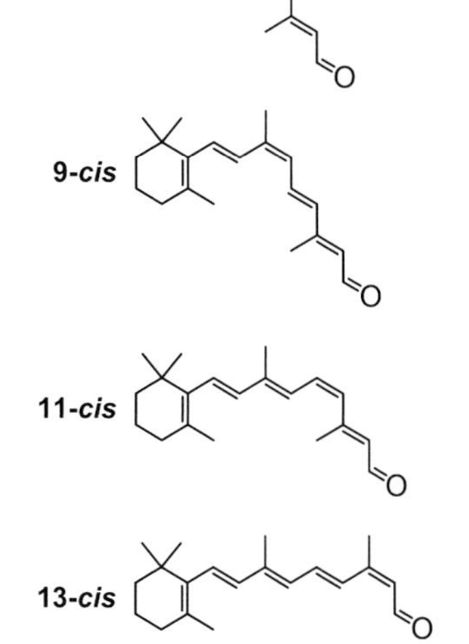

Scheme 3.1 The configurations of the all-*trans* and *cis* isomers of retinal used in the investigations described.

fied into peripheral-*cis* isomers, and the 9-*cis* and 11-*cis* isomers into central-*cis* isomers. Concerning the structure of the *cis*-bend, the 7-*cis* and 11-*cis* isomers can be classified into unmethylated *cis* isomers, while the 9-*cis* and 13-*cis* isomers are methylated *cis* isomers. Severe steric hindrance on the concave side is expected in unmethylated *cis* isomers, which must promote CTI upon triplet excitation.

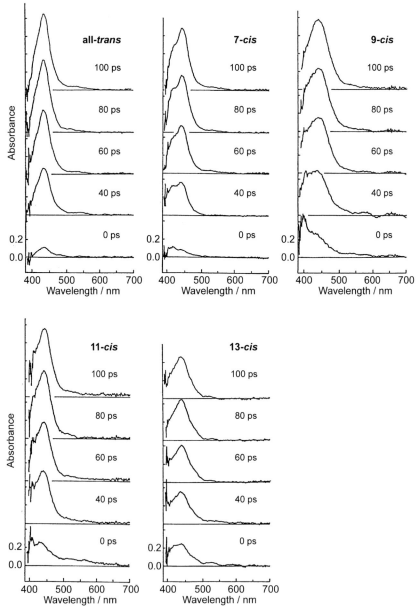

Fig. 3.2 Deca-picosecond time-resolved absorption spectra of the all-*trans* and *cis* isomers of retinal. The T₁ state was generated by direct excitation of isomeric retinal with the 355 nm, 20–25 ps pulses [5].

Figure 3.2 shows the deca-picosecond time-resolved absorption spectra of the all-*trans* and *cis* isomers of retinal [5]. Since retinal has a high quantum yield of intersystem crossing (0.50–0.61) [6], the T_1 absorption spectra can be recorded by 355 nm excitation to the $1B_u^+$ singlet state. In the all-*trans* isomer, a monotonical rise of the $T_n \leftarrow T_1$ absorption is seen within 100 ps. In the *cis* isomers, on the other hand, an intrinsic absorption peak appears immediately after excitation on the blue side of that of the all-*trans* isomer and then it decays with concomitant rise of the $T_n \leftarrow T_1$ absorption of the all-*trans* isomer. The spectral changes in the *cis* isomers must reflect CTI in the T_1 state. The convergence of the transient absorption to that of the *trans*-T_1 takes place in a similar time-scale among the 7-*cis*, 9-*cis*, and 11-*cis* isomers. In the 13-*cis* isomer, however, the sharpening of the $T_n \leftarrow T_1$ absorption takes place much more slowly.

Figure 3.3 shows that the transient absorptions, originating from the 7-*cis*, 9-*cis*, and 11-*cis* isomers, completely converge into that from the all-*trans* isomer at 5 ns after excitation, but the transient absorption from the 13-*cis* isomer still keeps its own contribution on the longer wavelength side. The above results of transient absorption spectroscopy suggest that the quantum yields of CTI, via the T_1 state, are similar among the 7-*cis*, 9-*cis*, and 11-*cis* isomers, but that of the 13-*cis* isomer is much smaller.

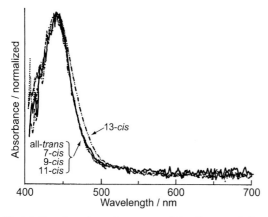

Fig. 3.3 Transient absorption spectra of the all-*trans* and *cis* isomers of retinal at 5 ns after excitation. The same conditions of excitation as described in Fig. 3.2 [5].

Figure 3.4 depicts the transient Raman spectra of the all-*trans* and *cis* isomers of retinal recorded at ~20 ns after excitation [7]. Consistent with the above results of time-resolved absorption spectroscopy, the 7-*cis*, 9-*cis*, and 11-*cis* isomers give rise to the same T_1 Raman spectra as that of the all-*trans* isomer. However, the difference Raman spectrum of "13-*cis* minus all-*trans*" exhibits another spectral pattern ascribable to the 13-*cis* T_1.

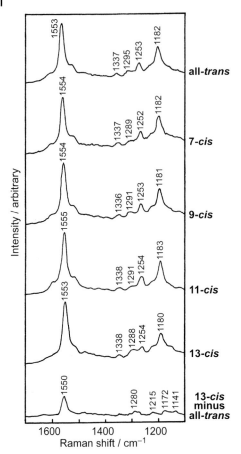

Fig. 3.4 Transient Raman spectra of the all-*trans* and *cis* isomers of retinal. The T_1 state was generated by excitation using 355 nm, 5 ns pulses, and the T_1 Raman spectra were recorded by the use of 532 nm, 5 ns pulses [7].

Figure 3.5 shows the processes of isomerization upon triplet excitation by the use of a sensitizer, Zn-tetraphenylporphyrin, starting from the all-*trans* and each *cis* isomer of retinal [8]. The all-*trans* isomer isomerizes very little. Concerning the *cis* isomers, on the other hand, 7-*cis* and 9-*cis* isomerize with a similar efficiency, 11-*cis* isomerizes with the highest efficiency, and 13-*cis* isomerizes with the lowest efficiency. The quantum yields of isomerization were determined by the use of the *initial* decrease of each starting isomer as follows: all-*trans*, 0.1; 7-*cis*, 0.8; 9-*cis*, 0.8; 11-*cis*, 0.9; and 13-*cis*, 0.2. The set of values was determined at the concentration of 1×10^{-4} mol L^{-1}. Those quantum yields in retinal can vary depending on its concentration due to the presence of a chain reaction, in which the resultant all-*trans* T_1 functions as an additional sensitizer [8]. The relative quantum yields of CTI are depicted in Scheme 3.2.

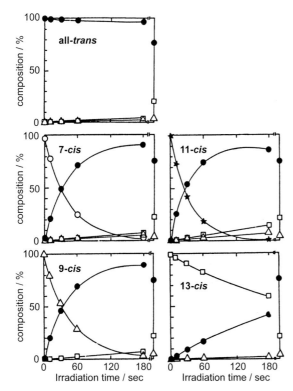

Fig. 3.5 Triplet-sensitized isomerization of retinal starting from the all-*trans* (filled circles), 7-*cis* (open circles), 9-*cis* (open triangles), 11-*cis* (stars) and 13-*cis* (open squares) isomers. The sensitizer, Zn-tetraphenylporphyrin, was irradiated with a 250 W halogen lamp after passing through water and a Toshiba V-Y52 filters [8].

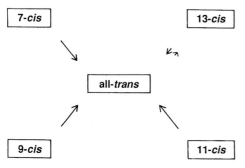

Scheme 3.2 A schematic presentation of triplet-state isomerization in retinal. The pathways of isomerization are shown by arrows, whose lengths are proportional to the values of quantum yield [8].

3.2.2
Triplet-Excited Region in All-*trans*-Retinal Shown in Terms of Stretching Force Constants Determined by Raman Spectroscopy and Normal Coordinate Analysis [9]

A set of carbon–carbon and carbon–oxygen stretching force constants is a useful measure of bond orders. Since the normal vibrations of a molecule are determined by a set of force constants (called "molecular force field") and the mass of atoms, isotopic substitution is a powerful technique to increase the number of observables (vibrational frequencies) keeping the molecular force field unchanged. Figures 3.6 and 3.7 show, respectively, the S_0 and T_1 Raman spectra of variously-deuterated species that played a key role in determining a set of stretching force constants of all-*trans*-retinal in each electronic state. (The two sets of assignments of Raman lines are shown in Ref. [9].)

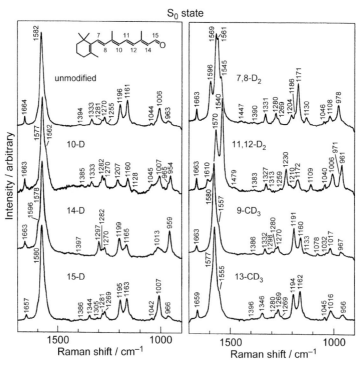

Fig. 3.6 The S_0 Raman spectra of undeuterated and various deuterio derivatives of all-*trans*-retinal. Raman spectra were recorded by the use of CW 488 nm line [9].

The strategy we have taken in the determination of the stretching force constants is as follows: First, we determined a complete set of force constants in the S_0 state, including the stretching, in-plane bending, out-of-plane wagging, and torsional force constants of Urey–Bradley–Shimanouchi (UBS) type, and also, non-UBS cross-terms, many of which were transferred from those determined by Saito and Tasumi [10]. Second, we transferred most of those S_0-state force constants, except for the carbon–carbon stretching force constants in the central part of the conjugated chain that were expected to change substantially upon triplet excitation. In the initial manual fitting of the C=C, C=O, and C–C stretching force constants, we used three different sets of force constants concerning the bond orders in the conjugated chain (as initial guess), that is, Set A: force constants reflecting the results of the PPP-SD-CI calculation including the inversion of bond orders (Figure 3.1, $n = 5$–7), Set B: those assuming a 1.5 bond order throughout the conjugated chain, and Set C: those determined in the S_0 state. Set A was not successful in the fitting at all, while Set B and Set C converged into a new set. Finally, we determined the force constants by least-square fitting.

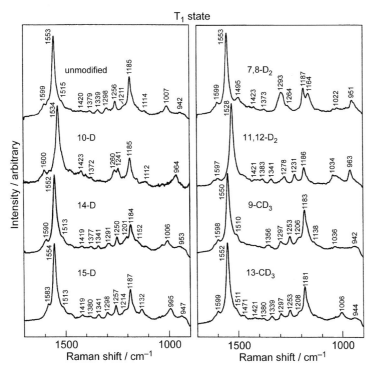

Fig. 3.7 The T_1 Raman spectra of undeuterated and various deuterio derivatives of all-*trans*-retinal. The T_1 state was generated by 355 nm, 5 ns pulses, and the T_1 Raman spectra were recorded by the use of 488 nm, 5 ns pulses [9].

Figure 3.8 shows the carbon–carbon stretching force constants thus determined for the S_0 and T_1 states. Changes in those force constants upon triplet excitation (from open circles to filled circles) can be characterized as follows: Concerning the conjugated polyene chain, the decrease in the C=C stretching force constant is in the order, C11=C12 > C9=C10 > C7=C8 > C13=C14, whereas the increase in the C–C stretching force constant is in the order, C8–C9 > C10–C11 > C12–C13.

The triplet-excited region can be defined, in terms of the carbon–carbon stretching force constants, to be a region from C6 or C7 to C13 or C14, although a large decrease in the stretching force constant is seen in the terminal C=O bond, as well.

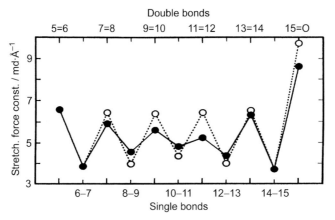

Fig. 3.8 Changes in the stretching force constants upon triplet excitation from the S_0 state (open circles) to the T_1 state (filled circles) of all-*trans*-retinal [9].

3.2.3
Dynamic Triplet-Excited Region in Retinal As Revealed by Deuteration Effects on the Quantum Yields of Isomerization via the T_1 State (Okumura, Koyama, unpublished results)

The concept of a triplet-excited region that causes changes in bond order in the conjugated chain and triggers the CTI is now experimentally established in retinal. However, the detailed process of atomic rearrangements in the T_1 excited state, resulting in the isomerization from the ground-state *cis* to *trans* configuration around a particular *cis* double bond, still remains to be determined. The rotational motion around a double bond must accompany sequential changes in the electronic structure, that is, $sp^2 \rightarrow sp^3 \rightarrow sp^2$, during which changes in the directions of a pair of the olefinic C–H groups are expected to take place. Ohmine and Morokuma [11,12] theoretically calculated the process of isomerization in butadiene, and predicted that the rotation around one of the double bonds should accompany both the flapping of the C–H bonds attached to it and the rotation around the neighboring double bond. Stimulated by this prediction, we have examined the effects of the 7,8-, 11,12-, and 14,15-deuterations on the quantum yields of triplet-sensitized isomerization of retinal starting from the 7-*cis*, 9-*cis*, 11-*cis*, and 13-*cis* isomers, although Waddell et al. [13] showed that the 14,15-deuteration did not affect the quantum yields of both the 13-*cis* → all-*trans* and all-*trans* → 13-*cis* isomerizations.

Figure 3.9 shows the effects of double deuteration of the C7=C8 or C11=C12 double bond and that of the C14–C15 single bond on the triplet-state CTI starting from the set of four *cis* isomers. The results can be summarized as follows: (1) The 7,8-deuteration (7,8-D_2) reduces the quantum yield of isomerization from the 7-*cis* to the all-*trans* isomer that includes rotation around the particular double bond to which deuterium substitution was made, and also, the quantum yield of isomerization from the 9-*cis* to the all-*trans* isomer around the neighboring double bond on the right-hand side of the retinal molecule (see Scheme 3.1). (2) The 11,12-deuteration (11,12-D_2) reduces the quantum yields of isomerization from the 7-*cis*, 9-*cis*, and 11-*cis* isomers to the all-*trans* isomer that include rotation around the particular *cis*-double bond to which deuterium substitution was made, and also, that around the neighboring double bonds on the left-hand side of the molecule. (3) The 14,15-deuteration (14,15-D_2) slightly reduces the quantum yield of isomerization from the 11-*cis* isomer. (4) Practically no deuteration effects on the quantum yields of isomerization are seen at all starting from the 13-*cis* isomer [13]. Table 3.1 lists the quantum yields of isomerization per triplet species generated for the undeuterated and variously deuterated retinal isomers.

Table 3.1 Quantum yields of isomerization per triplet species generated in undeuterated and deuterated retinals (Okumura, Koyama, unpublished results).

	D_0 retinal	7,8-D_2	7,8-D_2/D_0	11,12-D_2	11,12-D_2/D_0	14,15-D_2	14,15-D_2/D_0
7-*cis*	0.75 ± 0.02	0.65 ± 0.02[a]	0.87	0.70 ± 0.01	0.93	0.75 ± 0.01	1.00
9-*cis*	0.83 ± 0.01	0.70 ± 0.03	0.84	0.62 ± 0.03	0.75	0.80 ± 0.01	0.96
11-*cis*	0.83 ± 0.01	0.80 ± 0.03	0.96	0.65 ± 0.01	0.78	0.71 ± 0.02	0.86
13-*cis*	0.36 ± 0.01	0.38 ± 0.02	1.06	0.37 ± 0.03	1.03	0.36 ± 0.02	1.00

a Large decreases in the quantum yield are underlined.

It has been shown that double deuteration on an unmethylated double bond reduces the quantum yield of CTI in the triplet state not only around the particular double bond but also around the neighboring double bonds. The results are in agreement with the picture of structural changes in butadiene that was theoretically predicted by Ohmine and Morokuma [11,12]. The effect is the strongest in the central part of the conjugated chain. This "dynamic triplet-excited region" seems to be confined within the central double bonds including C7=C8, C9=C10, and C11=12.

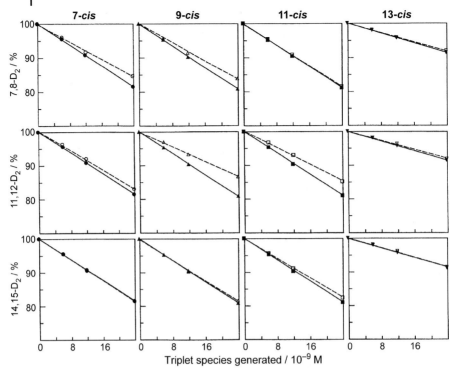

Fig. 3.9 Effects of deuteration on the decrease of the starting isomer of retinal
as a function of triplet species generated. The linear relations before deuteration
are shown in open symbols and broken lines, whereas those after deuteration
are shown in filled symbols and solid lines (Okumura, Koyama, unpublished results).

3.2.4
Summary and Future Trends

(1) The quantum yield of CTI via the T_1 state has been determined to be in the
order, 11-*cis* (0.9) > 9-*cis* (0.8) = 7-*cis* (0.8) > 13-*cis* (0.2). Time-resolved (transient)
electronic-absorption and Raman spectroscopies exhibited a clear difference between the 13-*cis* isomer and the rest of the isomers (7-*cis*, 9-*cis*, and 11-*cis*), reflecting the above values of quantum yield. (2) The triplet-excited region, predicted by
the PPP-SD-CI calculations of the π-bond order, has been spectroscopically evidenced in terms of carbon–carbon stretching force constants (Fig. 3.8). The results
support the idea that the quantum yield of CTI is determined by the elongation of
the *cis* double bond and the steric hindrance of the *cis*-bend structure to be
released. In the 11-*cis* and 9-*cis* isomers, the *cis*-bend structure is in the central
part of the triplet-excited region, enhancing the quantum yield of isomerization,
whereas in the 7-*cis* and 13-*cis* isomers, it is in the peripheral part, suppressing
the quantum of isomerization. However, the steric hindrance is much stronger in
the unmethylated 7-*cis* isomer than in the methylated 13-*cis* isomer; this is proba-

bly the reason why the quantum yield in the former becomes much higher. (3) The double deuteration on the unmethylated double bonds (C7=C8 and C11=C12) reduced not only the quantum yield of CTI around the particular double bond, but also around the neighboring double bonds. (4) Practically no deuteration effects are seen in the 13-*cis* to all-*trans* isomerization, most probably due to the fact that the 13-*cis*-bend is out of the triplet-excited region.

The deuteration effects have provided us with a unique insight into the atomic rearrangement during the process of isomerization in the triplet state. The results suggest the following: (1) The triplet-state isomerization may include the rearrangement of all the atoms in the triplet-excited region, which is to be described in terms of the rotations around the double bonds and the flapping of the C–H bonds as predicted by Ohmine and Morokuma [11,12]. (2) A pair of substitutions on an unmethylated double bond from C–H to C–D must substantially increase the momentum of its flapping motion, and as a result, slows down those atomic rearrangements. Obviously, those hypotheses need to be examined by further investigations.

3.3
Triplet-State Isomerization in β-Carotene and Spheroidene

3.3.1
Cis-Trans Isomerization in β-Carotene Studied by Electronic Absorption and Raman Spectroscopies and by HPLC Analysis

Scheme 3.3 shows the configurations of the all-*trans* and *cis* isomers of β-carotene. The all-*trans* isomer of this carotenoid has C_{2h} symmetry, and consists of an extended conjugated chain and a pair of β-ionone rings at both ends. Concerning the position of the *cis*-bend, the 7-*cis* and 9-*cis* isomers can be classified into peripheral-*cis* isomers, whereas the 13-*cis* and 15-*cis* isomers are central-*cis* isomers. Concerning the structure of the *cis*-bend, the 7-*cis* and 15-*cis* isomers can be classified as unmethylated *cis* isomers, whereas the 9-*cis* and 13-*cis* isomers are methylated *cis* isomers. Concerning the unmethylated *cis* isomers, severe steric hindrance is expected in the 7-*cis*-bend, but no severe steric hindrance is expected in the 15-*cis*-bend. Since the triplet-excited region is to be localized only in the central part of a long conjugated chain, the quantum yield of triplet-state isomerization is anticipated to be much higher in the central-*cis* isomers than in the peripheral-*cis* isomers. The 15-*cis* isomer of this carotenoid is bound to the reaction centers of photosystem I and photosystem II of plants and algae, and plays an important role in photo-protection [14].

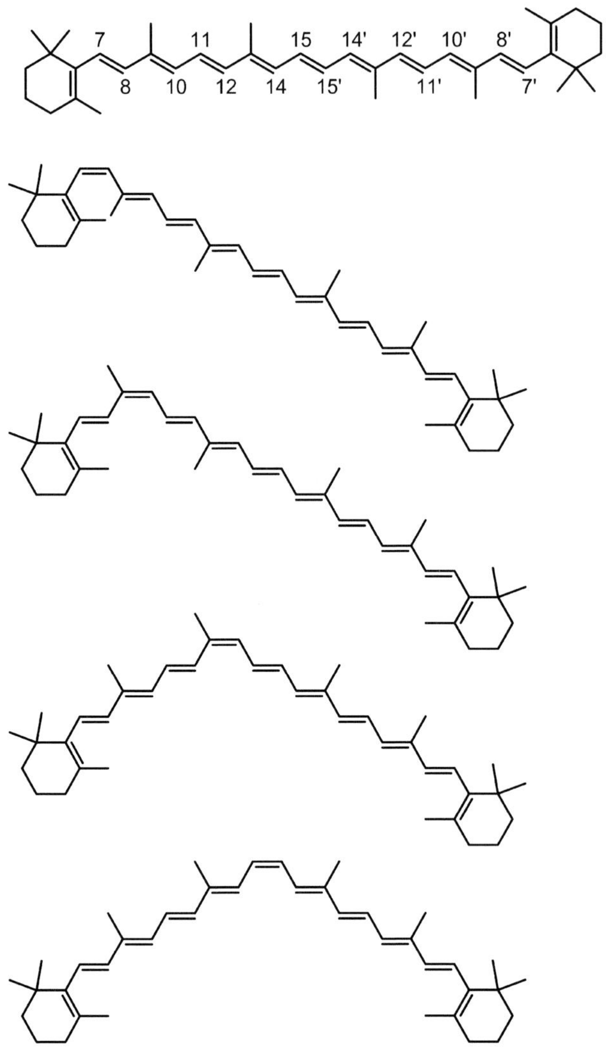

all-*trans*

7-*cis*

9-*cis*

13-*cis*

15-*cis*

Scheme 3.3 The configurations of the all-*trans* and *cis* isomers of *β*-carotene used in the investigations described.

Figure 3.10 shows the transient-absorption spectra of the set of all-*trans* and *cis* isomers recorded at 1.8 μs after excitation of the triplet sensitizer, anthracene [15]. The all-*trans*, 7-*cis*, 9-*cis*, and 13-*cis* isomers exhibit their unique spectral patterns in the T_1 state, whereas the 15-*cis* isomer exhibits the same spectral pattern as that of the all-*trans* isomer. Figure 3.11 shows the transient-Raman spectra of the same set of isomers (1.8 μs after excitation), which exhibits a very similar trend in the T_1 state [16]. Although the spectral patterns of the T_1 species are similar to one another, reflecting the triplet-excited region, unique profiles depending on the starting isomer are seen in the 1230–1130 cm^{-1} region. The all-*trans* isomer exhibits a peak at 1197 cm^{-1}; the 7-*cis* isomer at 1194 cm^{-1}; and the 9-*cis* isomer at 1180 cm^{-1}. However, the 13-*cis* isomer exhibits no peaks at all in this region. Most importantly, the spectral profile originating from the 15-*cis* isomer is identical to that from the all-*trans* isomer. The above results of electronic absorption and Raman spectroscopies indicate that the 15-*cis* to all-*trans* isomerization in the T$_1$ state is too rapid for the 15-*cis* T$_1$ to be detected, although other isomers have their own T$_1$ species whose lifetimes are long enough to be detected spectroscopically.

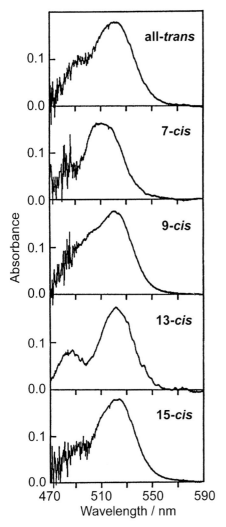

Fig. 3.10 The transient electronic absorption spectra of T$_1$ species generated from the all-*trans* and *cis* isomers (1.8 μs after excitation). The T$_1$ state was generated by excitation of the sensitizer anthracene, using 355 nm, 5–10 ns pulses [15].

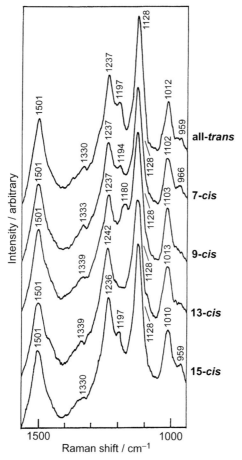

Fig. 3.11 The transient Raman spectra of T_1 species generated from the all-*trans* and *cis* isomers (1.8 μs after excitation). The T_1 state was generated by excitation of the sensitizer anthracene, using 337 nm pulses, and the T_1 Raman spectra were recorded by the use of 532 nm pulses [16].

The above idea was supported by the high-performance liquid chromatography (HPLC) analysis of triplet-sensitized isomerization [4]. Figure 3.12 shows the processes of triplet-sensitized isomerization starting from the set of *cis-trans* isomers of β-carotene. The quantum yields, defined as decrease in the starting isomer per T_1 species generated, were as follows: all-*trans*, 0.04; 7-*cis*, 0.12; 9-*cis*, 0.15; 13-*cis*, 0.87; and 15-*cis*, 0.98. Scheme 3.4 is a pictorial presentation of the isomerization pathways and the value of quantum yields; the length of each arrow is proportional to the quantum yield, and the length of a thicker arrow should be multiplied by 10 when compared with a thinner arrow. The 15-*cis* isomer exhibits almost complete one-way isomerization to the all-*trans* isomer with a quantum yield of almost unity.

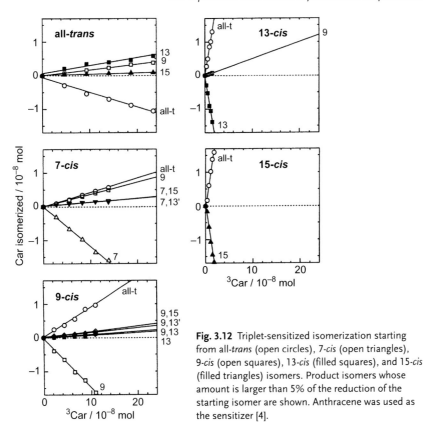

Fig. 3.12 Triplet-sensitized isomerization starting from all-*trans* (open circles), 7-*cis* (open triangles), 9-*cis* (open squares), 13-*cis* (filled squares), and 15-*cis* (filled triangles) isomers. Product isomers whose amount is larger than 5% of the reduction of the starting isomer are shown. Anthracene was used as the sensitizer [4].

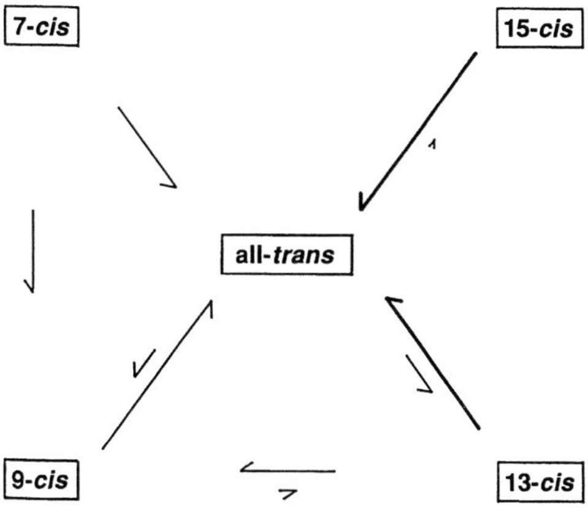

Scheme 3.4 A schematic presentation of the isomerization pathways and the quantum yields of isomerization via the T$_1$ state in β-carotene. The lengths of arrows are proportional to values of quantum yield. The lengths of thicker arrows should be multiplied by 10 to compare with those of thinner arrows [4].

3.3.2
Cis-Trans Isomerization in Spheroidene Studied by Time-Resolved Absorption Spectroscopy and by HPLC Analysis [17]

The T_1-state CTI in spheroidene was found to be much less efficient than that in β-carotene, which facilitated the examination of triplet-state dynamics by time-resolved absorption spectroscopy. Scheme 3.5 shows the configurations of the all-*trans* and *cis* isomers of spheroidene. The all-*trans* isomer consists of an open conjugated chain ($n = 10$) shifted to the left in the entire carbon skeleton, to which a pair of large peripheral groups are attached at both ends. Concerning the position of the *cis*-bend, the 13′-*cis* and 9-*cis* isomers can be classified into peripheral-*cis* isomers, whereas the 13-*cis* and 15-*cis* isomers are central-*cis* isomers. Concerning the structure of the *cis* bend, 13′-*cis*, 9-*cis*, and 13-*cis* isomers are methylated *cis* isomers, and only the 15-*cis* isomer is an umethylated *cis* isomer.

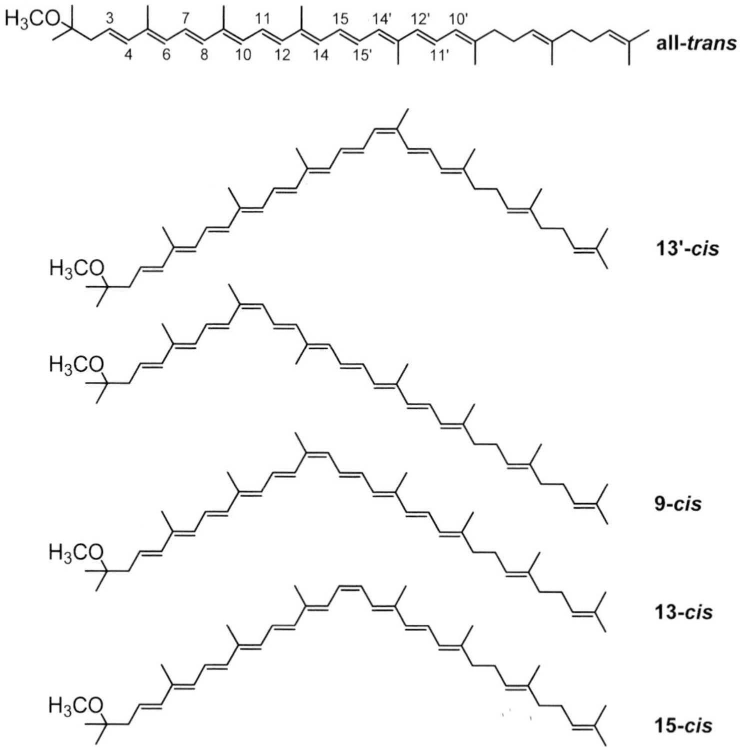

Scheme 3.5 The configurations of the all-*trans* and *cis* isomers of spheroidene used for time-resolved absorption spectroscopy.

Figure 3.13 shows the time-resolved absorption spectra of isomeric spheroidenes. In the all-*trans* isomer, concomitant with the decay of a sharp peak on the shorter wavelength side due to the sensitizer anthracene, the rise and decay of the

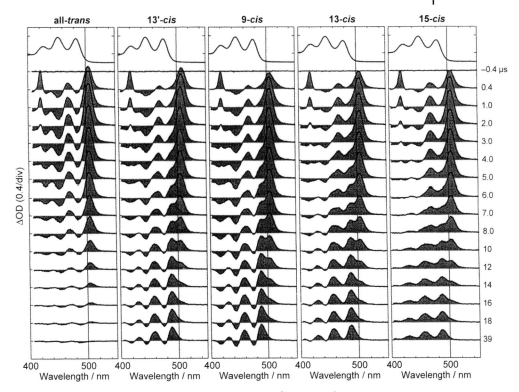

Fig. 3.13 Submicrosecond time-resolved absorption spectra of isomeric sphero-idenes. The sensitizer anthracene was excited at 355 nm to generate the T_1 state [17].

$T_n \leftarrow T_1$ absorption and the bleaching of the $1B_u^+ \leftarrow 1A_g^-$ absorption are clearly seen, the vibrational progression of the former being slightly shifted to the red in comparison with the latter. In the set of *cis* isomers, a difference spectral pattern of all-*trans* S_0 minus *cis* S_0 (a result of CTI) remains with different compositions after the complete decay of the $T_n \leftarrow T_1$ absorption (39 µs). By the use of singular-value decomposition (SVD) followed by three-component global fitting, we were able to extract the spectral patterns of each *cis* T_1 and the all-*trans* T_1 species together with that of (*trans–cis*) S_0, as shown in Fig. 3.14.

Table 3.2 lists the relevant spectral and kinetic parameters determined by the SVD and global-fitting analyses of time-resolved data matrices. The $T_n \leftarrow T_1$ absorption maxima of the set of triplet species are very similar to one another except for that of the 13'-*cis* T_1, a fact which caused difficulty in the analysis. The lifetimes of the *cis* T_1 species could be determined only as an average, although the lifetime of the all-*trans* T_1 could be easily determined. Fortunately, however, the time constants of CTI in the T_1 state could be determined successfully for the set of *cis* isomers. The ranking of the rate of triplet-state isomerization parallels that of the quantum yield of isomerization per triplet species generated, the latter

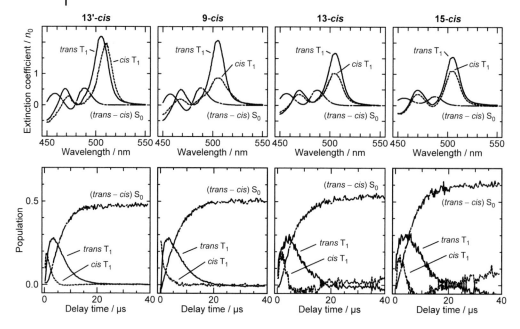

Fig. 3.14 The results of SVD followed by a three-component global fitting of the spectral data matrices, a part of which is shown in Fig. 3.13. The species-associated difference spectra (the upper panels) and time-dependent changes in population (the lower panels) are shown [17].

of which was independently determined by HPLC analysis of time-dependent changes in the isomeric composition. The rate and the quantum yield of triplet-state isomerization are in the order 15-*cis* > 13-*cis* > 9-*cis* > 13′-*cis*, and the most efficient isomerization of 15-*cis* $T_1 \rightarrow$ all-*trans* T_1 was evidenced.

Table 3.2 The $T_n \leftarrow T_1$ absorption maximum, the time constant of $T_1 \rightarrow S_0$ intersystem crossing (k_{trans}^{-1} or k_{cis}^{-1}), and the time constant (k_{iso}^{-1}) and the quantum yield (Φ_{iso}) of CTI for each T_1 species [17].

T_1 species	Absorption (nm)	k_{trans}^{-1} or k_{cis}^{-1} (µs)	k_{iso}^{-1} (µs)	Φ_{iso}
all-*trans*	505	4.76	–	–
13′-*cis*	510	0.83	0.91	0.48
9-*cis*	506	0.83	0.83	0.50
13-*cis*	505	0.83	0.77	0.52
15-*cis*	505	0.83	0.56	0.60

3.3.3
The Triplet-Excited Region of All-*trans*-Spheroidene in Solution and the Triplet-State Structure of 15-*cis*-Spheroidene Bound to the Bacterial Reaction Center Determined by Raman Spectroscopy and Normal Coordinate Analysis [18]

3.3.3.1 All-*trans*-Spheroidene in Solution

Figure 3.15 shows the S_0 and T_1 Raman spectra of a set of deuterio all-*trans*-spheroidene in *n*-hexane (see Ref. [18] for the assignments of Raman lines). The T_1 state was generated by excitation (at 355 nm) of the sensitizer anthracene, and subsequent triplet-energy transfer to spheroidene. The present deuterium substitution is focused on the olefinic protons in the central part of the conjugated chain. There are two reasons for this: (1) We are mainly interested in changes in bond order in the triplet-excited region that is located in the central part of the conjugated chain. (2) Those normal vibrations taking place in both peripherals are not in resonance with the $T_n \leftarrow T_1$ electronic transition that is confined to the triplet-excited region, and therefore, those vibrations were not resonance-enhanced enough to be detected.

In the T_1 Raman spectra we see Raman lines due to normal modes similar to those in the S_0 Raman spectra (see the spectra of D_0), which include the totally symmetric C=C and C–C stretchings (1500 and 1165 cm^{-1}), the C–H in-plane bendings (1268 cm^{-1}), the methyl in-plane rockings (1006 cm^{-1}), and the C–H out-of-plane waggings (944 cm^{-1}). This observation indicates that the conjugated chain takes a stretched planar all-*trans* configuration even in the T_1 state. For normal coordinate analysis of the T_1 Raman spectra we took the same strategy as in the case of retinal; most of the force constants in the S_0 state were transferred to the T_1 state, and then those carbon–carbon stretching force constants relevant to the central part of the conjugated chain were determined. In the calculation of normal vibrations, both peripheral parts beyond the conjugated chain were truncated and regarded as a ball.

Figure 3.16A presents the carbon–carbon stretching force constants in the S_0 (open circle) and T_1 (filled circle) states of all-*trans*-spheroidene that were determined by the normal-coordinate analysis: Decrease in the C=C stretching force constants are in the order, C13=C14 > C11=C12 > C9=C10 > C15=C15′, whereas increase in the C–C stretching force constants are in the order, C12–C13 > C14′–C15′ ≥ C14–C15. It is to be noted that the center of symmetry in those changes of the carbon–carbon stretching force constants is located *not* at the center of the entire skeleton (C15=C15′ bond) *but* at the center of the conjugated chain (C12–C13 bond). The exact span of the triplet-excited region could not be determined due to the limited number of observed frequencies as mentioned above.

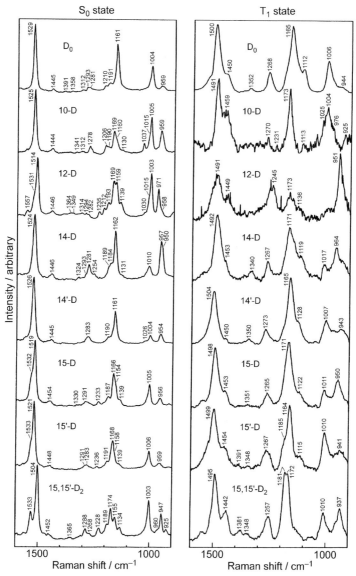

Fig. 3.15 The S_0 and T_1 Raman spectra of undeuterated and various deuterio derivatives of all-*trans*-spheroidene in *n*-hexane. The T_1 state was generated by 355 nm excitation of the sensitizer anthracene, and the T_1 Raman spectra were recorded by the use of the 532 nm pulses [18].

(a)   All-*trans*-spheroidene (in *n*-hexane)

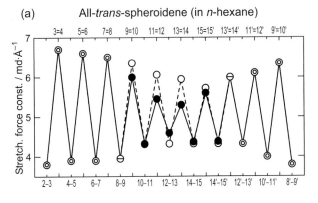

(b)   15-*cis*-spheroidene (reaction center-bound)

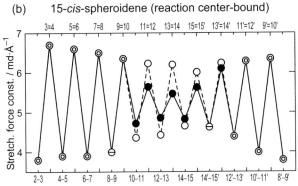

Fig. 3.16 Changes in the stretching force constants upon excitation from S_0 state (open circles) to the T_1 state (filled circles), showing the triplet-excited region for all-*trans*-spheroidene in solution (a) and for 15-*cis*-spheroidene bound to the reaction center (b). In both peripheral parts, some force constants (double circles) were assumed and could not be determined because of the limited number of deuterio species and the lack of observed Raman lines [18]. Circles with a horizontal line indicate that those force constants are the same in both the S_0 and T_1 states.

3.3.3.2 15-*cis*-Spheroidene Bound to the Reaction Center

The 15-*cis* isomer of spheroidene is bound to the reaction center from *Rhodobacter* (Rba.) *sphaeroides* 2.4.1 to give a photo-protective function, which involves the quenching of the triplet state of special-pair bacteriochlorophylls (^{3}P) to prevent the sensitized generation of harmful singlet oxygen and subsequent dissipation of the triplet energy transferred from ^{3}P (see Ref. [14] for a review). As will be described in the next section, the triplet-state isomerization of 15-*cis*-spheroidene bound to the reaction center plays the key role in dissipating the triplet energy.

Figure 3.17 shows the S_0 and T_1 Raman spectra of undeuterated 15-*cis*-spheroidene bound to the native reaction center from *Rba. sphaeroides* 2.4.1 (wild type) and variously deuterated spheroidenes incorporated into the reaction center from

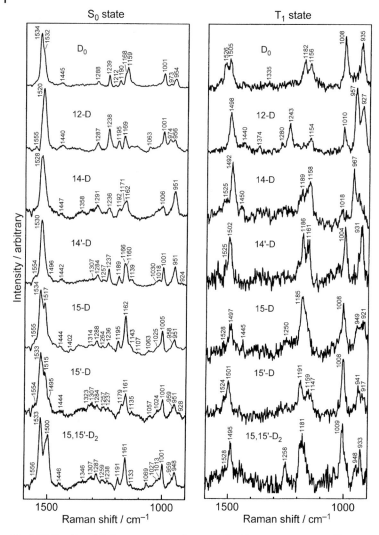

Fig. 3.17 The S_0 and T_1 Raman spectra of undeuterated 15-*cis*-spheroidene bound to the reaction center from *Rba. sphaeroides* 2.4.1 (wild type) and various deuterio derivatives incorporated into the reaction center from *Rba. sphaeroides* R26 (carotenoidless mutant). The T_1 state was generated by excitation of the bacteriochlorophyll Q_x absorption at 600 nm and subsequent triplet-energy transfer to the carotenoid, and the T_1 Raman spectra were recorded by the use of 532 nm pulses [18].

Rba. sphaeroides R26 (carotenoidless mutant). The reaction center-bound D_0 spheroidene exhibits unique spectral features in both the S_0 and T_1 states: In the S_0 state, it exhibits a key Raman line of the 15-*cis* configuration (at 1239 cm^{-1}), which is a coupled mode of the C15–H and C15′–H in-plane bendings. The Raman profiles of other normal modes are similar to the case of all-*trans*-spheroidene in solution, except for small splittings of the C=C and C–C stretching Raman lines. The results show that the reaction center-bound spheroidene takes a planar 15-*cis* configuration in the S_0 state. In the T_1 state, however, the C=C and C–C stretching Raman lines (~1520 cm^{-1} and ~1170 cm^{-1}) widely split into two components, the key Raman line of the 15-*cis* configuration (1239 cm^{-1}) disappears, and the methyl in-plane rocking (1008 cm^{-1}) and the C–H out-of-plane wagging (935 cm^{-1}) Raman lines are strongly enhanced in intensity. The results indicate the presence of *large twistings* of the conjugated chain in the T_1 state. The distorted structure of spheroidene is probably due to a result of compromise between the triplet-state isomerization from 15-*cis* toward all-*trans* and the barrier due to the wall of the binding pocket of the apo-complex preventing such isomerization.

After rough adjustment of the carbon–carbon stretching force constants we performed test calculations of normal vibrations to examine the effects of rotations around each of the C15=C15′, C11=C12, and C13=C14 bonds on the frequencies of the relevant normal modes. The rotations varied the coupling of vibrational modes among the C=C and C–C stretchings, the C–H in-plane bendings and out-of-plane waggings. Finally, the set of relevant force constants around the central part was determined by least-square fitting based on the most probable model.

Figure 3.16B shows the carbon–carbon stretching force constants thus determined. They are similar to, but different from, those determined for all-*trans*-spheroidene in solution. The decrease in the C=C stretching force constants is in the order C13=C14 > C11=C12 > C15=C15′ > C13′=C14′, whereas the increase in the carbon–carbon stretching force constants is in the order, C12–C13 > C10–C11 > C14–C15. The changes in the set of carbon–carbon stretching force constants in the reaction center-bound 15-*cis*-spheroidene, which are larger than those for all-*trans*-spheroidene free in solution, may reflect a substantial twisting of the conjugated chain. The rotational angles around the *cis* C15=C15′, *trans* C13=C14, and *trans* C11=C12 bonds in the reaction center-bound 15-*cis*-spheroidene in the T_1 state were determined to be (+45°, −30°, and +30°), respectively.

3.3.4
Conformational Changes and the Inversion of Spin-Polarization Identified by Low-Temperature Electron Paramagnetic Resonance Spectroscopy of the Reaction Center-Bound 15-*cis*-Spheroidene: A Hypothetical Mechanism of Triplet-Energy Dissipation [19]

Figure 3.18 shows the time-resolved Electron Paramagnetic Resonance (EPR) spectra of the reaction center from *Rba. sphaeroides* 2.4.1 at low temperatures. At 65 K, the initial spectrum (0.0 μs) can be ascribed to ^{3}P. After its decay, two different spectral patterns ascribable to the T_1 species of spheroidene appear; they are called ^{3}Car(I) and ^{3}Car(II). At higher temperatures, the contribution of ^{3}P becomes much smaller, but those triplet species of carotenoid exhibit basically the same time-resolved spectra.

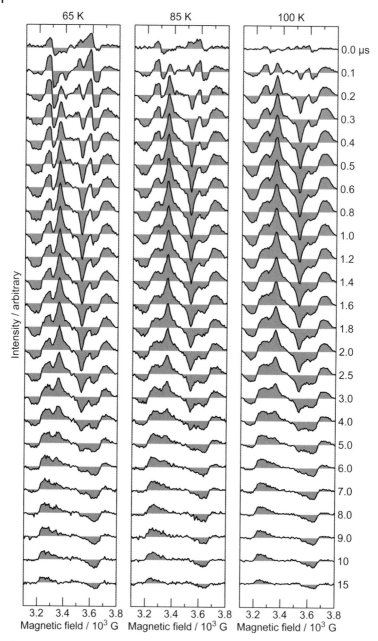

Fig. 3.18 Submicrosecond time-resolved EPR spectra of the reaction center from *Rba. sphaeroides* 2.4.1, recorded at low temperatures, after excitation of 15-*cis*-spheroidene by the use of 532 nm, ~6 ns pulses [19].

In the spectral analyses, first we applied SVD followed by a three-component global fitting to the data matrices, assuming a sequential transformation of $^3P \rightarrow$ $^3Car(I) \rightarrow {}^3Car(II)$; the results at 100 K are shown in Fig. 3.19A. The global fitting was successful, but we suspected that leakage of the triplet population may take

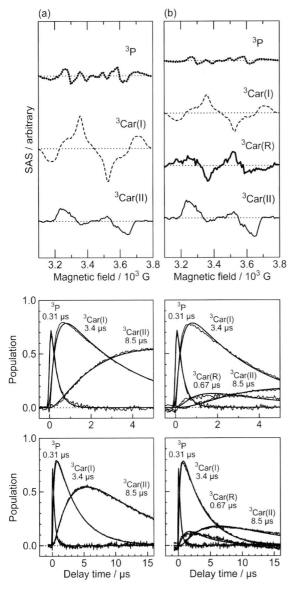

Fig. 3.19 The results of the SVD and global-fitting analysis of time-resolved EPR data matrix at 100 K. A three-component analysis (A) incorporated the special-pair bacterio-chlorophylls in the T_1 state (^3P) as well as two ($^3Car(I)$ and $^3Car(II)$) triplet species of spheroidene, whereas a four-component analysis (B) incorporated ^3P as well as three ($^3Car(I)$, $^3Car(R)$, and $^3Car(II)$) triplet species of spheroidene. The schemes used in the analyses concerning the triplet species of spheroidene are shown in Fig. 3.20 [19].

place in the process of ^{3}Car(I) → ^{3}Car(II) transformation, because the intensity of ^{3}Car(II) was reduced to approximately one-third when compared with that of ^{3}Car(I). Second, we tried a four-component global fitting, assuming an intermediate, ^{3}Car(R), that intervenes in the ^{3}Car(I) → ^{3}Car(II) transformation and functions as a leak channel. The results are shown in Fig. 3.19B. Then, another spectral pattern to be assigned to ^{3}Car(R) emerged, and the three triplet species of the carotenoid gave rise to similar intensities. The agreement between the observed and the simulated time profiles is almost equal between the three-component and four-component analyses. The results strongly suggest that the above pair of models are actually different expressions of the same physical phenomenon.

Figure 3.20 shows the schemes and the relevant time constants determined by the three-component (Fig. 3.20A) and the four-component analyses (Fig. 3.20B). In the former, the leakage of the triplet population takes place during the transformation from ^{3}Car(I) to ^{3}Car(II). The most important finding of leakage of triplet population is implicitly incorporated (see broken downward arrows) and reflected by the lowered intensity of ^{3}Car(II) (see Fig. 3.19A, the upper panel). In the latter, the leakage takes place from a representative intermediate, ^{3}Car(R). The dissipation of triplet energy is explicitly incorporated in terms of this leak channel, causing a fast decay of triplet population (a doubly downward arrow).

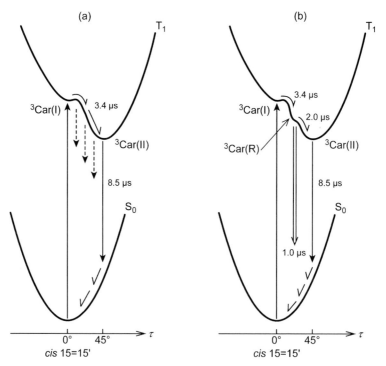

Fig. 3.20 Kinetic schemes used in the three-component (A) and four-component (B) global-fitting analyses, the results of which are shown in Fig. 3.19. See text and Ref. [19] for the details.

Third, we tried to determine the conformations of the conjugated chain in the three triplet species by the use of the zero-field splitting parameters, $|D|$ and $|E|$, that were derived from their spectral patterns depicted in Fig. 3.19B (upper panel) as species-associated spectra (SAS). The parameters $|D|$ and $|E|$ reflect the structure of a triplet species as deviation from spherical and spheroidal symmetry, respectively. We tried to determine the conformations of the conjugated chain in those triplet species, based on the theoretical calculation of zero-field splitting parameters. Table 3.3 compares the values of $|D|$ and $|E|$ spectroscopically determined for ^{3}Car(I), ^{3}Car(R), and ^{3}Car(II) with those simulated for the three different conformations around the *cis* C15=C15′, *trans* C13=C14, and *trans* C11=C12 double bonds. Assuming the conformation of ^{3}Car(I) generated immediately after excitation to be (0°, 0°, 0°), the conformations of ^{3}Car(R) and ^{3}Car(II) were determined to be (+20°, –20°, +20°) and (+45°, –40°, +40°), respectively. The fitting of the $|E|$ values is satisfactory, but the $|D|$ values tend to be higher in both models for some reason. These conformations are depicted in Fig. 3.21.

Table 3.3 Zero-field splitting parameters observed in ^{3}Car(I), ^{3}Car(R), and ^{3}Car(II) and those simulated in models with the rotational angles around (C15=C15′, C13=C14, C11=C12) bonds as specified for the reaction center-bound 15-*cis*-spheroidene [19].

	^{3}Car(I)	^{3}Car(R)	^{3}Car(II)
$|D|$	1[a]	0.90	0.73
$|E|$	1	0.89	0.50
	(0°, 0°, 0°)	(+20°, –20°, +20°)	(+45°, –40°, +40°)
$|D|$	1	0.98	0.82
$|E|$	1	0.89	0.50

a The $|D|$ and $|E|$ values in ^{3}Car(I) are normalized.

Finally, we tried to find the reason why such conformational changes can cause the dissipation of triplet population (energy) by examining the time-dependent changes in spin polarization (the difference in the populations on the spin sublevels, that is, $N_0 - N_{+1} (N_{-1})$). When the triplet energy is transferred from ^{3}P that has been generated by the radical-pair mechanism [20], the initial population on the spin sublevels should be $N_0 = 1$ and $N_{+1} = N_{-1} = 0$. Then, N_0 must decay to zero, keeping $N_{+1} = N_{-1}$ to be zero. Therefore, the spin polarization of $N_0 - N_{+1} (N_{-1})$ can never become negative in principle.

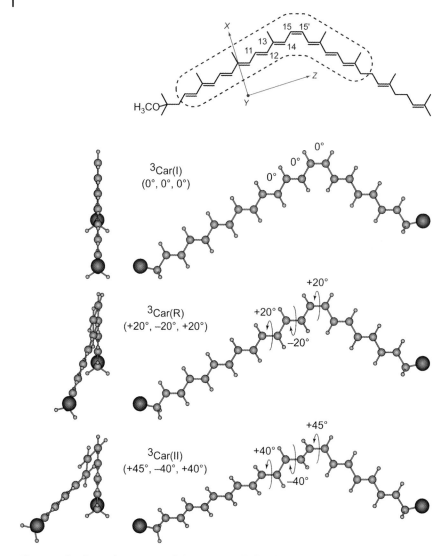

Fig. 3.21 The chemical structure and the principal axes of spheroidene (top), and the conformation of ³Car(I) assumed and those of ³Car(R) and ³Car(II) determined by comparison between the observed and the calculated zero-field splitting parameters, |D| and |E| (bottom 3); see Table 3.3 for the comparison of the zero-field splitting parameters [19].

The spectrum of each triplet species originates from an assembly of molecules in all the different spherical orientations. We decomposed the spectra of ^{3}Car(I), ^{3}Car(R), and ^{3}Car(II) into the $x–x'$, $y–y'$, and $z–z'$ components, taking a certain opening angle of a cone around the X, Y, and Z axes, and determined the scaling factors with the + or – sign to fit the observed spectra in reference to a set of simulated unit spectra along each direction. We used the scaling factors thus determined as a measure of spin polarization, and determined its time profile by the use of time-dependent changes in population of each triplet species (the lower panels of Fig. 3.19B). The results for each component and a sum of them are presented in Fig. 3.22.

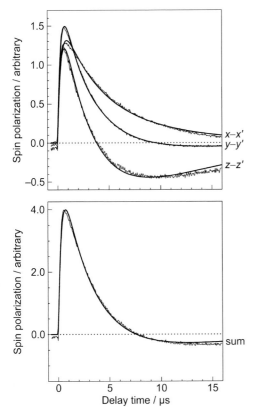

Fig. 3.22 Time-dependent changes in spin polarization, i.e. $N_0 – N_{+1}$ (N_{-1}), for the $x–x'$, $y–y'$, and $z–z'$ components and a sum of them [19].

We found something totally unexpected, but it turned out that the observation was along the line of our hypothetical mechanism of triplet-energy dissipation that accompanies the conformational changes of the triplet carotenoid. Figure 3.22 shows the time-dependent changes in spin polarization, which can be characterized as follows: (1) The time profile of a sum of the three components shows that the inversion of spin polarization *does take place* contrary to our expectation. The only mechanism we can think of to explain this observation is spin-orbit coupling. (2) Concerning the three components, the strongest inversion of the spin polarization takes place along the Z axis, which is approximately in the same direction as that of the C11=C12, C13=C14, and C15=C15′ double bonds (see Fig. 3.21), around which the rotational motions take place during the transformation of ^{3}Car(I) → ^{3}Car(R) → ^{3}Car(II). The results strongly suggest that the orbital angular momentum generated by the conformational changes is the origin of the unexpected change in spin polarization (or, in other words, in spin angular momentum). (3) The timing in the inversion of spin polarization along the Z axis (3.8 μs) approximately agrees with that in the generation of ^{3}Car(R) (3.4 μs as shown in Fig. 3.20B).

The results of this investigation, including (a) the conformational changes around the central double bonds, (b) the leak channel of triplet population, and (c) the inversion of the spin polarization during the conformational changes, have provided us with evidence for the hypothetical mechanism of triplet-energy dissipation we proposed previously [17,18,21,22]: "The rotational motions around the central double bonds cause a change in the orbital angular momentum, and through the spin-orbit coupling, a change in the spin angular momentum which facilitates the T_1 → S_0 intersystem crossing accompanying the triplet-energy dissipation."

3.3.5
Summary and Future Trends

In *β*-carotene, the quantum yield of CTI via the T_1 state has been determined to be in the order 15-*cis* (0.98) > 13-*cis* (0.87) > 9-*cis* (0.15) > 7-*cis* (0.12). The quantum yield of isomerization starting from the 15-*cis* isomer was extremely high so that 15-*cis* T_1 could be detected by neither electronic absorption nor Raman spectroscopy. The large difference in the quantum yield found between the central-*cis* and the peripheral-*cis* isomers supported the idea of the triplet-excited region that is localized in the central part of the conjugated chain. In spheroidene, on the other hand, the quantum yield of triplet-state isomerization was in the order, 15-*cis* (0.60) > 13-*cis* (0.52) > 9-*cis* (0.50) > 13′-*cis* (0.48). The lower quantum yields of CTI in spheroidene (in contrast to those in *β*-carotene) may be ascribable to the large peripheral groups hanging on both sides of the conjugated chain that give rise to a larger moment of inertia. The lower quantum yields facilitated the determination of time constants of CTI in the T_1 state of this particular carotenoid by electronic absorption spectroscopy (see Table 3.2); they were determined to be in the order, 15-*cis* (0.56 μs) < 13-*cis* (0.77 μs) < 9-*cis* (0.83 μs) < 13′-*cis* (0.91 μs). Importantly, the lifetimes of the *cis* T_1 species (0.83 μs) are in the same order as the time constants of CTI in the T_1 state, an observation that strongly supports the idea that the rotational motions trigger the T_1 → S_0 relaxation. Further, the lifetimes of the *cis* T_1 species are much shorter than the lifetime of the all-*trans* T_1 (4.76 μs), a big advantage in the dissipation of triplet energy. In both carotenoids, the quantum yield of isomerization is the highest in the 15-*cis* isomer; this is most probably the reason why this particular isomer has been widely selected by the reaction centers of photosynthetic bacteria and higher plants for photo-protective function [14].

The triplet-excited region has been spectroscopically identified in terms of carbon–carbon stretching force constants in all-*trans*-spheroidene in solution and in 15-*cis*-spheroidene bound to the bacterial reaction center (Fig. 3.16). Large changes in bond order are actually seen in the central part of the conjugated chain in both cases. The conformation of the reaction center-bound 15-*cis*-spheroidene in the T_1 state was determined to be (+45°, –30°, +30°) around the *cis* C15=C15′, *trans* C13=C14 and C11=C12 bonds.

A series of conformational changes was identified by EPR spectroscopy in the reaction center-bound 15-*cis*-spheroidene in the T_1 state, that is, ^{3}Car(I), (0°, 0°, 0°) → ^{3}Car(R), (+20°, –20°, +20°) → ^{3}Car(II), (+45°, –40°, +40°). The conformation of ^{3}Car(II) is similar to that of the T_1 species determined by Raman spectroscopy and normal coordinate analysis mentioned above. The intermediate, ^{3}Car(R), is a representative decay channel of triplet population. A large inversion of spin polarization, that is, $N_0 - N_{+1}$ (N_{-1}) < 0, was found in the Z direction of the principal axes. The direction and the timing in the inversion of spin polarization are in accord with those of rotations around the central double bonds. The results strongly support the long-standing hypothetical mechanism of triplet-energy dissipation. A theoretical proof of this mechanism is absolutely necessary.

3.4
Spectroscopic and Analytical Techniques for Studying *Cis-Trans* Isomerization in the T_1 State

3.4.1
Spectroscopic Techniques: Electronic Absorption, Raman, and Magnetic Resonance Spectroscopies

Transient absorption spectroscopy has been a conventional technique for studying the configurational (conformational) changes that are reflected in the $T_n \leftarrow T_1$ optical transition. Picosecond to microsecond time-resolved, pump and probe measurement has been widely used to examine the triplet-state isomerizations as far as each *cis*-T_1 and the all-*trans*-T_1 species are selectively observed. Even when their wavelengths are very similar to one another, the SVD and global-fitting analysis can successfully identify a set of triplet species appearing in different timescales, when a correct kinetic model is built.

Time-resolved (transient) Raman spectroscopy is unique in the sense that it provides structural information on the excited states. It is straightforward to identify a structural difference, although it is a tedious procedure to determine the structure. In order to increase the number of observed normal frequencies, isotope substitution is absolutely necessary; therefore, support by organic chemist(s) is inevitable. Normal coordinate analysis, including the empirical assignment of Raman lines and the adjustment of a set of force constants to reproduce the observed Raman frequencies and normal modes, needs some experience in this particular field. However, a combination of Raman spectra of a large number of isotope-substituted species and normal coordinate analysis provides us with reliable information concerning the bond order and the conformation in the triplet-excited region.

Time-resolved EPR spectroscopy is particularly useful for selective observation of triplet species. By the use of the zero-field splitting parameters $|D|$ and $|E|$ obtained from the spectral pattern of each triplet species, the actual conformation can be predicted reasonably well, when those parameters are simulated by theoretical calculation. The time-dependent changes in the spin polarization along the x, y, and z principal axes can provide information concerning the spin-orbit coupling in each direction.

In order to determine the quantum yields of isomerization via the T_1 state, the quantitative analysis, by HPLC, of time-dependent changes in the compositions of the starting and the product isomers becomes necessary. When a new isomer is found, its structure needs to be determined by ^1H and ^{13}C NMR spectroscopy using changes in the ^1H and ^{13}C chemical shifts in reference to those of the all-*trans* isomer ("isomerization shifts") and the $^1H–^1H$ NOE correlations.

3.4.2
A Useful Analytical Technique: Singular-Value Decomposition Followed by Global Fitting [23–25]

As exemplified in this chapter, the SVD and global-fitting analysis is a powerful method to extract, from time-resolved spectra, the major spectral patterns of the relevant triplet species as "species-associated (difference) spectra (SADS or SAS)" and their dynamics as "time-dependent changes in population." In the analysis, the noise level can be effectively removed as minor components. The analytical process consists of three steps: First, one collects time-resolved spectra consisting of a large number of data points along the spectral and time axes, or in other words, an intensity data matrix D as functions of wavelength and time. Second, one applies SVD to express it in a form $D = SVT^T$, where D is the $m \times n$ data matrix in which the number of rows m is greater than or equal to the number of column n, S is an $m \times n$ matrix called the "basis spectra", V is an $n \times n$ diagonal matrix ("a set of V_i values"), and T^T is the inverse of an $n \times n$ orthogonal matrix called "time profiles." Finally, a set of time constants is fit by the least-square method on the basis of a dynamical scheme to obtain SADS (SAS) and the time-dependent changes in population mentioned above. The most important point of this analysis is that not only the time profile (change in intensity as a function of time) but also the spectral profile (change in intensity as a function of wavelength) of each component are used. The pair of constraints once enabled the determination of the decay time constant even ~10 fs based on a data set that were collected by the use of the ~100 fs pump and probe pulses [2].

The principles of SVD followed by global fitting are briefly described below taking the case of electronic absorption spectroscopy as an example: The apparent optical density, $D(\lambda,t)$, of a system at wavelength λ and at time t can be given as

$$D(\lambda, t) = \sum_{i=1}^{p} [\varepsilon_i(\lambda) - \varepsilon_0(\lambda)] n_i(t) + R(\lambda, t) \tag{1}$$

where $\varepsilon_0(\lambda)$ is the molar extinction coefficient of the ground state, $\varepsilon_i(\lambda)$ is the molar extinction coefficient of the i-th excited state, and $R(\lambda,t)$ is noise due to environmental fluctuation. $\varepsilon_i(\lambda)$ is understood to include both effects of absorption and stimulated emission. Here, the optical pathlength is omitted for simplicity.

From the SVD theorem [23], it follows that

$$\sum_{i=1}^{p} [\varepsilon_i(\lambda_a) - \varepsilon_0(\lambda_a)] \tilde{n}_i(t_\beta) + \tilde{R}(\lambda_a, t_\beta) = \sum_{i} S_{ai} V_i T_{\beta i} \tag{2}$$

$$\sum_{a} S_{ai} S_{aj} = \delta_{ij}, \quad \sum_{\beta} T_{\beta i} T_{\beta j} = \delta_{ij} \tag{3}$$

where V_i are positive numbers arranged in the decreasing order ($V_1 > V_2 > ...$). When s_i is defined by $(S_{1i}, S_{2i}, ...)^T$ and t_i by $(T_{1i}, T_{2i}, ...)^T$, s_i and t_i are regarded as the eigenvectors of $\tilde{D}\tilde{D}^T$ and $\tilde{D}^T\tilde{D}$, respectively, the eigenvalues being V_i^2. Equation (3) shows that the sets of s_i ($i = 1, ..., p$) and t_i ($i = 1, ..., p$) constitute orthogonal basis sets in the N_λ and N_t-dimensional vector space, where N_λ and N_t denote the number of sampling points along the wavelength and the time axes. (Vectors s_i and t_i are called "basis spectra" and "time profiles.") In general, the time profile vector t_i originating from signals can be distinguished by autocorrelations or by power spectra from those originating from random noise. In most cases, the first p eigenvectors having larger eigenvalues ($V_1, ..., V_p$) can be regarded as signals. Then, we have

$$\sum_{i=1}^{p} [\varepsilon_i(\lambda_a) - \varepsilon_0(\lambda_a)] \tilde{n}_i(t_\beta) = \sum_{i=1}^{p} S_{ai} V_i T_{\beta i} \tag{4}$$

Thus, SVD serves to distinguish the signal from the noise components. With the help of the orthogonal relation of s_i shown in Eq. (3), Eq. (4) leads to

$$V_j T_{\beta j} = \sum_{i=1}^{p} \tilde{n}_i(t_\beta) C_{ij} \tag{5}$$

with

$$C_{ij} = \sum_{a} [\varepsilon_i(\lambda_a) - \varepsilon_0(\lambda_a)] S_{aj} \tag{6}$$

where C_{ij} is generally called "the C matrix." Equation (5) shows that weighted time profile $V_j t_i$ ($i = 1, ..., p$) can be represented by a linear combination of population $\tilde{n}_i [\equiv (\tilde{n}_i(t_1), \tilde{n}_i(t_2), ...)^T]$ and the C matrix; this provides us with a method to determine the decay time constants, k_i. The values of $k_1, k_2, ...$ and the components in the C matrix ($C_{11}, C_{12}, ...$) can be determined by minimizing a function, f, defined as follows:

$$f(k_1, k_2, ..., C_{11}, C_{12}, ...) = \sum_{j=1}^{p} \sum_{\beta} \left(V_j T_{\beta j} - \sum_{i=1}^{p} \tilde{n}_i(t_\beta) C_{ij} \right)^2 \tag{7}$$

This procedure is called "global fitting." The origin of t_β is also taken with a parameter, but for simplicity this parameter is omitted here. Since f is a nonlinear function of k_i values, it can have many false minima. In order to avoid such false minima and to choose the true minimum, we need to obtain f that is closest to zero; then Eq. (5) will be satisfied. Substituting Eq. (5) into the right-hand side of Eq. (4), and equating the coefficients of $\tilde{n}_i(t_\beta)$ of both sides, we obtain

$$\varepsilon_i(\lambda_a) - \varepsilon_0(\lambda_a) = \sum_{j=1}^{p} C_{ij} S_{aj} \tag{8}$$

The difference spectra thus obtained are called "species-associated difference spectra (SADS)."

When the decay time constants are determined by SVD and global fitting, Eq. (5) can be rewritten as

$$V_j t_j = \sum_{i=1}^{p} \tilde{n}_i(k_1, ..., k_p) C_{ij} \tag{9}$$

or equivalently as

$$\sum_{j=1}^{p} V_j t_j C_{ji}^{-1} = \tilde{n}_i(k_1, ..., k_p) \tag{10}$$

In the latter expression, C^{-1} is the inverse matrix of C, and the arguments $k_1, k_2, ...$ of $\tilde{n}_i$ are added to explicitly show the dependence of $\tilde{n}_i$ on those time constants. This equation means that all the vectors $\tilde{n}_i(k_1, ..., k_p)$ ($i = 1, ..., p$) should be on the hyperplane spanned by the vectors, $t_1, ..., t_p$. Thus, the procedure of global fitting is to find a set of values of $k_1, ..., k_p$, so that all the $\tilde{n}_i$ vectors are on the particular hyperplane. Generally speaking, since each vector component of $\tilde{n}_i$ is a nonlinear function of k_i, SVD and global fitting will be able to find a set of satisfactory values of k_i only when the assumed model is correct.

The following are the necessary conditions to make the SVD and global fitting analysis successful: (1) A very high signal-to-noise ratio of the spectral data for analysis is a prerequisite to apply this method successfully; this determines the number of signal components clearly distinguished from the noise components after the SVD of the time-resolved spectra. (2) A correct kinetic model is absolutely essential for complete global fitting; in other words, the complete fitting will *never* be achieved based on a wrong model. (3) The correct choice of the initial-guess values (k_i and C_{ij}) is another key factor in order to obtain the f value closest to zero, or in other words, to achieve a complete global fitting. (4) Shorter durations of the pump and probe pulses, relative to the decay time constant of the shortest-lived species in question, enhance its spectral contribution in the time-resolved spectra, and strongly facilitates the SVD and global-fitting analysis. However, it should be mentioned that a temporally short (spectrally broad) pump pulse can cause nonspecific excitation.

References

1 Y. Koyama, Y. Mukai, In *Biomolecular Spectroscopy Part B*, R. J. H. Clark, R. E. Hester, Eds. John Wiley & Sons, Chichester, **1993**, 49–137.

2 Y. Koyama, F. S. Rondonuwu, R. Fujii, Y. Watanabe, *Biopolymers* **2004**, *74*, 2–18.

3 Y. Koyama, Y. Mukai, M. Kuki, In *Laser Spectroscopy of Biomolecules*, J. E. Korppi-Tommola, Ed. SPIE – The International Society for Optical Engineering, Bellingham, **1993**, *1921*, 191–202.

4 M. Kuki, Y. Koyama, H. Nagae, *J. Phys. Chem.* **1991**, *95*, 7171–7180.

5 Y. Mukai, Y. Koyama, Y. Hirata, N. Mataga, *J. Phys. Chem.* **1988**, *92*, 4649–4653.

6 M. M. Fisher, K. Weiss, *Photochem. Photobiol.* **1974**, *20*, 423–432.

7 H. Hamaguchi, H. Okamoto, M. Tasumi, Y. Mukai, Y. Koyama, *Chem. Phys. Lett.* **1984**, *107*, 355–359.

8 Y. Mukai, H. Hashimoto, Y. Koyama, *J. Phys. Chem.* **1990**, *94*, 4042–4051.

9 Y. Mukai, M. Abe, Y. Katsuta, S. Tomozoe, M. Ito, Y. Koyama, *J. Phys. Chem.* **1995**, *99*, 7160–7171.

10 S. Saito, M. Tasumi, *J. Raman Spectrosc.* **1983**, *14*, 236–245.

11 I. Ohmine, K. Morokuma, *J. Chem. Phys.* **1980**, *73*, 1907–1917.

12 I. Ohmine, K. Morokuma, *J. Chem. Phys.* **1981**, *74*, 564–569.

13 W. H. Waddell, R. Crouch, K. Nakanishi, N. J. Turro, *J. Am. Chem. Soc.* **1976**, *98*, 4189–4192.

14 Y. Koyama, R. Fujii, In *The Photochemistry of Carotenoids*, H. A. Frank, A. J. Young, G. Britton, R. J. Cogdell, Eds. Kluwer Academic Publishers, Dordrecht, **1999**, 161–188.

15 H. Hashimoto, Y. Koyama, K. Ichimura, T. Kobayashi, *Chem. Phys. Lett.* **1989**, *162*, 517–522.

16 H. Hashimoto, Y. Koyama, *J. Phys. Chem.* **1988**, *92*, 2101–2108.

17 R. Fujii, K. Furuichi, J.-P. Zhang, H. Nagae, H. Hashimoto, Y. Koyama, *J. Phys. Chem. A* **2002**, *106*, 2410–2421.

18 Y. Mukai-Kuroda, R. Fujii, N. Ko-chi, T. Sashima, Y. Koyama, M. Abe, R. Gebhard, I. van der Hoef, J. Lugtenburg, *J. Phys. Chem. A* **2002**, *106*, 3566–3579.

19 Y. Kakitani, R. Fujii, Y. Koyama, H. Nagae, L. Walker, B. Salter, A. Angerhofer, *Biochemistry* **2006**, *45*, 2053–2062.

20 M. C. Thurnauer, J. J. Katz, J. R. Norris, *Proc. Natl Acad. Sci. USA* **1975**, *72*, 3270–3274.

21 Y. Koyama, *J. Photochem. Photobiol. B: Biol.* **1991**, *9*, 265–280.

22 N. Ohashi, N. Ko-chi, M. Kuki, T. Shimamura, R. J. Cogdell, Y. Koyama, *Biospectroscopy* **1996**, *2*, 59–69.

23 F. Chatelin, *Eigenvalues of Matrices*, John Wiley & Sons, Chichester, **1993**.

24 E. R. Henry, J. Hofrichter, *Methods Enzymol.* **1992**, *210*, 129–192.

25 R. A. Goldbeck, D. S. Kliger, *Methods Enzymol.* **1993**, *226*, 147–177.

4

Retinal Binding Proteins

Hideki Kandori

4.1
Retinal Chromophore in Rhodopsins

4.1.1
Specific Color Regulation of the Retinal Chromophore in Protein

The human eye (Fig. 4.1A) is an excellent light sensor, possessing high photosensitivity, low noise, wide dynamic range, high spatial and time resolution, etc. Visual excitation is initiated by light absorption through visual pigments present in our eyes [1–10]. Therefore, the excellent light-sensing ability of our vision is due to the properties of the light-sensor proteins in the retina. Protonated 11-*cis*-retinal Schiff base is a chromophore molecule in visual rhodopsins (Fig. 4.1B). Protonated retinal Schiff base of the all-*trans* form is a chromophore of archaeal rhodopsins that acts as an ion pump or light sensor in some archaea (bacteria) (Fig. 4.2) [11–17]. Visual and archaeal rhodopsins are retinal biding proteins that convert light into signals or energy.

   In rhodopsins, the retinal (vitamin A aldehyde) molecule forms the Schiff base linkage with a side-chain of a lysine residue, which is a covalent bond between the chromophore and protein [2]. The Schiff base is protonated in most rhodopsins, and the protonation plays a crucial role in color regulation. It should be noted that the retinal Schiff base absorbs in the UV region (λ_{max} ~360 nm), and the absorption is not very sensitive to the environment. In contrast, the protonated Schiff base of retinal exhibits a large variation of absorption that covers the visible region. In other words, the "visible" region has been determined by the color tuning of our visual rhodopsins. Humans have four photoreceptive proteins: one works for twilight vision, and the other three work for color vision [18]. The former is present in rod cells, and its photoreceptive protein is called "rhodopsin" (λ_{max} ~500 nm). The latter are present in cone cells, and are named after the colors they absorb, such as "human blue" (λ_{max} ~425 nm), "human green" (λ_{max} ~530 nm), and "human red" (λ_{max} ~560 nm). Other animals have additional proteins. For exmaple, chicken has a violet-sensitive protein, and some insects have UV-sensitive proteins [6]. Although color pigments are not "rhodopsin," the word rhodopsin is often used to represent general visual pigments.

cis-trans Isomerization in Biochemistry. Edited by Christophe Dugave
Copyright © 2006 WILEY-VCH Verlag GmbH & Co. KGaA, Weinheim
ISBN: 3-527-31304-4

(a)

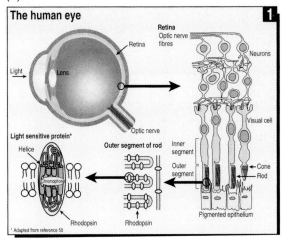

(b)

Fig. 4.1 (A) Structure of human eye. (B) The chromophore of visual rhodopsins, protonated Schiff base of 11-*cis*-retinal. This figure is modified from Kandori [5].

Wide color tuning in rhodopsins implies that the $\pi-\pi$ transition of the retinal chromophore is highly sensitive to the protein environment. It should be noted, however, that such wide color tuning is not the case for the protonated Schiff base of retinal in solution. The chromophore in solution has been used so far as a standard model system of rhodopsins, and an acid such as HCl is normally added to stabilize the protonation state of the Schiff base. The absorption spectra of the protonated Schiff base of retinal in solution are limited (λ_{max} 430–480 nm) under various counterions and solvents tested. This indicates that wide color tuning only occurs in proteins. So what is the molecular mechanism of color tuning in proteins? A working hypothesis on color tuning mechanism is as follows: "In rhodopsins, the protonation of the chromophore is stabilized by the presence of a negatively charged counterion. The mechanism of color tuning is primarily a result of the distance between the ion pair (i.e. the longer the distance, the longer the wavelength). The protein moiety regulates the distance through a specific chromophore–protein interaction, and hence specific colors are given." [5].

(a)

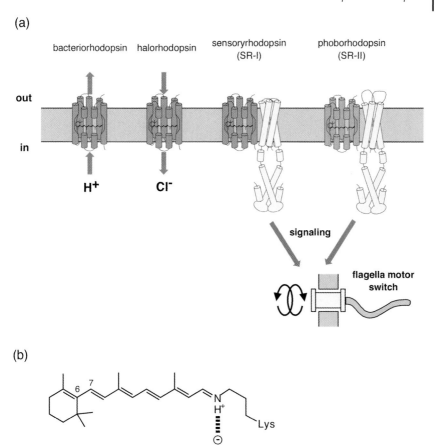

(b)

Fig. 4.2 (A) Archaeal rhodopsins. (B) The chromophore of archaeal rhodopsins, protonated Schiff base of all-*trans*-retinal.

Color originates from the energy gap of the π–π transition between electronically excited and ground states. The positive charge originally located at the Schiff base is more delocalized in the excited state, and more or less charge delocalization results in a smaller or larger energy gap, respectively. Accordingly, the presence of a counterion that localizes a positive charge gives a large energy gap (i.e. spectral blue shift). This was first tested experimentally by measuring the absorption spectra of the chromophore in various counterions and solvents, leading to the conclusion that the proximity of the anion regulates the λ_{max} of the chromophore [19]. Thus, as an intrinsic property of the chromophore, the absorption spectrum is now known to be determined by the distance between the ion pair (Figs 4.1B and 4.2B).

Figures 4.1B and 4.2B show that the isomeric compositions between visual and archaeal rhodopsins differ not only for the C11=C12 double bond, but also for the C6–C7 single bond. In visual rhodopsins, the C6–C7 bond is in a *cis* form, and the polyene and β-ionone ring is not planer because of steric hindrance between the C5-methyl group and the C8-hydrogen [9]. Consequently, the conjugation of π-electrons is not extended to the β-ionone ring. In contrast, the C6–C7 bond is in a *trans* form for archaeal rhodopsins, and the polyene and β-ionone ring is planer. Extended conjugation of π-electrons presumably contributes to the red-shifted absorption spectra in archaeal rhodopsins. In fact, bacteriorhodopsin, halorhodopsin, and sensoryrhodopsin have a λ_{max} at 570–590 nm.

4.1.2
Unique Photochemistry of the Retinal Chromophore in Protein

The photochemistry of the rhodopsin chromophore is also unique in proteins. For instance, the quantum yield of the photoreaction of rhodopsin is known to be high, which forms the molecular basis of the high sensitivity of human vision [2]. The quantum yield is essentially independent of temperature and excitation wavelength. Its fluorescence quantum yield was found to be very low ($\phi \sim 10^{-5}$), and an ultrafast reaction was inferred through barrierless excited-state potential surfaces [2,10,12]. Product formation also takes place for rhodopsins at low temperatures such as liquid nitrogen (77 K) or helium (4 K) temperatures, where molecular motions are frozen [3,13]. In the early stage of investigation, these observations questioned *cis-trans* isomerization (CTI) as the primary event in vision, because isomerization needs certain molecular motion of the chromophore. In fact, the first ultrafast spectroscopy of bovine rhodopsin led to a conclusion that favors a reaction mechanism other than isomerization as the primary reaction of rhodopsin. The history will be reviewed in the following sections, where ultrafast spectroscopy revealed that *cis-trans* photoisomerization is indeed a primary event in rhodopsins.

Figure 4.3 shows photochemical reactions in visual (Fig. 4.3A) and archaeal (Fig. 4.3B) rhodopsins. In visual rhodopsins, the 11-*cis*-retinal is isomerized into the all-*trans* form. The selectivity is 100%, and the quantum yield is 0.67 for bovine rhodopsin [20]. In archaeal rhodopsins, the all-*trans*-retinal is isomerized into the 13-*cis* form. The selectivity is 100%, and the quantum yield is 0.64 for bacteriorhodopsin [21]. Squid and octopus possess a photoisomerase called retinochrome, which supplies the 11-*cis*-retinal for their rhodopsins through the specific photoreaction. Retinochrome possesses all-*trans*-retinal as the chromophore, and the all-*trans*-retinal is isomerized into the 11-*cis* form with a selectivity of 100% [22]. Thus, the photoproduct is different between archaeal rhodopsins and retinochrome, the all-*trans* form being converted into the 13-*cis* and 11-*cis* forms, respectively. This fact implies that protein environment determines the reaction pathways of photoisomerization in their excited states.

Previous HPLC analysis revealed that the protonated Schiff base of 11-*cis*-retinal in solution is isomerized into the all-*trans* form almost predominantly, indicating that the reaction pathway in visual rhodopsins is the nature of the chromophore itself [23]. On the other hand, the quantum yield was found to be 0.15 in methanol solution [23]. Therefore, the isomerization reaction is 4–5 times more efficient in protein than in solution. Ultrafast spectroscopy directly captured the excited state dynamics of the rhodopsin chromophore, and the kinetic difference between protein and solution discovered will be presented in Section 4.2.

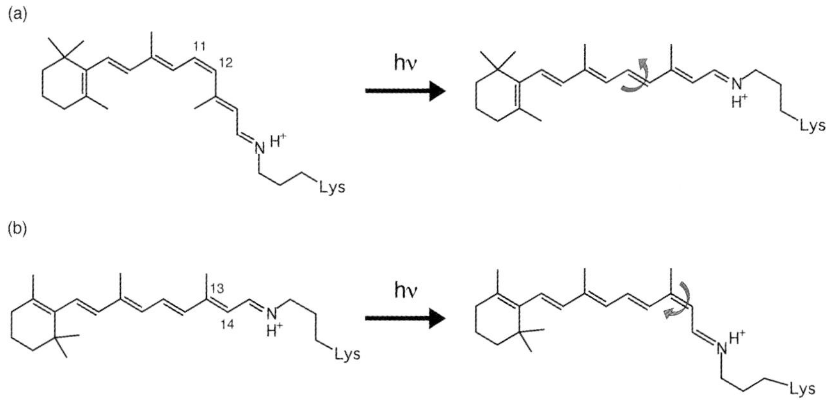

(a)

(b)

Fig. 4.3 Photochemical reactions in (A) visual and (B) archaeal rhodopsins. *Cis-trans* photoisomerization is a common reaction.

HPLC analysis also revealed that the protonated Schiff base of all-*trans*-retinal in solution is isomerized predominantly into the 11-*cis* form (82% 11-*cis*, 12% 9-*cis*, and 6% 13-*cis* in methanol) [23]. The 11-*cis* form as a photoproduct is the nature of retinochrome, not those of archaeal rhodopsins. This suggests that the protein environment of retinochrome serves as the intrinsic property of the photo-isomerization of the retinal chromophore. In contrast, it seems that the protein environment of archaeal rhodopsins forces the reaction pathway of the isomeriza-tion to change into the 13-*cis* form. In this regard, it is interesting that the quan-tum yield of bacteriorhodopsin (0.64) is 4–5 times higher than that in solution (~0.15) [21,23]. The altered excited state reaction pathways in archaeal rhodopsins never reduce the efficiency. Rather, archaeal rhodopsins discover the reaction pathway from the all-*trans* to 13-*cis* form efficiently. Consequently, the system of efficient isomerization reaction is achieved as well as in visual rhodopsins. Struc-tural and spectroscopic studies on archaeal rhodopsins are also reviewed in Sec-tion 4.3.

4.2
Photoisomerization in Visual Rhodopsins

4.2.1
Structure and Function of Visual Rhodopsins

The role of visual rhodopsins is to activate transducin, a heterotrimeric G protein, in the signal transduction cascade of vision [6,7,24]. Rhodopsin, a member of the

G protein-coupled receptor (GPCR) family, is composed of seven transmembrane helices. The 11-*cis*-retinal forms the Schiff base linkage with a lysine residue of the seventh helix (Lys296 in the case of bovine rhodopsin), and the Schiff base is protonated, which is stabilized by a negatively charged carboxylate (Glu113 in the case of bovine rhodopsin) [25,26]. The *β*-ionone ring of the retinal is coupled with hydrophobic region of opsin through hydrophobic interactions. Thus, the retinal chromophore is fixed in the retinal binding pocket of rhodopsin by three kinds of chemical bonds: covalent bond, hydrogen bond, and hydrophobic interaction.

Although the structure of the protein environment had been unknown for a long time, Okada et al. crystallized bovine rhodopsin, and its three-dimensional structure was determined with atomic resolution [27,28]. It was the first (and even now only) crystal structure of a GPCR. Figure 4.4 shows the retinal molecule and surrounding amino acids according to the crystallographic structure of bovine rhodopsin [28]. All 18 amino acid residues and a water molecule within 4.0 Å from the retinal are shown, including Lys296 that forms the Schiff base with 11-*cis*-retinal. Bulky side-chains, Ile189, Tyr191, and Trp265, sandwich the retinal vertically, while Thr118 and Tyr268 are located at both sides of the polyene chain (Fig. 4.4).

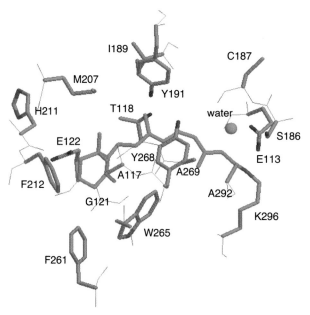

Fig. 4.4 Crystallographic structure of the retinal binding pocket of bovine rhodopsin (PDB: 1L9H) [28]. In total, 18 amino acids and a water are shown that have atoms within 4.0 Å from the retinal chromophore.

Previous site-directed mutagenesis revealed glutamate at position 113 (E113) to be the counterion in bovine rhodopsin (Fig. 4.4), which was determined by the fact that the substitution of Glu113 dramatically decreases the pK_a of the Schiff base [25,26]. Since the glutamate at this position is completely conserved among vertebrate photoreceptive pigments, the electrostatic interaction between the Schiff base (NH^+) and the glutamate (COO^-) is likely to be the common feature. It has been also suggested that water molecules may stabilize the ion-pair state at hydrophobic environment. One possible model is that a water molecule bridges the Schiff base and E113. Solid-state NMR studies reported the distance between the Schiff base nitrogen and an oxygen of E113 to be >4 Å [29,30], which favors the presence of a bridged water molecule. Nevertheless, the crystal structure of bovine rhodopsin reported the absence of the bridged water molecule, though a water molecule is present near the Schiff base (Fig. 4.4) [27,28]. The distance is 3.1 Å between the Schiff base nitrogen and the corresponding oxygen of E113 in the crystal structure. Our Fourier transform infrared (FTIR) spectroscopic study also suggested the absence of the bridged water molecule between the Schiff base and the oxygen of E113 by measuring stretching vibrations of water directly [31].

4.2.2
Primary Process in Vision Studied by Ultrafast Spectroscopy

Absorption of a photon by the chromophore induces primary photoreaction, followed by conformational changes of protein, and eventually activates transducin. This is called the "bleaching process" because rhodopsin loses its color. To investigate the primary photoreaction processes in rhodopsin, two spectroscopic approaches have been applied; low-temperature and time-resolved spectroscopies [3,10]. They can detect primary photointermediate states by reducing the thermal reaction rate at low temperature (low-temperate spectroscopy) or directly probing dynamic processes at physiological temperature (time-resolved spectroscopy). Historically, the former technique was in advance of the latter because generation of ultrashort pulses was necessary to detect primary intermediates of rhodopsin at physiological temperature, and picosecond pulses only emerged in the 1970s [3,10]. After the discovery of the chromophore (vitamin A) by George Wald in the 1930s, Yoshizawa and Kito first found a red-shifted photoproduct of bovine rhodopsin at low temperature (−186 °C), which reverted to rhodopsin by illumination [32]. On warming above −140 °C, this photoproduct (now called bathorhodopsin [3]) is converted to lumirhodopsin and finally decomposed to all-*trans*-retinal and opsin through several intermediates. Based on low-temperature spectrophotometric experiments, Yoshizawa and Wald proposed in 1963 that bathorhodopsin has a "highly constrained and distorted" all-*trans*-retinal as its chromophore and is on a higher potential energy level than rhodopsin and subsequent intermediates [33]. According to their prediction, the process of rhodopsin to bathorhodopsin should be a CTI of the chromophore.

The first challenge to the isomerization model was made in 1972 by picosecond laser photolysis [34]. The first picosecond laser photolysis observed formation of bathorhodopsin within 6 ps after excitation of bovine rhodopsin at room temperature, and interpreted that its formation would be too fast to be attributed to such a conformational change as the CTI of the retinal chromophore [34]. On the basis of both isotope effect and non-Arrhenius temperature dependence of the formation time of bathorhodopsin, the same group proposed a model that the formation of bathorhodopsin is accompanied by proton translocation [35]. Which reaction takes place in visual rhodopsins – isomerization or proton translocation? The isomerization model was favored by various low-temperature spectroscopic results [3], such as considerable angle change (26°) of the transition dipole moment between rhodopsin and bathorhodopsin [36], and no formation of bathorhodopsin for an 11-*cis*-locked rhodopsin analog [37]. Low-temperature resonance Raman spectroscopy revealed that a chromophore isomerization occurs in the rhodopsin–bathorhodopsin transformation [38]. To obtain more direct information on the reaction mechanism at physiological temperature, we studied primary processes of the rhodopsin analogs possessing 11-*cis*-locked ring retinals such as 5-, 7-, and 8-membered analogs.

In the case of 5-membered rhodopsin, only a long-lived excited state ($\tau = 85$ ps) was formed without any ground-state photoproduct (Fig. 4.5D), giving direct evidence that the CTI is the primary event in vision [39]. Excitation of 7-membered rhodopsin, on the other hand, yielded a ground-state photoproduct with a spectrum similar to photorhodopsin (Fig. 4.5C). These different results were interpreted in terms of the rotational flexibility along the C11–C12 double bond [39]. This hypothesis was further supported by the results with an 8-membered rhodopsin that possesses a more flexible ring. Upon excitation of 8-membered rhodopsin with a 21 ps pulse, two photoproducts – photorhodopsin-like and bathorhodopsin-like products – were observed (Fig. 4.5B) [40]. Photorhodopsin is a precursor of bathorhodopsin found by picosecond transient absorption spectroscopy [41]. Thus, the picosecond absorption studies directly elucidated the correlation between the primary processes of rhodopsin and the flexibility of the C11–C12 double bond of the chromophore, and we eventually concluded that the respective potential surfaces were as shown in Fig. 4.5 [10,40].

The structure of the intermediate states in Rh7 and Rh8 has been studied recently by theoretical investigation [42]. Regarding the proton translocation model, it should be noted that the excitation photon density was extremely high in the low-temperature picosecond experiments [10,35]. Therefore, the non-Arrhenius dependence of the formation rate of bathorhodopsin on temperature and the deuterium isotope effect may be results which could be detected only under intense excitation conditions. In fact, a deuterium isotope effect was not observed in the process from photorhodopsin to bathorhodopsin under weak excitation conditions [43].

Direct observation of the CTI process in real time has been performed by use of femtosecond pulses. In 1991, two groups first reported femtosecond transient absorption spectroscopy of bovine rhodopsin [44,45], but their conclusions were remarkably divergent. One group excited bovine rhodopsin with a 35 fs pulse and probed with a 10 fs pulse, and concluded that product formation completed within 200 fs [44]. In contrast, the other group measured transient absorption of bovine

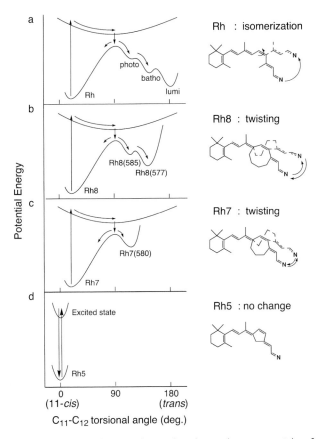

Rh : isomerization

Rh8 : twisting

Rh7 : twisting

Rh5 : no change

Potential Energy

Rh8(585)
Rh8(577)
Rh8

Rh7(580)
Rh7

Excited state

Rh5

0   90   180
(11-*cis*)   (*trans*)
C_{11}-C_{12} torsional angle (deg.)

Fig. 4.5 Schematic drawing of ground- and excited-state potential surfaces along the 11-ene torsional coordinates of the chromophore of rhodopsin (A), 8-membered rhodopsin (B), 7-membered rhodopsin (C) and 5-membered rhodopsin (D). This figure is modified from Mizukami et al. [40].

rhodopsin with 300 fs resolution, and concluded that the primary isomerized photointermediate appears in 3 ps [45]. In the following year, a different group applied femtosecond transient absorption spectroscopy to octopus rhodopsin, and reported that there are two time constants for the formation of the primary photointermediates, 400 fs and 2 ps [46].

Thus, the first trials provided rather confusing results on the primary processes of rhodopsin photoisomerization. The first group then reported several femtosecond pump probe studies, one after another, which involved the measurement of bovine rhodopsin with a wide spectral window [47] and the measurements of 9-*cis* rhodopsin [48] and 13-demethyl rhodopsin [49]. In addition, they observed oscillatory features with a period of 550 fs (60 cm^{-1}) on the kinetics at probe wavelengths within the photoproduct absorption band of rhodopsin, whose phase and amplitude demonstrate that they are the result of nonstationary vibrational motion in the ground state of photorhodopsin [50]. The observation of coherent vibrational motion in photorhodopsin supports the idea that the primary step in vision is a vibrationally coherent process and that the high quantum yield of the CTI in rhodopsin is a consequence of the extreme speed of the excited-state torsional motion [8].

All these studies with femtosecond pulses on the primary photochemical processes of rhodopsin were done by means of transient absorption (pump probe) spectroscopy [10]. However, absorption spectroscopy may not be the best way to probe the excited-state dynamics of rhodopsin, because other spectral features, such as ground-state depletion and product absorption, are possibly superimposed on the excited-state spectral features (absorption and stimulated emission) in the obtained data. Each spectral feature may even vary in the femtosecond time domain, which provides further difficulty in analyzing the data. In contrast, fluorescence spectroscopy focuses only on the excited-state processes, so that the excited-state dynamics can be observed more directly.

Thus, we attempted to apply another experimental approach, femtosecond fluorescence up-conversion spectroscopy, to detect the excited-state dynamics of bovine rhodopsin in real time [51,52]. In addition, we also measured the protonated Schiff bases of 11-*cis*-retinal and 11-*cis*-locked 5-membered ring retinal in methanol solution, which provided experimental evidence about how protein contributes to efficient isomerization [53].

Figure 4.6A shows typical fluorescence decays of bovine rhodopsin at 580 and 680 nm. Although it was known that the fluorescence quantum yield of rhodopsin is very low, accurate fluorescence detection with femtosecond time resolution enabled us to capture the excited state directly. The linear relationship between the excitation laser power and the fluorescence intensity indicates that the observed fluorescence decay indeed originated from the excited state of rhodopsin [51]. The kinetics were nonexponential throughout the observed wavelengths between 530 and 780 nm, and fitted by the fast 100–300 fs and slow 1.0–2.5 ps components [52]. While the origin of these decay components was not fully understood, the average amplitude of the femtosecond components over six wavelengths was about 70%, which is very close to the quantum yield of photoisomerization of rhodopsin (0.67). We thus concluded that the slow components (~30%) originate from the nonreactive excited state of rhodopsin, while the fast components (~70%) come from the coherent isomerization observed in the femtosecond transient absorption spectroscopy [52].

Figure 4.6B shows typical fluorescence decays of the rhodopsin chromophore, protonated Schiff base of 11-*cis*-retinal, in methanol solution at 605 and 695 nm. The kinetic features are very similar to those of rhodopsin in terms of ultrafast and nonexponential components (Fig. 4.6A), but the kinetics are considerably slower. The fluorescence lifetimes for five wavelengths obtained in the study [53] were classified by two features: the fast femtosecond (90–600 fs) and the slow picosecond (2–3 ps) components. The populations of fast and slow components were 25 and 75%, respectively. Figure 4.6C shows typical fluorescence decays of protonated Schiff base of 11-*cis*-locked 5-membered retinal in methanol solution

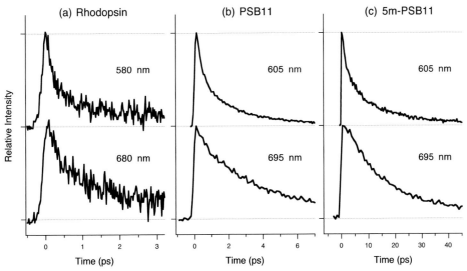

Fig. 4.6 The fluorescence decay kinetics of bovine rhodopsin (A), protonated Schiff base of 11-*cis*-retinal (PSB11) in methanol (B), and protonated Schiff base of 5-membered locked 11-*cis* retinal (5m-PSB11) in methanol (C). The data in (A) are from Kandori et al. [52], while those in (B) and (C) are from Kandori et al. [53]. Fitting parameters are as follows: (A) 0.14 (80%) and 1.5 (14%) ps at 580 nm and 0.33 (60%) and 2.0 (31%) ps at 680 nm, (B) 0.5 (59%) and 2.0 (39%) ps at 605 nm and 3.1 ps at 695 nm, (C) 2.0 (51%) and 12.6 (47%) ps at 605 nm and 15.8 ps at 695 nm.

at 605 and 695 nm. Again, the kinetic features are similar to those shown in Figs 4.6A and B, whereas the lifetimes were about 5 times longer than in Fig. 4.6B. The estimated population was about 25 and 75% for the fast and slow components, respectively, being identical to those in the protonated Schiff base of 11-*cis* retinal. We therefore interpreted both components of PSB11 to become ~5 times longer by locking the 11-*cis* form, suggesting that both fast and slow components are due to the rotational relaxation around the C11–C12 double bond.

Identification of the fast and slow components was apparently different in bovine rhodopsin (Fig. 4.6A) and the protonated Schiff base of 11-*cis*-retinal in solution (Fig. 4.6B). In the case of rhodopsin, we interpreted the results as showing that only the femtosecond component is coupled to isomerization. In contrast, both femtosecond and picosecond components are coupled to the relaxation processes of the C11–C12 double bond. It is noted, however, that the excited state of the protonated Schiff base of 11-*cis*-locked 5-membered retinal possesses a very short lifetime (<15 ps), though the 11-*cis* form being locked. This suggests that torsional relaxation of the polyene chain plays a significant role in the decay of the excited state, and torsional modes are highly coupled with each other. In solution, locking of the 11-*cis* form with a 5-membered ring probably changes such coupling dramatically, because other torsional modes are used for the decay of the excited state (Fig. 4.6C). In this context, it is interesting to compare the processes between solution and protein. In methanol, the fluorescence lifetime of the

5-membered PSB11 is 5 times longer than that of PSB11 [53], whereas in protein, the fluorescence lifetime of a rhodopsin analog possessing 5-membered retinal (85 ps) is 2 orders of magnitude longer than that of the native rhodopsin [39,51,52]. These observations suggest that the protein moiety enhances the isomerization rates of the chromophore.

4.2.3
Structural Changes of the Chromophore and Protein upon Retinal Photoisomerization

It is well known that rhodopsin is an excellent molecular switch for converting a light signal to an electric response in the photoreceptor cell. As mentioned in the preceding chapter, the highly efficient photoisomerization of rhodopsin (quantum yield 0.67) is assured by the extremely fast CTI of the chromophore that is facilitated by the protein environment [10]. How is such a reaction achieved in protein? What is the role of protein environment? To understand the role of protein, we need to understand how retinal and protein change their shapes during isomerization. One intriguing question is which side of the C11=C12 bond rotates. Figure 4.3A shows that the Schiff base side rotates, while there is no change on the side of the β-ionone ring. This makes sense because β-ionone ring is bulky, being hardly moved. However, the molecular motion of the Schiff base in Fig. 4.3A is also significant, and entire rotation must be not the case for bathorhodopsin that is formed in an ultrafast time scale.

Vibrational spectroscopy provided important information about the hydrogen-bonding strength of the Schiff base from C=N stretching vibration [54–56]. The C=N stretch in H_2O is considerably upshifted from the intrinsic stretching mode by coupling with the N–H in-plane bending vibration. The intrinsic C=N stretch can be measured in D_2O, where the coupling of the N–H bending vibration is removed. Since the frequency of the N–H in-plane bending increases if the hydrogen bond of the Schiff base is strengthened, the difference in frequency of the C=N stretch in H_2O and D_2O has been regarded as the marker of hydrogen-bonding strength of the Schiff base. A large difference corresponds to a strong hydrogen bond [54–56]. The differences in the C=NH and C=ND frequencies were reported to be identical (32 cm^{-1}) between bovine rhodopsin and bathorhodopsin, implying that there was no change of the Schiff base hydrogen bond upon retinal isomerization [57]. Therefore, the chromophore structure of the photoproduct in Fig. 4.3A is not true for bathorhodopsin. It is likely that the Schiff base is not moved between rhodopsin and bathorhodopsin.

It should be noted that the hydrogen-bonding strength is not the only determinant of N–H in-plane bending frequency, and the frequency upshift of the C=N stretch by coupling with the N–H bending is complex, so that the frequency difference of the C=N stretch in H_2O and D_2O is not necessarily a direct marker of hydrogen-bonding strength of the Schiff base. In this regard, the N–H stretch is a more direct probe of hydrogen-bonding strength of the Schiff base, where the strong hydrogen bond causes lowered frequency. Nevertheless, analysis by C=N stretches has successfully explained the Schiff base hydrogen bond so far. It is also supported by the results for N–D stretches in D_2O for bovine rhodopsin [31].

Photoisomerization of rhodopsin does not alter the Schiff base environment. Motion of the β-ionone ring side is also unlikely. How does the retinal chromophore change its shape? Low-temperature photocalorimetric studies have revealed that about 60% of light energy ($\sim$35 kcal mol^{-1}) is stored in the structure of bathorhodopsin [58]. This fact suggests that the chromophore structure of bathorhodopsin is significantly distorted in a high-energy state. Such structural deformation of the polyene chain has been studied by the analysis of hydrogen-out-of-plane (HOOP) vibrational modes at 1000–800 cm^{-1} [59,60]. It is known that conformational distortion leads to the enhancement of specific HOOP modes in both resonance Raman and infrared spectroscopy. Figure 4.7A shows the bathorhodopsin minus rhodopsin Fourier transform infrared (FTIR) spectra at 77 K, in which the stronger HOOP bands in bathorhodopsin (positive signal) is a marker of chromophore distortion. Previous resonance Raman study assigned the bands at 967 (–), 921 (+), 875 (+), and 850 (+) cm^{-1} as in-phase C11–C12, C11, C10, and C14 HOOP modes, respectively [59]. Among the observed HOOP modes, only the 850 cm^{-1} band due to C14 HOOP is H–D exchangeable. This is in clear contrast to the case in bacteriorhodopsin (Fig. 4.7B), as discussed in Section 4.3. The strong band at 921 cm^{-1} due to C11 HOOP is a clear monitor of the structural deformation at this moiety of bathorhodopsin. The positive band at 858 cm^{-1} in D_2O originates from the N–D HOOP mode, suggesting that the Schiff base is somewhat perturbed. In fact, N–D stretch in D_2O is also perturbed between rhodopsin and bathorhodopsin, though no frequency changes were observed [31].

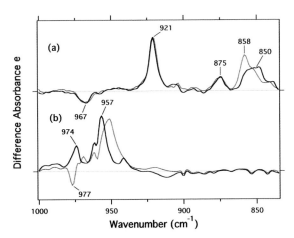

Fig. 4.7 Bathorhodopsin minus bovine rhodopsin (A) and K minus bacteriorhodopsin (B) difference infrared spectra measured at 77 K. Solid and dotted lines represent the measurements in H_2O and D_2O, respectively.

It is thus likely that light energy storage in bathorhodopsin ($\sim$35 kcal mol^{-1}) predominantly originates from the distorted chromophore structure. This is of course achieved because of the protein environment. Although protein structural changes are very limited in bathorhodopsin as discussed above, some energy storage is also caused by the structural constraint of protein. In fact, low-temperature FTIR spectroscopy observed such protein changes by monitoring peptide backbone vibrations (amide-I, -II, and -A), several side-chains, and internal water molecules [31,61,62].

4.3
Photoisomerization in Archaeal Rhodopsins

4.3.1
Structure and Function of Archaeal Rhodopsin

Halobacteria contain four rhodopsins: bacteriorhodopsin, halorhodopsin, sensoryrhodopsin, and phoborhodopsin (Fig. 4.2A) [11–17]. Bacteriorhodopsin and halorhodopsin are light-driven ion pumps, which act as an outward proton pump and an inward Cl$^-$ pump, respectively. Sensoryrhodopsin and phoborhodopsin are photoreceptors that act to produce attractant and repellent responses in phototaxis, respectively. These four archaeal rhodopsins have similar structures: seven helices constitute the transmembrane portion of the protein, and a retinal chromophore is bound to a lysine residue of the seventh helix via a protonated Schiff base linkage (Fig. 4.1). A negatively charged counterion is present to stabilize the positive charge inside the protein; the counterion is an aspartate except for in halorhodopsin, which possesses a chloride ion. In sensoryrhodopsin, interaction with a transmembrane transducer protein raises the pK_a of the aspartate, so that the aspartate is protonated at neutral pH.

X-ray crystallographic structures of three of the four archaeal rhodopsins – bacteriorhodopsin [63,64], halorhodopsin [65], and phoborhodopsin from *Natronobacterium pharaonis* [66,67] – have been determined. Figure 4.8 shows the structure of the retinal chromophore and the Schiff base region. In the case of bacteriorhodopsin, the Schiff base region has a quadrupolar structure with positive charges located at the protonated Schiff base and Arg82, and counterbalancing negative charges located at Asp85 and Asp212 (Fig. 4.8). One of the aspartates is replaced by threonine, and chloride ion is bound to the region of the chloride pump halorhodopsin [65]. While *N. pharaonis* phoborhodopsin is a light-sensor protein, the Schiff base region is similar to that of bacteriorhodopsin. In these archaeal rhodopsins, the quadrupole inside the protein is stabilized by three water molecules (water401, 402, and 406 for bacteriorhodopsin and *N. pharaonis* phoborhodopsin).

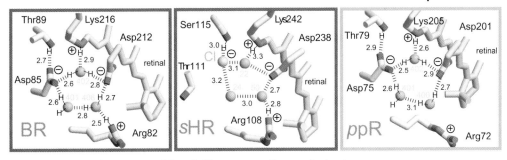

Fig. 4.8 Crystallographic structure of the Schiff base region of bacteriorhodopsin (A, PDB: 1C3W) [64], halorhodopsin of *Halobacterium salinarum* (B, PDB: 1E12) [65], and phoborhodopsin of *Natronobacterium pharaonis* (C, PDB: 1JGJ) [66]. Green spheres represent oxygen atoms of water. Dotted lines are putative hydrogen bonds, whose distances are shown in angstroms.

It should be noted that unlike visual rhodopsins, bridged water molecules are present for archaeal rhodopsins. In the case of bacteriorhodopsin, water402 is a hydrogen-bonding acceptor for the protonated Schiff base, while the same water (water402) is a hydrogen-bonding donor for Asp85. A notable structural feature is that Asp85 and Asp212 are located at similar distances from the retinal Schiff base (Fig. 4.8A), whereas the Schiff base proton is transferred only to Asp85 in microseconds. This suggests that the water molecules in the Schiff base region play important roles in the proton transfer reaction.

Figure 4.9 illustrates another structural feature of bacteriorhodopsin. The linear polyene chain of the all-*trans*-retinal is sandwiched vertically by two tryptophans, Trp86 and Trp182, while the phenol ring of Tyr185 is located parallel to the polyene chain of the retinal chromophore. Thr89 is located at the other side (not shown). The presence of three bulky groups, Trp86, Trp182, and Tyr185, presum-

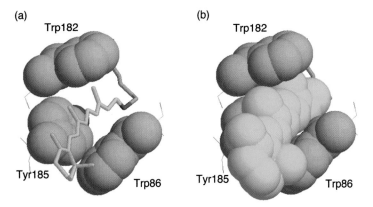

Fig. 4.9 Crystallographic structure of bacteriorhodopsin (A, PDB: 1C3W) [64]. Trp86, Trp182, and Tyr185 are shown in space-filling model together with the retinal chromophore in stick (A) or space-filling (B) drawing.

ably determines the isomerization pathway from all-*trans* to 13-*cis* form. As described above, the 11-*cis* form is the main photoproduct of the protonated Schiff base of all-*trans*-retinal in solution. Therefore, the reaction pathway is altered in the protein environment of bacteriorhodopsin, whereas the quantum yield of isomerization is 4–5 times higher in bacteriorhodopsin (0.64) than that in solution (~0.15) [21,23].

4.3.2
Primary Process in Bacterial Photosynthesis and Light Sensor Studied by Ultrafast Spectroscopy

Unlike visual rhodopsins that bleach upon illumination, archaeal rhodopsins exhibit photocycle. This is highly advantageous in ultrafast spectroscopic studies and these techniques have been extensively applied in addition to low-temperature spectroscopy [2,12,13]. In particular, bacteriorhodopsin has been regarded historically as the model system to test new spectroscopic methods. As in visual rhodopsins, the light absorption of archaeal rhodopsins causes formation of red-shifted primary intermediates [68]. The primary K intermediate can be stabilized at 77 K. Time-resolved visible spectroscopy of bacteriorhodopsin reveals the presence of the precursor, called the J intermediate [12,13]. The J intermediate is more red-shifted (λ_{max} ~625 nm) than the K intermediate (λ_{max} ~590 nm). The excited state of bacteriorhodopsin possesses blue-shifted absorption, which decays nonexponentially. The two components of the stimulated emission decay at about 200 and 500 fs [69]. The J intermediate is formed in <500 fs, and converted to the K intermediate within 3 ps [12,69].

So when does isomerization take place from the all-*trans* to 13-*cis* form? To answer this question, as for visual rhodopsin, all-*trans*-locked 5-membered retinal was incorporated into bacteriorhodopsin [70–72]. In experiments with a picosecond time resolution, an intermediate was found with properties similar to those of the J intermediate [70]. Together with the ultrafast pump probe [71] and coherent anti-Stokes Raman [72] spectroscopic results, it was concluded that isomerization around C13=C14 is not a prerequisite for producing the J intermediate. More importantly, since the J intermediate is a ground-state species, isomerization does not take place in the excited state of bacteriorhodopsin according to their interpretation [70–72]. However, other experimental data favor a common mechanism between visual and archaeal rhodopsins; namely, isomerization taking place in the excited state. Femtosecond visible-pump and infrared-probe spectroscopy showed the 13-*cis* characteristic vibrational band at 1190 cm^{-1} appearing with a time constant of ~0.5 ps, indicating that the all-*trans* to 13-*cis* isomerization takes place in femtoseconds [73]. This time scale is coincident with formation of the J intermediate. Fourier transform of the transient absorption data with <5 fs resolution also showed the appearance of the 13-*cis* form in <1 ps, supporting the suggestion that the all-*trans* to 13-*cis* isomerization takes place in femtoseconds [74]. Previous anti-Stokes resonance Raman spectroscopy proposed that the J intermediate is a vibrational hot state of the K intermediate [75]. Thus, many experimental results are consistent with the isomerization model in the excited state.

Comparative investigation of mutant proteins of bacteriorhodopsin, other archaeal rhodopsins, and the protonated Schiff base of all-*trans*-retinal in solution is useful for better understanding of the primary photoisomerization mechanism. Ultrafast spectroscopy of various bacteriorhodopsin mutants revealed that only the replacements of the charged residues reduced the photoisomerization rate, leading to less efficient photoisomerization [76]. This observation explains an important role of the electrostatic interaction of the counterion complex in the primary photoisomerization mechanism (Fig. 4.8A). The excited state is more long lived in the chloride pump halorhodopsin [77–79] and light-sensor *N. pharaonis* phoborhodopsin [80], and hence less efficient for photoisomerization. These observations suggest that bacteriorhodopsin possesses the optimized structure for the primary photoisomerization mechanism, though the structures in Fig. 4.8 look essentially similar.

As for visual rhodopsins, spectroscopic studies of the protonated Schiff base of all-*trans*-retinal in solution are important for understanding the isomerization mechanism. We first reported the excited state dynamics of the protonated Schiff base of all-*trans*-retinal in methanol solution [81], and found that the kinetics is very similar to that of the 11-*cis* form (Fig. 4.6B). The only difference was that the lifetimes are 1.2–1.4 times longer in the all-*trans* form than in the 11-*cis* form [53,81]. Slightly faster decay of the 11-*cis* form may be reflected by their molecular structures, namely the initial steric hindrance between C_{10}-H and C_{13}-CH$_3$ in the 11-*cis* form (Fig. 4.3) that accelerates the fluorescence decay. Interestingly, it was found that the all-*trans*-locked 5-membered system, which prohibits both $C11=C12$ and $C13=C14$ isomerizations, exhibits similar kinetics to those of the all-*trans* form in solution [82]. These results are entirely different from those of the 11-*cis*-locked 5-membered system, in which the excited-state lifetime is 5-times longer (Fig. 4.6B,C) [53]. This suggests more complex excited-state dynamics for the all-*trans* form. Observation of the J-like state in protein [70–72] might be correlated with such properties of the protonated Schiff base of the all-*trans* form.

4.3.3
Structural Changes of the Chromophore and Protein upon Retinal Photoisomerization

The light energy storage in archaeal rhodopsin is lower than that in bovine rhodopsin. Low-temperature photocalorimetric studies reported the energy stored in the primary intermediates to be ~35 kcal mol^{-1} and ~16 kcal mol^{-1} for bovine rhodopsin [58] and bacteriorhodopsin [83], respectively. This indicates that only 30% of light energy is stored in the structure of the primary K intermediate, which is about half that in visual rhodopsins. In other words, about 70% of light energy is dissipated in the formation of the K intermediate. Nevertheless, such energy loss may not be serious from the functional point of view, because bacteriorhodopsin pumps a single proton by use of one photon, and the free energy gain by pumping a proton is about 6 kcal mol^{-1} [2]. Interestingly, quantum yields of photoisomerization are not so different between bovine rhodopsin (0.67) and bacteriorhodopsin (0.64). Therefore, such difference in energy storage must be correlated with the structures of the primary intermediates.

In the case of archaeal rhodopsins, X-ray crystallographic structures of the primary K intermediates has been reported for bacteriorhodopsin [84–86] and *N. pharaonis* phoborhodopsin [87]. As expected, protein structures are little changed before and after isomerization. One group concluded that the energy storage in the K state of bacteriorhodopsin is almost completely explained by the distortion at position C13 [85]. However, it would be difficult to determine the bond angle accurately under the current resolution (>2.0 Å). In fact, three reported structures of the K intermediate of bacteriorhodopsin are considerably different among groups [84–86].

As described in the section on visual rhodopsin, vibrational spectroscopy, such as resonance Raman and infrared, is a powerful technique for monitoring detailed structural changes. In particular, HOOP modes are good probe of chromophore distortion. Figure 4.7B shows the K minus bacteriorhodopsin difference FTIR spectra, in which only positive bands appear in H_2O. A negative band at 977 cm^{-1} in D_2O originates from the N–D in-plane bending vibration. The most intense peak at 957 cm^{-1} exhibits isotope shift in D_2O, which was assigned as the C15 HOOP that also contains the N–H wagging mode [88]. It is interesting to note that the band at 921 cm^{-1} in bovine rhodopsin is not sensitive to H–D exchange, while the band at 957 cm^{-1} in bacteriorhodopsin downshifts in D_2O. Such comparison clearly illustrates which part in retinal is changing upon isomerization. In bacteriorhodopsin, chromophore distortion is localized at the Schiff base moiety. Bulky side-chains shown in Fig. 4.9 may lead to localization of the reaction near the Schiff base region accompanying the isomerization from the all-*trans* to 13-*cis* form.

So, is the hydrogen bond of the Schiff base changed upon photoisomerization of bacteriorhodopsin? While such analysis has been performed through the C=N stretch, we have also succeeded in measuring a more direct probe: the N–H (N–D in D_2O) stretch. In normal infrared spectroscopy of rhodopsins, measured frequencies are limited at <1800 cm^{-1}. It was difficult to obtain accurate spectra at the higher frequencies even for light-induced spectral changes. By optimizing the measuring conditions, we were able to obtain the difference spectra in the entire mid-infrared region [89]. In addition, by hydrating the sample with D_2O, H–D exchangeable X–H stretches (X=N or O) are downshifted into the 2700–2000 cm^{-1} region (Fig. 4.10A). By use of [ζ-^{15}N]lysine-labeled samples, we assigned the N–D stretches of bacteriorhodopsin and the K intermediate to about 2150 and 2490 cm^{-1}, respectively [90]. Significant upshift of the N–D stretch and parallel dipolar orientation to the membrane indicates that the hydrogen bond of the Schiff base is broken upon retinal isomerization. Similar observation was gained for *N. pharaonis* phoborhodopsin [91], whereas more complex frequency change was observed for halorhodopsin [92]. Thus, in the case of archaeal rhodopsins, retinal isomerization accompanies rotational motions of the Schiff base side, so that the hydrogen bond is significantly altered, unlike visual rhodopsins.

Figure 4.8 shows that the hydrogen-bonding acceptor of the Schiff base is a water molecule, which bridges the ion pair. Motion of the Schiff base probably enforces rearrangement of the water-containing hydrogen-bonding network in the Schiff base region. This was indeed probed by FTIR spectroscopy [93–96]. Figure 4.10A shows assignment of water O–D stretches in the K minus bacteriorhodopsin spectra by use of isotope water, in which six water bands in bacteriorhodopsin (negative side) were observed. By use of systematic mutations, we were able to conclude that these vibrations originate from three water molecules in the Schiff base region (Fig. 4.10B) [95]. Among them, it is particularly noted that the band at 2171 cm^{-1} is unusually low as water O–D stretch [94,95]. Water402 is located at a similar distance from two aspartates, Asp85 and 212, whereas the hydrogen-bonding interaction is much stronger with Asp85. An FTIR study of subsequent intermediates showed that such strong hydrogen-bonding interaction and transient changes are essential for the proton transfer reaction in bacteriorhodopsin [97,98]. It is noteworthy that strongly hydrogen-bonded water molecules in bacteriorhodopsin are generally weakenend upon retinal isomerization, as is demonstrated by the frequency upshift (Fig. 4.10A). This observation strongly suggests that hydrogen-bonding interaction is highly destabilized in the K state, which possibly contributes to the high energy state. In fact, quantum chemical/molecular mechanics calculation of the K intermediate concluded that 11 kcal mol^{-1} in the stored energy of 16 kcal mol^{-1} originates from such hydrogen-bonding interaction [99,100]. Thus, in the case of archaeal rhodopsins, both chromophore distortion and hydrogen-bonding alteration contribute to the energy storage. The higher loss of energy in the primary reaction of bacteriorhodopsin than in bovine rhodopsin may be related to the different mechanisms of light energy storage.

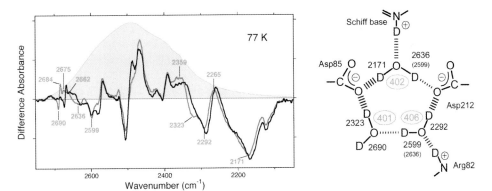

Fig. 4.10 (A) The K minus bacteriorhodopsin difference infrared spectra measured at 77 K in the 2750–1930 cm^{-1} region. The spectra are compared between hydration with D$_2$O (red lines) and D$_2$^{18}O (blue lines). The gray curve in the 2700–2000 cm^{-1} region represents O–D stretching vibrations of D$_2$O. Green-labeled frequencies correspond to those identified as O–D stretching vibrations of water. This figure is modified from Kandori [98]. (A) Schematic drawing of the deuterated pentagonal cluster structure in the Schiff base. The numbers are the O–D stretching frequencies of water molecules in cm^{-1} assigned in the FTIR study. This figure is reproduced from Shibata and Kandori [95].

4.4
Summary and Prospects

This chapter has gathered together the current understanding of retinal photoisomerization in visual and archaeal rhodopsins mainly from the experimental
point of view. Extensive studies by means of ultrafast spectroscopy of visual and
archaeal rhodopsins have provided an answer to the question, "What is the primary reaction in vision?" We now know that it is isomerization: from 11-*cis* to all-
trans form in visual rhodopsins and from all-*trans* to 13-*cis* form in archaeal rhodopsin. Femtosecond spectroscopy of visual and archaeal rhodopsins eventually
captured their excited states and, as a consequence, we now know that this unique
photochemistry takes place in our eyes and in archaea. Such unique reactions are
facilitated in the protein environment, and recent structural determinations have
further improved our understanding on the basis of structure. In parallel, vibrational analysis of primary intermediates, such as resonance Raman and infrared
spectroscopies, have provided insight into the isomerization mechanism.

In solution, photochemical properties are similar between the protonated Schiff
bases of 11-*cis*- and all-*trans*-retinal. Nevertheless, we have learned that there are
considerable differences in the photochemistries of 11-*cis* (visual) and all-*trans*
(archaeal) forms in protein (rhodopsin). In visual rhodopsins, conformational distortion takes place at the center of the chromophore, whereas no changes occur at
the Schiff base. Such changes may lead to the coherent product formation in femtoseconds. In contrast, structural changes take place only at the Schiff base region
of archaeal rhodopsins, which accompanies changes in the hydrogen-bonding network. Hydrogen-bonding alteration also plays an important role in the function of
ion pumps.

Due to constraints of space, I could not introduce many important theoretical
studies here. Various important models have been proposed on the primary isomerization mechanism in rhodopsins, including the bicycle pedal model [101], sudden polarization [102], and the hula-twist model [103]. The finding of a conical
intersection between the excited and ground states is also an important contribution [104]. Since the atomic structures of visual and archaeal rhodopsins are now
available, theoretical investigations will become more important in the future.
The combination of three methods – diffraction, spectroscopy, and theory – will
lead to a real understanding of the isomerization mechanism in rhodopsins.

Acknowledgments

The author thanks many collaborators in the references.

References

1 G. Wald, *Science* **1968**, *162*, 230–239.

2 R. R. Birge, *Biochim. Biophys. Acta* **1990**, *1016*, 293–327.

3 T. Yoshizawa, H. Kandori, in *Progress in Retinal Research*, N. Osborne, G. Chader, Ed. Pergamon Press, Oxford, **1992**, 33–55,.

4 H. G. Khorana, *J. Biol. Chem.* **1992**, *267*, 1–4.

5 H. Kandori, *Chem. Industr. (Lond.)* **1995**, *18*, 735–739.

6 Y. Shichida, H. Imai, *Cell. Mol. Life Sci.* **1998**, *54*, 1299–1315.

7 T. P. Sakmar, *Prog. Nucleic Acid Res. Mol. Biol.* **1998**, *59*, 1–33.

8 R. A. Mathies, J. Lugtenburg, in *Handbook of Biological Physics*, D. G. Stavenga, W. J. de Grip, E. N. Pugh, Ed. Elsevier, Amsterdam, **2000**, Vol. 3, 1197–1209,.

9 T. Okada, O. P. Ernst, K. Palczewski, K. P. Hofmann, *Trends Biochem. Sci.* **2001**, *26*, 318–324.

10 H. Kandori, Y. Shichida, T. Yoshizawa, *Biochemistry (Moscow)* **2000**, *66*, 1197–1209.

11 W. Stoeckenius, *Acc. Chem. Res.* **1980**, *13*, 337–344.

12 R. A. Mathies, S. W. Lin, J. B. Amea, W. T. Pollard, *Annu. Rev. Biophys. Biophys. Chem.* **1991**, *20*, 491–518.

13 T. G. Ebrey, in *Thermodynamics of Membranes, Receptors and Channels*, M. Jackson, Ed. CRC Press, New York, **1993**, 353–387.

14 J. L. Spudich, J. K. Lanyi, *Curr. Opin. Cell. Biol.* **1996**, *8*, 452–457.

15 U. Haupts, J. Tittor, D. Oesterhelt, *Annu. Rev. Biophys. Biomol. Struct.* **1999**, *28*, 367–399.

16 J. K. Lanyi, *J. Phys. Chem. B* **2000**, *104*, 11441–11448.

17 J. L. Spudich, C.-S. Yang, K.-H. Jung, E. N. Spudich, *Annu. Rev. Cell Dev. Biol.* **2000**, *16*, 365–392.

18 J. Nathans, D. Thomas, D. S. Hogness, *Science* **1986**, *232*, 193–202.

19 P. E. Blatz, J. H. Mohler, H. V. Navangul, *Biochemistry* **1972**, *11*, 848–855.

20 H. J. A. Dartnall, *Vision Res.* **1967**, *8*, 339–358.

21 J. Tittor, D. Oesterhelt, *FEBS Lett.* **1990**, *263*, 269–273.

22 Y. Furutani, A. Terakita, Y. Shichida, H. Kandori, *Biochemistry* **2005**, *44*, 7988–7997.

23 Y. Koyama, K. Kubo, M. Komori, H.; Yasuda, Y. Mukai, *Photochem. Photobiol.* **1991**, *54*, 433–443.

24 K.-P.; Hofmann, E. J. M. Helmreich, *Biochim. Biophys. Acta* **1996**, *1286*, 285–322.

25 E. A. Zhukovsky, D. D. Oprian, *Science* **1989**, *246*, 928–930.

26 T. P. Sakmar, R. R. Franke, H. G. Khorana, *Proc. Natl Acad. Sci. USA* **1989**, *86*, 8309–8313.

27 K. Palczewski, T. Kumasaka, T. Hori, C. A. Behnke, H. Motoshima, B. A. Fox, I. Le Trong, D. C. Teller, T. Okada, R. E. Stenkamp, M. Yamamoto, M. Miyano, *Science* **2000**, *289*, 739–745.

28 T. Okada, Y. Fujiyoshi, M. Silow, J. Navarro, E. M. Landau, Y. Shichida, *Proc. Natl Acad. Sci. USA* **2002**, *99*, 5982–5987.

29 M. Eilers, P. J. Reeves, W. Ying, H. G. Khorana, S. O. Smith, *Proc. Natl Acad. Sci. USA* **1999**, *96*, 487–492.

30 A. F. L. Creemers, C. H. W. Klaassen, P. H. M. Bovee-Geurts, R. Kelle, R. Kragl, J.; Raap, W. J. De Grip, J. Lugtenburg, H. J. M. de Groot, *Biochemistry* **1999**, *38*, 7195–7199.

31 Y. Furutani, Y. Shichida, H. Kandori, *Biochemistry* **2003**, *42*, 9619–9625.

32 T. Yoshizawa, Y. Kito, *Nature* **1958**, *182*, 1604–1605.

33 T. Yoshizawa, G. Wald, *Nature* **1963**, *197*, 1279–1286.

34 G. E. Busch, M. L. Applebury, A. A. Lamola, P. M. Rentzepis, *Proc. Natl Acad. Sci. USA* **1972**, *69*, 2802–2806.

35 K. Peters, M. L. Applebury, P. M. Rentzepis, *Proc. Natl Acad. Sci. USA* **1977**, *74*, 3119–3123.

36 S. Kawamura, F. Tokunaga, T. Yoshizawa, A. Sarai, T. Kakitani, *Vision Res.* **1979**, *19*, 879–884.

37 Y. Fukada, Y. Shichida, T. Yoshizawa, M. Ito, A. Kodama, K. Tsukida, *Biochemistry* **1984**, *23*, 5826–5832.

38 R. Callender, *Methods Enzymol.* **1982**, *88*, 625–633.

39 H. Kandori, S. Matuoka, Y. Shichida, T. Yoshizawa, M. Ito, K. Tsukida, V. Balogh-Nair, K. Nakanishi, *Biochemistry* **1989**, *28*, 6460–6467.

40 T. Mizukami, H. Kandori, Y. Shichida, A.-H. Chen, F. Derguini, C. G. Caldwell, C. Bigge, K. Nakanishi, T. Yoshizawa, *Proc. Natl Acad. Sci. USA* **1993**, *90*, 4072–4076.

41 Y. Shichida, S. Matuoka, T. Yoshizawa, *Photobiochem. Photobiophys.* **1984**, *7*, 221–228.

42 L. De Vico, M. Garavelli, F. Bernardi, M. Olivucci, *J. Am. Chem. Soc.* **2005**, *127*, 2433–2442.

43 H. Kandori, S. Matuoka, Y. Shichida, T. Yoshizawa, *Photochem. Photobiol.* **1989**, *49*, 181–184.

44 R. W. Schoenlein, L. A. Peteanu, R. A. Mathies, C. V. Shank, *Science* **1991**, *254*, 412–415.

45 M. Yan, D. Manor, G. Weng, H. Chao, L. Rothberg, T. M. Jedju, R.R. Alfano, C. H. Callender, *Proc. Natl Acad. Sci. USA* **1991**, *88*, 9809–9812.

46 M. Taiji, K. Bryl, M. Nakagawa, M. Tsuda, T. Kobayashi, *Photochem. Photobiol.* **1992**, *56*, 1003–1011.

47 L. A. Peteanu, R. W. Schoenlein, Q. Wang, R. A. Mathies, C. V. Shank, *Proc. Natl Acad. Sci. USA* **1990**, *90*, 11762–11766.

48 R. W. Schoenlein, L. A. Peteanu, Q. Wang, R. A. Mathies, C. V. Shank, *J. Phys. Chem.* **1993**, *97*, 12087–12092.

49 Q. Wang, G. G. Kochendoerfer, R. W. Schoenlein, P. J. E. Verdegem, J. Lugtenburg, R. A. Mathies, C. V. Shank, *J. Phys. Chem.* **1996**, *100*, 17388–17394.

50 Q. Wang, R. W. Schoenlein, L. A. Peteanu, R. A. Mathies, C. V. Shank, *Science* **1994**, *266*, 422–424.

51 H. Chosrowjan, N. Mataga, Y. Shibata, S. Tachibanaki, H. Kandori, Y. Shichida, T. Okada, T. Kouyama, *J. Am. Chem. Soc.* **1998**, *120*, 9706–9707.

52 H. Kandori, Y. Furutani, S. Nishimura, Y. Shichida, H. Chosrowjan, Y. Shibata, N. Mataga, *Chem. Phys. Lett.* **2001**, *334*, 271–276.

53 H. Kandori, Y. Katsuta, M. Ito, H. Sasabe, *J. Am. Chem. Soc.* **1995**, *117*, 2669–2670.

54 B. Aton, A. G. Doukas, R. H. Callender, U. Dinur, B. Honig, *Biophys. J.* **1980**, *29*, 79–94.

55 H. S. Rodman-Gilson, B. Honig, A. Croteau, G. Zarrilli, K. Nakanishi, *Biophys. J.* **1988**, *53*, 261–269.

56 T. Baasov, N. Friedman, M. Sheves, *Biochemistry* **1987**, *26*, 3210–3217.

57 G. Eyring, R. A. Mathies, *Proc. Natl Acad. Sci. USA* **1979**, *76*, 33–37.

58 A. Cooper, *Nature* **1979**, *282*, 531–533.

59 G. Eyring, B. Curry, A. Broek, J. Lugtenburg, R. A. Mathies, *Biochemistry* **1982**, *21*, 384–393.

60 I. Palings, E. M. M. van den Berg, J. Lugtenburg, R. A. Mathies, *Biochemistry* **1989**, *28*, 1498–1507.

61 H. Kandori, A. Maeda, *Biochemistry* **1995**, *34*, 14220–14229.

62 T. Nagata, T. Oura, A. Terakita, H. Kandori, Y. Shichida, *J. Phys. Chem. A* **2002**, *106*, 1969–1975.

63 H. Belrhali, P. Nollert, A. Royant, C. Menzel, J. P. Rosenbusch, E. M. Landau, E. Pebay-Peyroula, *Struct. Fold. Des.* **1999**, *7*, 909–917.

64 H. Luecke, B. Schobert, H.-T. Richter, J. P. Cartailler, J. K. Lanyi, *J. Mol. Biol.* **1999**, *291*, 899–911.

65 M. Kolbe, H. Besir, L-O. Essen, D. Oesterhelt, *Science* **2000**, *288*, 1390–1396.

66 H. Luecke, B. Schobert, J. K. Lanyi, E. N. Spudich, J. L. Spudich, *Science* **2001**, *293*, 1499–1503.

67 Royant, A.; Nollert, P.; Edman, K.; Neutze, R.; Landau, E. M.; Pebay-Peyroula, E.; Navarro, J. (**2001**) *Proc. Natl Acad. Sci. USA* 98, 10131–10136.

68 R. A. Mathies, C. H. Brito Cruz, W. T. Pollard, C. V. Shank, *Science* **1988**, *240*, 777–779.

69 J. Dobler, W. Zinth, W. Kaiser, *Chem Phys. Lett.* **1988**, *144*, 215–220.

70 J. K. Delaney, T. L. Brack, G. H. Atkinson, M. Ottolenghi, G. Steinberg, M. Sheves, *Proc. Natl Acad. Sci. USA* **1995**, *92*, 2101–2105.

71 Q. Zhong, S. Ruhman, M. Ottolenghi, M. Sheves, N. Friedman, *J. Am. Chem. Soc.* **1996**, *118*, 12828–12829.

72 G. H. Atkinson, L. Ujj, Y. Zhou, *J. Phys. Chem. A* **2000**, *104*, 4130–4139.

73 J. Herbst, K. Heyne, R. Diller, *Science* **2002**, *297*, 822–825.

74 T. Kobayashi, T. Saito, H. Ohtani, *Nature* **2001**, *414*, 531–534.

75 S. J. Doig, P. J. Reid, R. A. Mathies, *J. Phys. Chem.* **1991**, *95*, 6372–6379.

76 L. Song, M. A. El-Sayed, J. K. Lanyi, *Science* **1993**, *261*, 891–894.

77 H. Kandori, K. Yoshihara, H. Tomioka, H. Sasabe, *J. Phys. Chem.* **1992**, *96*, 6066–6071.

78 H. Kandori, K. Yoshihara, H. Tomioka, H. Sasabe, Y. Shichida, *Chem. Phys. Lett.* **1993**, *211*, 385–391.

79 T. Arlt, S. Schmidt, W. Zinth, U. Haupts, D. Oesterhelt, *Chem. Phys. Lett.* **1995**, *241*, 559–565.

80 H. Kandori, H Tomioka, H. Sasabe, *J. Phys. Chem. A* **2002**, *106*, 2091–2095.

81 H. Kandori, H. Sasabe, *Chem. Phys. Lett.* **1993**, *216*, 126–172.

82 B. Hou, N. Friedman, S. Ruhman, M. Sheves, M. Ottolenghi, *J. Phys. Chem. B* **2001**, *105*, 7042–7048.

83 R. R. Birge, T. M. Cooper, *Biophys. J.* **1983**, *42*, 61–69.

84 K. Edman, P. Nollert, A. Royant, H. Belrhali, E. Pebey-Peyroula, J. Hajdu, R. Neutze, E. M. Landau, *Nature* **1999**, *401*, 822–826.

85 B. Schobert, J. Cupp-Vickery, V. Hornak, S. O. Smith, J. K. Lanyi, *J. Mol. Biol.* **2002**, *321*, 715–726.

86 Y. Matsui, K. Sakai, M. Murakami, Y. Shiro, S. Adachi, H. Okumura, T. Kouyama, *J. Mol. Biol.* **2002**, *324*, 469–481.

87 K. Edman, A. Royant, P. Nollert, C. A. Maxwell, E. Pebey-Peyroula, J. Navarro, R. Neutze, E. M. Landau, *Structure* **2002**, *10*, 473–482.

88 A. Maeda, *Israel J. Chem.* **1995**, *35*, 387–400.

89 H. Kandori, N. Kinoshita, Y. Shichida, A. Maeda, *J. Phys. Chem. B* **1998**, *102*, 7899–7905.

90 H. Kandori, M. Belenky, J. Herzfeld, *Biochemistry* **2002**, *41*, 6026–6031.

91 K. Shimono, Y. Furutani, N. Kamo, H. Kandori, *Biochemistry* **2003**, *42*, 7801–7806.

92 M. Shibata, N. Muneda, T. Sasaki, K. Shimono, N. Kamo, M. Demura, H. Kandori, *Biochemistry* **2005**, *44*, 12279–12286.

93 H. Kandori, Y. Shichida, *J. Am. Chem. Soc.* **2000**, *122*, 11745–11746.

94 M. Shibata, T. Tanimoto, H. Kandori, *J. Am. Chem. Soc.* **2003**, *125*, 13312–13313.

95 M. Shibata, H. Kandori, *Biochemistry* **2005**, *44*, 7406–7413.

96 H. Kandori, *Biochim. Biophys. Acta* **2000**, *1460*, 177–191.

97 T. Tanimoto, Y. Furutani, H. Kandori, *Biochemistry* **2003**, *42*, 2300–2306.

98 H. Kandori, *Biochim. Biophys. Acta* **2004**, *1658*, 72–79.

99 S. Hayashi, E. Tajkhorshid, K. Schulten, *Biophys. J.* **2002**, *83*, 1281–1297.

100 S. Hayashi, E. Tajkhorshid, H. Kandori, K. Schulten, *J. Am. Chem. Soc.* **2004**, *126*, 10516–10517.

101 A. Warshel, *Nature* **1976**, *260*, 679–683.

102 L. Salem, P. Bruckmann, *Nature* **1975**, *258*, 526–528.

103 R. S. H. Liu, A. E. Asato, *Proc. Natl Acad. Sci. USA* **1985**, *82*, 259–263.

104 M. Garavelli, P. Celani, F. Bernardi, M. A. Robb, M. Olivucci, *J. Am. Chem. Soc.* **1997**, *119*, 6891–6901.

5
Non-Retinal Chromophoric Proteins

Marc Zimmer

5.1
Introduction

Although they are not related, photoactive yellow protein (PYP), green fluorescent protein (GFP), and phytochromes are all discussed in this chapter that deals with the *cis-trans* isomerization (CTI) of non-retinal binding proteins. All three systems contain a chromophore that is attached to the protein and that undergoes CTI. In this chapter we attempt to provide the reader with a brief introduction to PYP, GFP, and phytochromes, and present a summary of the current understanding of the CTIs that are central to the photochemistry observed in these systems.

5.2
Photoactive Yellow Protein

Photoactive yellow protein (PYP) was discovered 20 year ago in *Halorhodospira halophila*, then known as *Ectothiorhodospira halophila* [1,2]. In several halophilic purple bacteria it has a vital role in the avoidance response to blue light (phototaxis). It has been thoroughly studied as a model photoreceptor system and as the structural prototype for the PAS class of signal transduction proteins. PYP has 125 amino acid residues in an a/β-fold with six antiparallel β-sheets and several helices (see Fig. 5.1). The covalently bound *p*-coumaric acid chromophore is linked to the only cysteine in the protein (Cys69) (see Fig. 5.1). Hellingwerf has published an excellent review of the photophysical behavior of PYP [1].

   Figure 5.1 shows a model of the PYP photocycle with all its distinguishable steps (with transient UV/Vis spectroscopy). In its resting dark state, pG (also referred to as P or PYP), the PYP chromophore is in an anionic *trans* conformation. The charge on the chromophore is stabilized by hydrogen bonding interactions with Tyr42, Glu46, and Thr50. Blue light induces a *trans* to *cis* isomerization of the C7–C8 bond, which is completed within a few picoseconds. Time-dependent density functional theory has been used to suggest that the first thermally stable photocycle intermediate, the *cis* I_0 state, is formed from the resting pG state

cis-trans Isomerization in Biochemistry. Edited by Christophe Dugave
Copyright © 2006 WILEY-VCH Verlag GmbH & Co. KGaA, Weinheim
ISBN: 3-527-31304-4

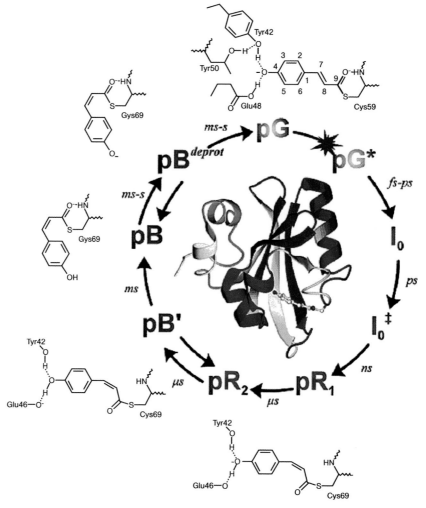

Fig. 5.1 Detailed model of all distinguishable steps (with transient UV/Vis spectroscopy) in the photocycle of PYP [1]. The relevant changes in the configuration of the chromophore and in the surrounding functional groups, at the various steps, are indicated in the structural diagrams. The inner part of the figure shows the structure of PYP, color-coded according to the extent of structural change in pB, as measured with NMR [10]. Atom numbering of the carbon atoms of the chromophore is given in the inset corresponding to pG. From Hellingwerf, K. J., Hendriks, J., Gensch, T. *J. Phys. Chem. A* 2003, 107, 1082–1094. Copyright (2003) American Chemical Society, USA.

by crossing of the transition state barrier, and relaxation from the excited state to the ground state by internal conversion [3]. This intermediate has a red-shifted absorption spectrum and still retains its hydrogen bonds with Tyr42 and Glu46. The C1–C7–C8–C9 dihedral angle (see Fig. 5.1 for numbering) of the I_0 intermediate has been calculated to be –80° [3], which corresponds with that observed in the crystal structure of a cryogenically trapped early intermediate of the photocycle [4]. It has been suggested that the light energy that drives the photocycle is stored in this highly strained *cis* conformation [3,5]. Within nanoseconds several relaxation processes of the strained chromophore and the hydrogen bonding network lead to the relaxed, pR, intermediate (also known as I_1 or PYP_L). Vibrational spectroscopy has been used to show that the hydrogen bond between Glu46 and the anionic chromophore is the same in the dark *trans* pG state as it is in the *cis* pR state [6]. Protonation of the *cis* p-coumaric acid chromophore in the pR state through intramolecular proton transfer leads to conformational changes in the protein. The proton transfer to the blue-shifted intermediate, pB (also referred to as I_2 or PYP_M) occurs from either Glu46 or the solvent. In the solid state these structural changes are fairly minor [4,7–9], but in solution there are major differences [10].

It has been proposed that the structural changes associated with the protonation of the chromophoric phenol are responsible for the signal transduction by the pB state. The pB→pR conversion is photoreversible and its kinetics have been examined [11]. Calculations [12] have shown that proton transfer is much more likely in the *cis* pR state than in the initial dark *trans* pG state. The PYP photocycle is completed by the pB state relaxing back to the pG dark state. This deprotonation and reisomerization is pH-dependent ($\tau = 140$ ms). It has been shown that isomerization of the deprotonated chromophore is faster than for the protonated form, therefore it has been suggested that protonation precedes isomerization.

Since PYP is one of the most well-characterized photosensors it has attracted a lot of attention from computational chemists. Robb et al. have used a QM/MM molecular dynamics strategy to examine the complete photocycle of PYP [12]. By comparing the behavior of the chromophore in vacuo with that of the chromophore within the protein they were able to determine the "chemical role" of the protein cavity. They found that CTI of the chromophore does not occur in vacuo, however in the protein the isomerization is facilitated by electrostatic stabilization of the chromophore's excited state with the guanidium group of Arg52 that lies just above the negatively charged phenolate of the chromophore.

A similar study by Yamada et al. [13] concluded that the protein prevents the chromophore from adopting a completely planar structure. Based on their calculations they proposed that the efficiency of photoisomerization in PYP is due to the asymmetric protein–chromophore interaction that can serve as the initial accelerant for the light-induced photocycle. They also found that the C4–C7–C8–C9 dihedral always twists counterclockwise.

PYP is a perfect example of how a CTI of a small prostetic group can lead to large conformational changes within a protein.

5.3
Green Fluorescent Protein and Other GFP-like Proteins

Green fluorescent protein (GFP), its mutants and homologs are widely used as biological markers [14–17]. They are particularly useful due to their stability, and the fact that the chromophore is formed in an autocatalytic cyclization that does not require a cofactor. This means that unlike most other bioluminescent reporters GFP fluoresces in the absence of any other proteins, substrates, or cofactors. Furthermore, it appears that fusion of GFP to a protein does not alter the function or location of the protein. This has enabled researchers to use GFP as a tracer in living systems and it has led to GFP's widespread use in many areas of modern science, such as cell dynamics and development studies.

GFP from the jellyfish *Aequorea victoria* has 238 residues, which form 11 β-sheets arranged in a barrel shape (Fig. 5.2). The barrel is a nearly perfect cylinder with a height of 42 Å and a radius of 12 Å. The chromophore is located in the middle of the so-called β-barrel and deletion mapping experiments have shown that nearly the entire structure (residues 2–232) is required for chromophore formation and/or fluorescence [18].

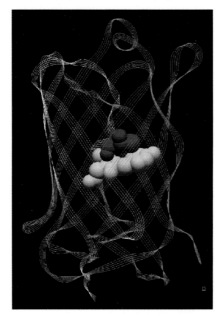

Fig. 5.2 Crystal structure of green fluorescent protein [70] with the chromophore shown as light CPK spheres. Thr62 and Phe165 are largely responsible for restricting the *cis-trans* isomerization of the chromophore.

The autocatalytic chromophore formation in GFP (Scheme 5.1) is comprised of three steps: formation of the 5-membered imidazole ring by nucleophilic attack of the Gly67 amide nitrogen on the Ser65 carbonyl carbon, dehydration of the Ser65 carbonyl oxygen, and oxidation of the C_a–C_β bond to form the conjugated chromophore [19–23]. The initial precyclized GFP is referred to as immature GFP and is nonfluorescent. Two different mechanisms have been proposed for the chromophore maturation: a cyclization–oxidation–dehydration mechanism [24] and a conjugation trapping mechanism [25].

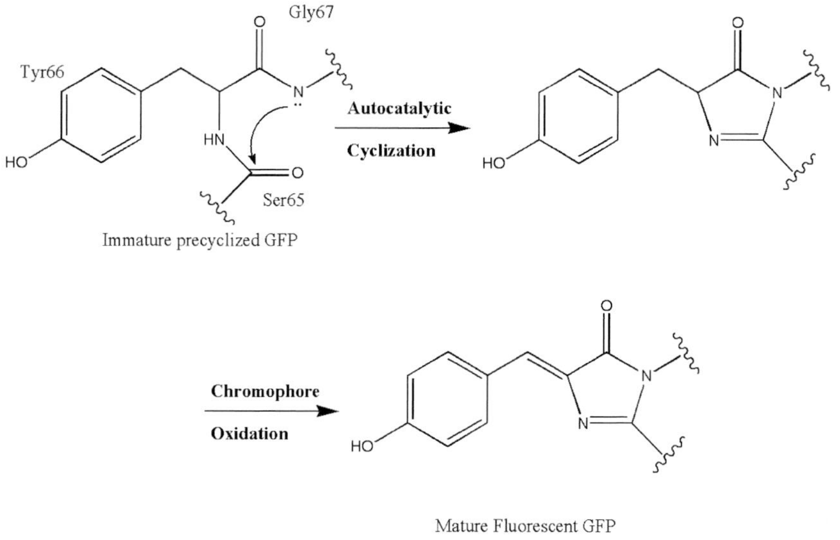

Tyr66 Gly67 Ser65 Immature precyclized GFP

Autocatalytic
Cyclization

Chromophore
Oxidation

Mature Fluorescent GFP

Scheme 5.1 Chromophore formation in green fluorescent protein.

Wild-type GFP has a major absorption at 398 nm [26] and a minor absorption at 475 nm with a shoulder on the red edge [27,28]. Excitation at 398 nm results in an emission maximum at 508 nm, while irradiation at 475 nm produces an emission with a maximum at 503 nm [20]. It is generally accepted that the two absorptions are due to GFP existing in two different substates, one in which the chromophore is in the neutral phenolic form (A state) and the other in which it is in the anionic phenolate form (B state). Based on the observation that the Y66H mutant only absorbs at 384 nm [20], changes in the absorption and fluorescence spectra that accompany other mutations, and the crystal structure of the Y66H mutant [29], a mechanism shown in Fig. 5.3 was proposed for the photoisomerization of wild-type GFP [20,29,30]. The neutral form of the chromophore can convert to the anionic species (B) by going through an intermediate state (I). In going from the neutral chromophore (species A) to the charged chromophore (B) the Tyr66 phenolic proton is shuttled through an extensive hydrogen bonding network to the carboxylate oxygen of Glu222. It has been proposed that the B state can also be irreversibly formed by decarboxylation of Glu222 [31,32].

The change from forms A to I is solely a protonation change, while the change from I to B is a conformational change with most changes occurring at Thr203. The A and B states are populated in a 3:1 ratio at equilibrium. Spectral hole-burning experiments have shown that the ground state of form I is higher in energy than both the ground states of A and B, and that it is separated from them by energy barriers of several hundred wavenumbers [28]. Ultrafast multipulse control spectroscopy has been used to show that there are in fact two distinct anionic groundstate intermediates, I1 and I2 [33].

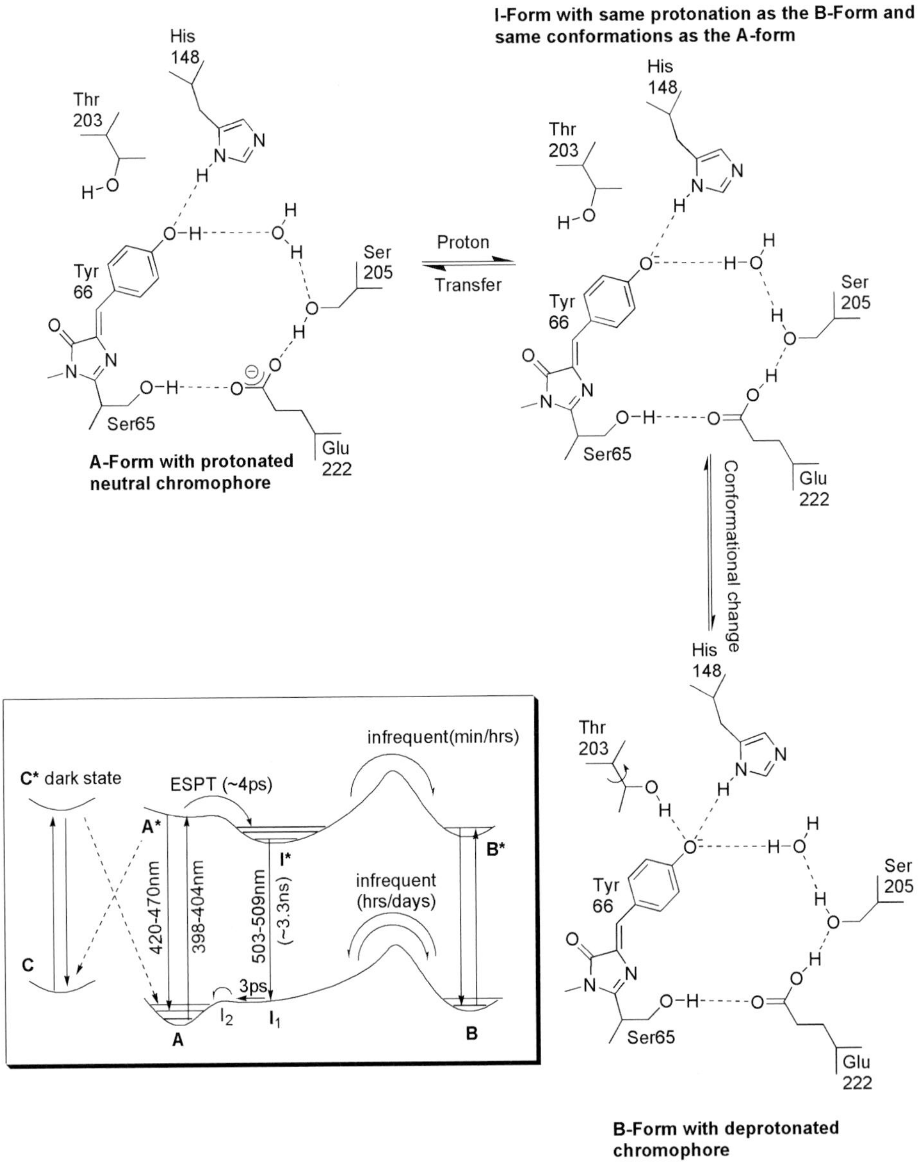
I-Form with same protonation as the B-Form and same conformations as the A-form
His 148
Thr 203
Tyr 66
Ser 205
Ser65
Glu 222
Proton
Transfer
A-Form with protonated neutral chromophore
Conformational change
B-Form with deprotonated chromophore
C* dark state
ESPT (~4ps)
infrequent(min/hrs)
A*
I*
B*
420-470nm
398-404nm
503-509nm
(~3.3ns)
infrequent (hrs/days)
C
3ps
I2
I1
A
B

◀ **Fig. 5.3** Proposed mechanism for the photo-isomerization of wild-type green fluorescent protein. The neutral form of the chromophore (A) can convert to the anionic species (B) by going through the intermediate state (I). In going from the neutral species (A) to the charged species (B) the Tyr66 phenolic proton is shuttled through an extensive hydrogen bonding network to the carboxylate oxygen of Glu222. The change from forms A to I is solely a protonation change, while the change from I to B is a conformational change with most changes occurring at Thr203. The inset shows the fluorescence mechanism [15]. Upon excitation of the A state an excited-state proton transfer (ESPT) occurs in which the proton is transferred from the chromophore to Glu222 in a time-scale of the order of pico-seconds. Following radiative relaxation from the excited state intermediate (I*) the system returns to the ground state A through the ground state intermediates I₁ and I₂ [33]. Excitation of the anionic B state results in direct emission from the excited state (B*) at 482 nm. Recently a nonfluorescent dark state, state C, has been observed that is distinct from states A and B and absorbs at higher energies [52]. The C state, perhaps the neutral *trans* form of the chromophore, may be populated by nonradiative decay from A* and it maybe depopulated by excitation to the excited C* state with *trans-cis* isomerization to repopulate state A.

In the excited state the barrier between A* and I* is low whereas that between I* and B* is at least 2000 cm^{-1}. Upon excitation of the A state an excited-state proton transfer occurs in which the proton is transferred from the chromophore to Glu222 in a time-scale of the order of picoseconds. Following radiative relaxation from the excited-state intermediate (I*) the systems returns to the ground state A through the ground state intermediates I$_1$ and I$_2$. Excitation of the anionic B state results in direct emission from the excited state (B*) at 482 nm. This description of the photophysical behavior of GFP is shown in the inset of Fig. 5.3, and has been partially validated by the interpretation of the absorption and Stark spectra of the wild type, the S65T and Y66H/Y145F GFP mutants [34]. The electronic spectra show that the excitation of species A only involves a small charge displacement, while excitation of species B involves a significant change from the ground state. Since the intermediate state (I) is structurally similar to both the ground and excited state of species A and electronically similar to the ground and excited state of species B, the protein has to undergo structural changes in going from state A to state B. Additional evidence comes from Raman studies [35] and the X-ray structure of S65T at low pH, which shows that there is no hydrogen bonding interaction between the side-chain of Thr203 and the phenolic oxygen of the chromophore at low pH, while the side chain χ_1 dihedral of Thr203 rotates by 100° to form a hydrogen bond in the high pH structure [36].

The photophysical behavior of GFP described above and summarized in the inset in Fig. 5.3 is further complicated by transitions between bright and dark fluorescent states. At the single molecule level these transitions are responsible for the reversible fast blinking and photobleaching that has been observed in single protein experiments [37–39]. The most commonly accepted models used to explain these observations are based on nonradiative relaxation pathways between the excited and ground state that involve torsional changes of the φ and τ dihedrals of the chromophore shown in Fig. 5.4.

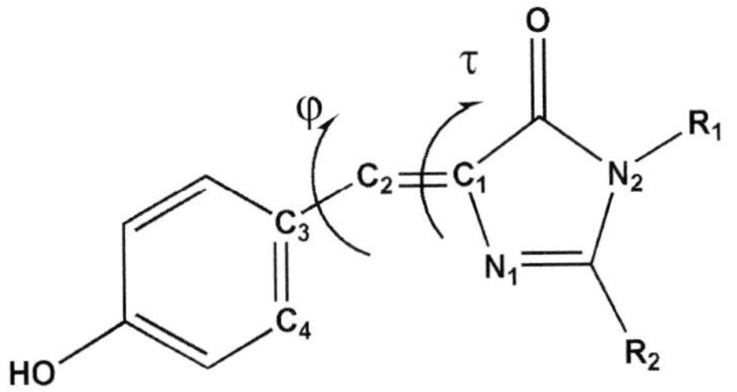

Fig. 5.4 The τ (N1–C1–C2–C3) and φ (C1–C2–C3–C4) dihedral angles of the green fluorescent protein chromophore. In the protein R_1 is Gly67 and R_2 is Ser65, and in HBDI, an often used model compound, $R_1 = R_2 = CH_3$. In τ one-bond flips (τ-OBF) the dihedral rotation occurs around the τ torsional angle, in a φ-OBF it is around the φ dihedral angle, in a hula twist (HT) the φ and τ dihedral angles concertedly rotate.

A model for the light/dark behavior of GFP has been proposed [40]. It is based on quantum mechanical calculations of the energy barriers for the φ and τ one-bond flips (OBF) and the φ/τ hula twists (HTs) that were calculated in the ground and first singlet excited states for a small nonpeptide model compound. Figure 5.5 shows the calculated energy profiles.

While the planar geometries are the ground state minima this is not necessarily so for the excited state, in fact in some cases the excited state has an energy minimum with a perpendicularly twisted chromophore. According to the calculations the ground and excited states for the τ-OBF and HT in the neutral form (A) and the φ-OBF in the zwitterionic form come very close to each other. It has been proposed that this can lead to fluorescence quenching nonadiabatic crossing (NAC).

Figure 5.6 shows a model for the photophysical behavior of GFP [41]. According to the model the excited A state of the chromophore can undergo an excited-state proton transfer (ESPT) to form the excited intermediate form, I* (pathway A2), which emits radiation at 505 nm (pathway I1), or the excited A state can undergo fluorescence quenching NAC by means of a τ-OBF or a HT (pathway A3). Excitation of the anionic B state leads to excited B* which fluoresces (B1). The model also contains a zwitterionic form that has not gained much acceptance in the literature.

Model compounds of the chromophore do not fluoresce in solution. This is presumably due to the lack of constraints imposed by the protein environment. The excited state of the model compounds can freely rotate around their φ and τ dihedral angles, which allows NAC to occur, resulting in fluorescence quenching. Fluorescence can, however, be obtained by lowering the temperature to 77K, this presumably freezes the solution and imposes steric barriers to rotation. Similar behavior is observed when the protein is denatured – the fluorescence yield decreases by at least three orders of magnitude [42]. Furthermore, chromophore model compounds that are non- or minorly substituted emit minimal fluorescence, while sterically bulky substituents modify the equilibrium between radiative and nonradiative deexcitation pathways, making the sterically hindered compounds more fluorescent [43].

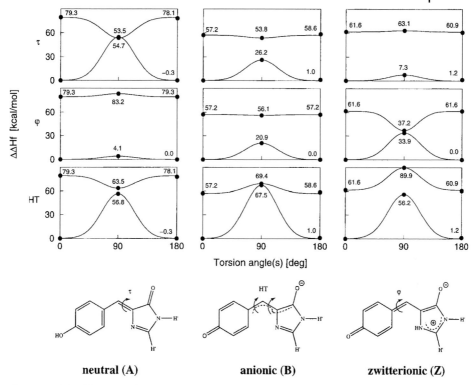

Fig. 5.5 Models of the green fluorescent protein chromophore in the neutral, anionic, and zwitterionic forms used in the quantum chemical calculations, shown in those resonance structures that best represent the calculated bond orders. Rotation by 180° around φ leaves the structure unchanged. The configurations displayed represent $\tau = 0°$ and are referred to as *cis* configurations. The upper panels show energy profiles for rotation around the dihedral angles τ and φ and for the hula twist (HT) motion in the ground and first singlet excited states. Calculated values are marked by dots, and the Gaussian profiles are shown as visual aid. Note that, for these calculations the dihedral angles were fixed while relaxing all other degrees of freedom. From W. Weber, V. Helms, J.A. McCammon, P.W. Langhoff, *Proc. Natl Acad. Sci. USA* 1999, 96, 6177–6182. Copyright (2000) National Academy of Sciences, USA.

The radiationless decay has been investigated by ultrafast polarization spectroscopy [44] and time-resolved fluorescence [45,46]. The results confirm that the radiationless decay occurs by an ultrafast internal conversion, due to intramolecular motion about the bridging bond of the chromophore in the excited state, that the isomerization is nearly barrierless, and that there is only a very weak dependence on medium viscosity, thereby implying that the isomerization occurs by a volume-conserving motion such as a hula twist [47].

A volume analysis of the τ and φ 90° OBFs and HTs in a GFP chromophore model compound, revealed that the τ-OBF displaces a larger volume than both the HT and the φ-OBF. However, the HT and φ-OBF processes displace the same vol-

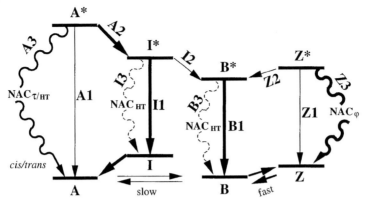

Fig. 5.6 Model for the photophysical behavior of green fluorescent protein [40]. Excited states are labeled by asterisks. Barriers may exist for processes of types 2 and 3. Excitation arrows have been omitted for simplicity. The relative free energies of the ground state forms A (neutral), B (anionic), I (intermediate) and Z (zwitterionic) depend on the protein environment. From W. Weber, V. Helms, J.A. McCammon, P.W. Langhoff, *Proc. Natl Acad. Sci. USA* 1999, 96, 6177–6182. Copyright (2000) National Academy of Sciences, USA.

ume, and therefore the volume-conserving property of the HT is not a sufficient reason for the excited chromophore to undergo a HT [48].

Interestingly ab initio calculations on a chromophore model compound have shown that a correlated torsion about a double bond and its adjacent single bond exists in both the gas and solution phases, implying that a HT may occur even in the absence of volume-conserving requirements [49].

According to molecular mechanics calculations the protein matrix of wild-type GFP forms a cavity around the chromophore that is complementary to an excited state conformation in which the phenol and imidazolidinone rings are perpendicular to each other – a conformation that was obtained by a concerted τ and φ 45° HT [48]. Therefore even though an HT motion in wild-type GFP may not be more volume conserving than the φ-OBF, it still occurs as it is intrinsically favoured [49] and is complementary with the protein matrix surrounding the chromophore. Similar behavior has been proposed for PYP [13].

Ramachandran-type plots in which the τ and φ dihedrals of the chromophore within the GFP protein matrix were systematically varied [50] showed that there are two minima for all protonation states, one at $\tau = 60 \pm 30°$ and $\varphi = 120 \pm 30°$, and the other at $\tau = 60 \pm 30°$ and $\varphi = -60 \pm 30°$, and that the protein environment of GFP allows the chromophore some rotational freedom, especially by a HT or in the φ dihedral angle (Fig. 5.7). There is a significant energy barrier for $\tau = 180$–270°, therefore a *cis-trans* photoisomerization cannot occur by a 180° rotation of the φ dihedral angle. The protein exerts some strain on the chromophore when it is planar, and the only reason planar chromophores are found in GFP is due to their delocalized π-electrons. These results have been confirmed by molecular dynamics simulations of the chromophore with freely rotating τ and φ dihedral angles [50]. We have recently shown that wild-type GFP is not an anomaly, most of the GFP and GFP-like proteins in the protein databank have a protein matrix that is not complementary with a planar chromophore [51]. When the π-conjugation across the ethylenic bridge of the chromophore is removed, the protein matrix will significantly twist the freely rotating chromophore from the planar structures found in the crystal structure. These calculations [51] were done by minimizing, with freely rotating τ and φ dihedral angles, the crystal structure of 38 GFP analogs and mutants found in the protein databank.

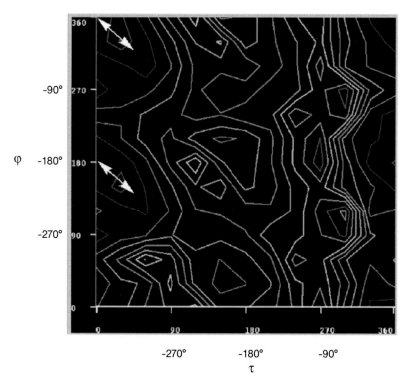

Fig. 5.7 Ramachandran plot of the τ vs. φ dihedral angles of the chromophore in GFP with a protonation state of $-O,N,Glu-$. See Fig. 5.4 for the definition of the dihedral angles. Energy contours increase in energy from purple through blue and yellow to red, each contour line corresponds to an energy difference of 28 kJ/mol. The two lowest energy minima are at $\tau = 27°$ and $\varphi = -27°$, and $\tau = 31°$ and $\varphi = 150°$. The white two-headed arrows show the low-energy hula-twist pathway from the planar chromophore to the perpendicularly twisted chromophore.

According to the calculations for a chromophore model in state A, summarized in Fig. 5.6, NAC can occur when the two rings making up the chromophore are twisted perpendicularly to each other. In Figs 5.3 and 5.6 we saw that the excited state A* could undergo ESPT to the I* state or that the phenol and imidazolinone rings of the chromophore can twist relative to one another before undergoing NAC and relaxing back to the A state. However it also possible that the two rings' systems could continue their rotation passed the perpendicular form which undergoes NAC, and form the *trans* isomer. Recently a nonfluorescent dark state, state C, has been observed that is distinct from states A and B and absorbs at higher energies [52]. Based on spectroscopic and computational studies of a yellow mutant of GFP it has been suggested that the dark C state is a neutral form of the chromophore that is less hydrogen-bonded than the neutral A state – perhaps the neutral *trans* form of the chromophore. The C state maybe populated by nonradiative decay from A* (switching off fluorescence) or it maybe depopulated by excitation to the excited C* state with *trans-cis* isomerization to repopulate state A (see inset in Fig. 5.3) [52]. It has been suggested that this on/off switching of GFP can be used as optical memory elements [15]. To the best of my knowledge, the only examination of the protein barrier to *cis-trans* isomerization are the Ramachandran type plots mentioned above which showed that there was a sizable barrier to a 180° rotation in the τ dihedral angle of wild-type GFP. However, it is quite possible that the barrier to rotation is much smaller in the yellow GFP variants than in wild-type GFP [40].

More than 50 crystal structures of GFP, GFP mutants, and GFP analogs have been deposited in the protein databank [53], and of these two have their chromophore's oriented in a *trans* configuration. They are the intensely colored blue nonfluorescent pocilloporin pigment [54], which has a distinctly nonplanar *trans* configuration with most of the deviation from planarity occurring due to φ rotation, and a GFP-like protein from the sea anemone, *Entacmaea quadricolor* (eqFP611) [55].

Lukyanov et al. [56] have also proposed that CTI can occur in some GFP-like proteins, where it leads to a dark nonfluorescent state. They based their CTI model on some GFP-like proteins they have isolated. The majority of GFP-like proteins, such as DsRed, are fluorescent and have been isolated from corals. However, there are some nonfluorescent proteins that are in the so-called "chromo" state ("The chromo state indicates that the protein has a high extinction coefficient but a low quantum yield, whereas in the fluorescent state the protein is characterized by a high quantum yield.") [56]. Most interesting of these is asCP, a unique nonfluorescent GFP-like protein discovered in the sea anemone *Anemonia sulcata* [57]. Initially nonfluorescent, asCP can be made to fluoresce (kindled) by intense green light irradiation. After kindling the protein relaxes back to its nonfluorescent state, or it can be quenched instantly by short blue light irradiation.

Using site-directed mutagenesis Lukyanov et al. were able to create asCP mutants that were always fluorescent, and some that were nonfluorescent and could not be kindled. On the basis of their findings they proposed that GFP-like molecules with the chromophore in the *cis* conformation were fluorescent, while those with a *trans* conformation were nonfluorescent. In asCP the initial state has a *trans* chromophore – it is nonfluorescent. However, upon kindling the chromophore adopts the *cis* conformation and becomes fluorescent. Table 5.1 summarizes this hypothesis.

Table 5.1 Possible chromophore conformations in GFP-like proteins and their photophysical consequences.

Proteins	*cis*-Chromophore	*trans*-Chromophore
GFP	Initial state, fluorescent	Dark state, nonfluorescent
Known fluorescent coral proteins	Initial state, fluorescent	
Known chromophoric coral proteins		Initial state, chromo, nonfluorescent
asCP	Kindled state, fluorescent	Initial state, chromo, nonfluorescent

The crystal structure of the dark state of asCp has recently been released [58], and as predicted it is in the *trans* conformation. However the chromophore has only one covalent link to the protein. Fragmentation of the protein has occurred – this has been shown to be an intrinsic step in the maturation of the asCP chromophore. The cleavage of the Cys62–chromophore bond (asCP numbering) may provide the chromophore freedom of movement not observed in GFP and other GFP-like proteins – by lowering the activation barriers for *cis/trans* conformational transitions it may be responsible for asCP kindling abilities.

5.4
Phytochromes

Light perception in plants is governed by a series of photoreceptors that can be classified into three groups – the phytochromes, cryptochromes, and phototropins. Originally phytochromes were defined as the receptors that were responsible for the red and far-red reversible, plant responses [59]. However they have also been found in bacteria (even in nonphotosynthetic bacteria). They are typically homodimers consisting of two polypeptides with a molecular weight of ~125 kDa, each containing a linear tetrapyrrole (bilin) chromophore that is covalently linked to a conserved cysteine by a thioether bond. Attachment of the bilin prosthetic group is autocatalytic. Phytochromes typically form covalent adducts with phytochromobilin (PΦB) but can also bind other phycobilin analogs (see Fig. 5.8). The bilins adopt cyclic porphyrin-like conformations in aqueous solution. Upon association with proteins they form more extended conformations that alter the pathways for light deexcitation [60]. The open chain bilin has 64 isomers that differ in their methine bridge configurations (*Z/E*) and conformations (*syn*(s)/*anti*(a)). Phytochromes undergo a *cis/trans* photoisomerization of the bilin, which leads to substantial changes in the conformation that presumably results in the photosignaling response that regulates plant growth and development [61].

One of the distinguishing characteristics of phytochromes is a reversible photoisomerization between a red light-absorbing form known as P$_r$ (λ_{max} = 666 nm), and the far red-absorbing form termed the P_{fr} form ($\lambda_{max} = 730$ nm) [62]. Depending on the species, either P_r or P_{fr} can be the active form. Photoreversible reactions such as this one play a key role in signal transduction; besides being found in phytochromes they also have an important function in sensory rhodopsin I and photoactive yellow protein [11]. The conformational differences between P_r and P_{fr} have been observed using several techniques, including limited proteolysis, cysteine labeling, circular dichroism, and chromatography [63]. Gartner and Braslavsky have recently written a review of the molecular basis of bilin photochemistry and the role of the phytochrome protein [64].

Although there is no crystal structure of phytochromes, they have been the focus of numerous spectroscopic studies [65,66]. Several intermediates have been trapped by time-resolved experiments in both the P_r:P_{fr} and the P_{fr}:P_r interconversions. The P_r:P_{fr} steady-state ratio is determined by the incident light. It is commonly accepted that the bilin adopts a ZZZ/asa configuration/conformation in P_r that converts to ZZE/ass in P_{fr}. Conformational changes of the protein backbone are required to maintain the high-energy P_{fr} state. The first step(s) in the photoconversion is a rapid isomerization (picoseconds) around the C15=C16 double bond (see Fig. 5.8). This is followed by a series of slower conformational changes (micro- and milliseconds) that occur in the dark phase. In oat phytochrome (PhyA) three intermediates have been found in the P_rgP_{fr} conversion (lumi-R, meta-Ra, and meta-Rc) and two were found in the reverse reaction (lumi-F and meta-F) [67,68].

Recently a combination of resonance Raman spectroscopy and density functional calculations has been used to examine the conformations of the phytochromobilin structure of phytochrome phyA (oat) [69]. They conclude that the chromophore is in the ZZZasa configuration, and that the reaction cycle is initiated by a ZZZasa (P_r) → ZZEasa (Lumi-R) photoisomerization followed by thermal relaxation steps that include at least a partial a to s single bond rotation at the methane bridge A–B. This may explain the change in hydrogen bonding of the C=O group of ring A that occurs with the formation of the P_{fr} precursor meta-Rc [65].

If phytochromobilin (PΦB) is replaced with phycocyanobilin (PCB) the photochemistry remains the same as described above. However if the D ring of the chromophore is modified then differences in the time-course of the photoreaction are observed [68].

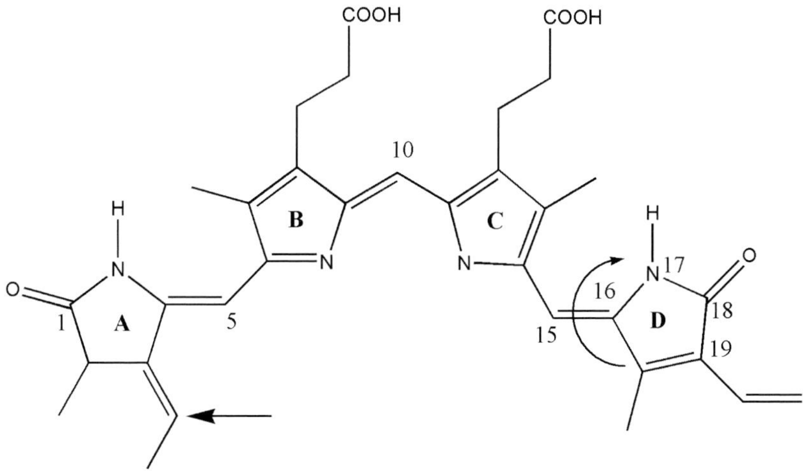

3E-phytochromobilin (PΦB)

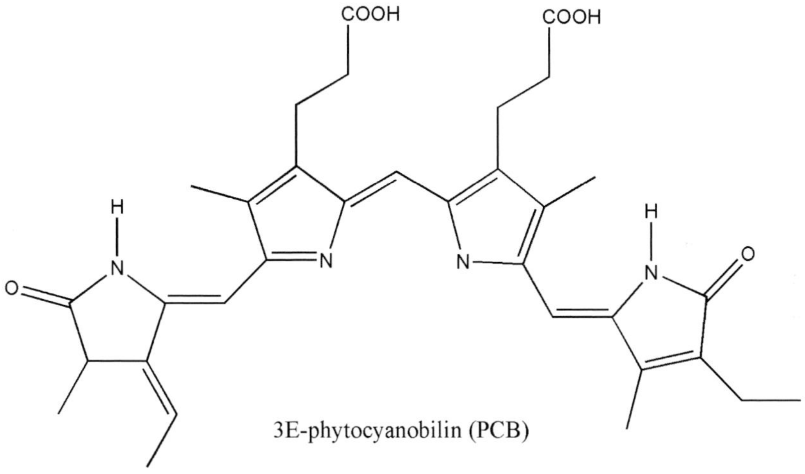

3E-phytocyanobilin (PCB)

Fig. 5.8 Phytochrome autocatalytically forms adducts with phytochromobilin (PΦB) as well phycobilin analogs, such as phytocyanobilin (PCB). The bilins are covalently attached by means of a thioester bond through the site, indicated with the arrow, and photoisomerize around the C15=C16 double bond.

References

1 K. J. Hellingwerf, J. Hendriks, T. Gensch, *J. Phys. Chem. A* **2003**, *107*, 1082–1094.

2 T. E. Meyer, *Biochim. Biophys. Acta* **1985**, *806*, 175–183.

3 M. J. Thompson, D. Bashford, L. Noodleman, E. D. Getzoff, *J. Am. Chem. Soc.* **2003**, *125*, 8186–8194.

4 U. K. Genick, S. M. Soltis, P. Kuhn, I. L. Canestrelli, E. D. Getzoff, *Nature* **1998**, *392*, 206–209.

5 R. Kort, K. J. Hellingwerf, R. B. G. Ravelli, *J. Biol. Chem.* **2004**, *279*, 26417–26424.

6 R. Brudler, R. Rammelsberg, T. T. Woo, E. D. Getzoff, K. Gerwert, *Nat. Struct. Biol.* **2001**, *8*, 265–270.

7 U. K. Genick, G. E. O. Borgstahl, K. Ng, Z. Ren, C. Pradervand, P. M. Burke, V. Srajer, T. Y. Teng, W. Schildkamp, D. E. McRee, K. Moffat, E. D. Getzoff, *Science* **1997**, *275*, 1471–1475.

8 B. Perman, V. Srajer, Z. Ren, T. Y. Teng, C. Pradervand, T. Ursby, D. Bourgeois, F. Schotte, M. Wulff, R.;Kort, K. Hellingwerf, K. Moffat, *Science* **1998**, *279*, 1946–1950.

9 Z. Ren, B. Perman, V. Srajer, T. Y. Teng, C. Pradervand, D. Bourgeois, F. Schotte, T. Ursby, R. Kort, M. Wulff, K. Moffat, *Biochemistry* **2001**, *40*, 13788–13801.

10 C. J. Craven, N. M. Derix, J. Hendriks, R. Boelens, K. J. Hellingwerf, R. Kaptein, *Biochemistry* **2000**, *39*, 14392–14399.

11 C. P. Joshi, B. Borucki, H. Otto, T. E. Meyer, M. A. Cusanovich, M. P. Heyn, *Biochemistry* **2005**, *44*, 656–665.

12 G. Groenhof, M. Bouxin-Cademartory, B. Hess, S. P. De Visser, H. J. C. Berendsen, M. Olivucci, A. E. Mark, M. A. Robb, *J. Am. Chem. Soc.* **2004**, *126*, 4228–4233.

13 A. Yamada, T. Ishikura, T. Yamato, *Proteins* **2004**, *55*, 1063–1069.

14 M. Zimmer, *Glowing Genes: A Revolution in Biotechnology.* Prometheus Books, Amherst, NY, **2005**.

15 V. Tozzini, V. Pellegrini, F. Beltram, *CRC Handbook of Organic Photochemistry and Photobiology*, 2nd edn. CRC Press, Boca Raton, FL, **2004**.

16 M. Zimmer, *Chem. Rev. 102*, **2002**, 759–781.

17 R. Y. Tsien, *Annu. Rev. Biochem.* **1998**, *67*, 509–544.

18 J. Dopf, T. M. Horiagon, *Gene* **1996**, *173*, 39–44.

19 A. B. Cubitt, R. Heim, S. R. Adams, A. E. Boyd, L. A. Gross, R. Y. Tsien, *Trends Biochem Sci* **1995**, *20*, 448–455.

20 R. Heim, D. C. Prasher, R. Y. Tsien, *Proc. Natl Acad. Sci. USA* **1994**, *91*, 12501–12504.

21 B. G. Reid, , G. C. Flynn, *Biochemistry* **1997**, *36*, 6786–6791.

22 B. R. Branchini, A. R. Nemser, M. Zimmer, *J. Am. Chem. Soc.* **1998**, *120*, 1–6.

23 M. Donnelly, F. Fedeles, M. Wirstam, P. E. M. Siegbahn, M. Zimmer, *J. Am. Chem. Soc.* **2001**, *123*, 4679–4686.

24 M. A. Rosenow, H. A. Huffman, M. E. Phail, R. M. Wachter, *Biochemistry* **2004**, *43*, 4464–4472.

25 D. P. Barondeau, C. D. Putnam, C. J. Kassmann, J. A. Tainer, E. D. Getzoff, *Proc. Natl Acad. Sci. USA* **2003**, *100*, 12111–12116.

26 H. Morise, O. Shimomura, F. H. Johnson, J. Winant, *Biochemistry* **1974**, *13*, 2656–2662.

27 W. W. B. Ward, *Biochemistry* **1982**, *21*, 4535–4540.

28 T. M. H. Creemers, A. J. Lock, V. Subramaniam, T. M. Jovin, S. Volker, *Nature Struct Biol* **1999**, *6*, 557–560.

29 R. M. Wachter, B. A. King, R. Heim, K. Kallio, R. Y. Tsien, S. G. Boxer, S. J. Remington, *Biochemistry* **1997**, *36*, 9759–9765.

30 M. Chattoraj, B. A. King, G. U. Bublitz, S. G. Boxer, *Proc. Natl Acad. Sci. USA* **1996**, *93*, 8362–8367.

31 J. J. van Thor, T. Gensch, K. J. Hellingwerf, L. N. Johnson, *Nat. Struct. Biol.* **2002**, *9*, 37–41.

32 A. F. Bell, D. Stoner-Ma, R. M. Wachter, P. J. Tonge, *J. Am. Chem. Soc.* **2003**, *125*, 6919–6926.

33 J. T. M. Kennis, D. S. Larsen, I. H. M. van Stokkum, M. Vengris, J. J. van Thor, R. van Grondelle, *Proc. Natl Acad. Sci. USA* **2004**, *101*, 17988–17993.

34 G. Bublitz, B. King, S. Boxer, *J. Am. Chem. Soc.* **1998**, *120*, 9370–9371.

35 A. F. Bell, X. He, R. M. Wachter, P. J. Tonge, *Biochemistry* **2000**, *39*, 4423–4431.

36 M. A. Elsliger, R. M. Wachter, G. T. Hanson, K. Kallio, S. J. Remington, *Biochemistry* **1999**, *38*, 5296–5301.

37 M. F. Garcia-Parajo, G. M. J. Segers-Nolten, J.-A. Veerman, J. Greeve, N. F. V. Hulst, *Proc. Natl Acad. Sci. USA* **2000**, *97*, 7237–7242.

38 G. Chirico, F. Cannone, A. Diaspro, *J. Phys. D Appl. Phys.* **2003**, *36*, 1682–1688.

39 G. Chirico, F. Cannone, A. Diaspro, S. Bologna, V. Pellegrini, R. Nifosi, F. Beltram, *Phys. Rev. E* **2004**, *7003*, 901–901.

40 W. Weber, V. Helms, J. McCammon, P. Langhoff, *Proc. Natl Acad. Sci. USA* **1999**, *96*, 6177–6182.

41 A. A. Voityuk, M.-E. Michel-Beyerele, N. Roesch, *Chem. Phys.* **1998**, *231*, 13–25.

42 H. Niwa, S. Inouye, T. Hirano, T. Matsuno, S. Kojima, M. Kubota, M. Ohashi, F. I. Tsuji, *Proc. Natl Acad. Sci. USA* **1996**, *93*, 13617–13622.

43 A. Follenius-Wund, M. Bourotte, M. Schmitt, F. Iyice, H. Lami, J. J. Bourguignon, J. Haiech, C. Pigault, *Biophys. J.* **2003**, *85*, 1839–1850.

44 N. M. Webber, K. L. Litvinenko, S. R. Meech, *J. Phys. Chem. B* **2001**, *105*, 8036–8039.

45 D. Mandal, T. Tahara, N. M. Webber, S. R. Meech, *Chem. Phys. Lett.* **2002**, *358*, 495–501.

46 D. Mandal, T. Tahara, S. R. Meech, *J. Phys. Chem. B* **2004**, *108*, 1102–1108.

47 K. L. Litvinenko, S. R. Meech, *Phys. Chem. Chem. Phys.* **2004**, *6*, 2012–2014.

48 N. Y. A. Baffour-Awuah, M. Zimmer, *Chem. Phys.* **2004**, *303*, 7–11.

49 A. Toniolo, S. Olsen, L. Manohar, T. J. Martinez, *Faraday Disc.* **2004**, *127*, 149–163.

50 M. C. Chen, C. R. Lambert, J. D. Urgitis, M. Zimmer, *Chem. Phys.* **2001**, *270*, 157–164.

51 S. L. Maddalo, M. Zimmer, *Photochem. Photobiol.* **2006**, *82*, 367–372.

52 R. Nifosi, A. Ferrari, C. Arcangeli, V. Tozzini, V. Pellegrini, F. Beltram, *J. Phys. Chem. B* **2003**, *107*, 1679–1684.

53 H. M. Berman, J. Westbrook, Z. Feng, G. Gilliland, T. N. Bhat, H. Weissig, I. N. Shindyalov, P. E. Bourne, *Nucleic Acids Res.* **2000**, *28*, 235–242.

54 M. Prescott, M. Ling, T. Beddoe, A. J. Oakley, S. Dove, O. Hoegh-Guldberg, R. J. Devenish, Rossjohn, J. *Structure* **2003**, *11*, 275–284.

55 J. Petersen, P. G. Wilmann, T. Beddoe, A. J. Oakley, R. J. Devenish, M. Prescott, J. Rossjohn, *J. Biol. Chem.* **2003**, *278*, 44626–44631.

56 D. M. Chudakov, A. V. Feofanov, N. N. Mudriku, S. Lukyanov, K. A. Lukyanov, *J. Biol. Chem.* **2003**, *278*, 7215–7219.

57 K. A. Lukyanov, A. F. Fradkov, N. G. Gurskaya, M. V. Matz, Y. A. Labas, A. P. Savitsky, M. L. Markelov, A. G. Zaraisky, X. N. Zhao, Y. Fang, W. Y. Tan, S. A. Lukyanov, *J. Biol. Chem.* **2000**, *275*, 25879–25882.

58 P. G. Wilmann, J. Petersen, R. J. Devenish, M. Prescott, J. Rossjohn, *J. Biol. Chem.* **2005**, *280*, 2401–2404.

59 C. Fankhauser, *J. Biol. Chem.* **2001**, *276*, 11453–11456.

60 A. J. Fischer, J. C. Lagarias, *Proc. Natl Acad. Sci. USA* **2004**, *101*, 17334–17339.

61 P. H. Quail, *Nat. Rev. Mol. Cell Biol.* **2002**, *3*, 85–93.

62 S. H. Wu, J. C. Lagarias, *Biochemistry* **2000**, *39*, 13487–13495.

63 B. Esteban, M. Carrascal, J. Abian, T. Lamparter, *Biochemistry* **2005**, *44*, 450–461.

64 W. Gartner, S. E. Braslavsky, In *Photoreceptors and Light Signalling*, A. Batschauer, Ed., Royal Society of Chemistry, Cambridge, UK, **2004**, *3*, 136–180.

65 H. Foerstendorf, C. Benda, W. Gartner, M. Storf, H. Scheer, F. Siebert, *Biochemistry* **2001**, *40*, 14952–14959.

66 B. Borucki, H. Otto, G. Rottwinkel,
J. Hughes, M. P. Heyn, T. Lamparter,
Biochemistry **2003**, *42*, 13684–13697.

67 P. Eilfeld, J. Vogel, R. Maurer, *Photo-chem. Photobiol.* **1987**, *45*, 825–830.

68 P. Eilfeld, W. Rudiger, *Z. Naturforsch.
C-a J. Biosci.* **1985**, *40*, 109–114.

69 M. A. Mroginski, D. H. Murgida, D. von
Stetten, C. Kneip, F. Mark,
P. Hildebrandt, *J. Am. Chem. Soc.* **2004**,
126, 16734–16735.

70 F. Yang, L. G. Moss, G. N. J. Phillips,
Nat. Biotechnol. **1996**, *14*, 1246–1251.

6

Fatty Acids and Phospholipids

Chryssostomos Chatgilialoglu and Carla Ferreri

6.1
Introduction

Lipids are a group of molecules with a wide structural diversity, classified together for their insolubility in water [1]. The primary building blocks of most cell membranes are glycerol-phosphate-containing lipids, generally referred to as phospholipids. The general structure of an L-α-phosphatidylcholine is shown in Scheme 6.1, with two hydrophobic fatty acid chains in the positions *sn*-1 and *sn*-2 of L-glycerol and the phosphorous-containing polar headgroup in *sn*-3 position.

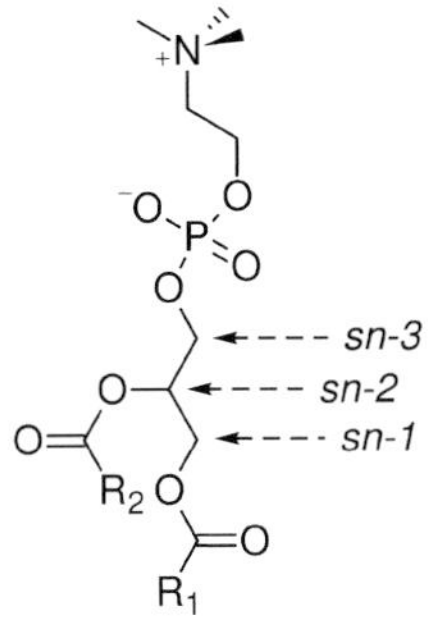

Scheme 6.1 Structure of L-α-phosphatidylcholine (PC). R_1 and R_2 are fatty acid residues.

The hydrophobic part consists of fatty acid residues that are carboxylic acids with a long hydrocarbon chain (up to 26 carbon atoms), saturated or unsaturated with up to six double bonds. Some of the most common mono- and polyunsaturated fatty acid (MUFA and PUFA) structures are shown in Scheme 6.2, with their common names and the abbreviations describing the position and geometry of the double bonds (e.g. 9-*cis*), as well as the notation of the carbon chain length and total number of unsaturations (e.g. C18:1). Naturally occurring MUFA and PUFA residues of glycerol-based phospholipids in eukaryotes generally have the *cis* double bond geometry, and PUFA double bonds have the characteristic methylene-interrupted motif. Being the *cis* geometry connected with biological activities

cis-trans Isomerization in Biochemistry. Edited by Christophe Dugave
Copyright © 2006 WILEY-VCH Verlag GmbH & Co. KGaA, Weinheim
ISBN: 3-527-31304-4

then, during MUFA and PUFA biosynthesis this feature is strictly controlled by the regiospecific and stereoselective enzymatic activity of desaturases [2].

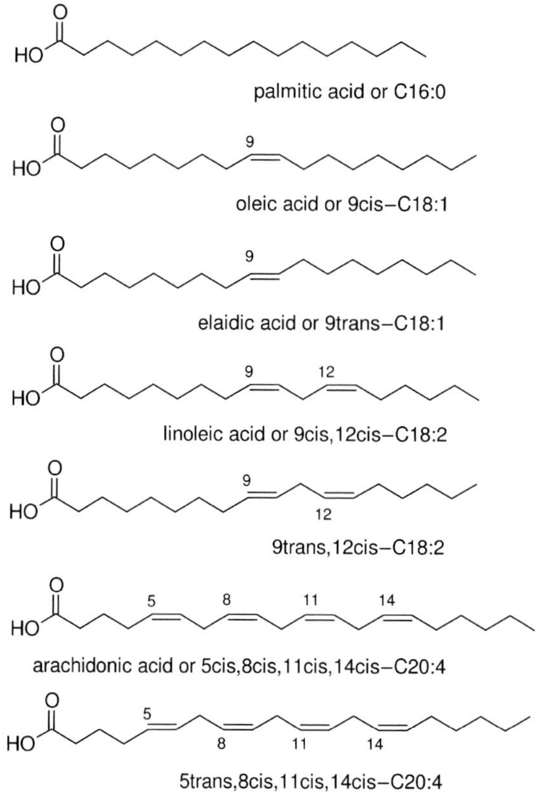

palmitic acid or C16:0

oleic acid or 9cis–C18:1

elaidic acid or 9trans–C18:1

linoleic acid or 9cis,12cis–C18:2

9trans,12cis–C18:2

arachidonic acid or 5cis,8cis,11cis,14cis–C20:4

5trans,8cis,11cis,14cis–C20:4

Scheme 6.2 Common names and numerical abbreviations of some natural fatty acids and examples of geometrical isomers.

Research on *cis-trans* isomerization (CTI) of lipid double bonds focused both on the conversion that occurs in some bacteria enzymatically and on *trans* isomers that are present in mammalian cells after a dietary supplementation of chemically modified fats [3,4]. It is known that *cis/trans* isomeric mixtures of fats result from vegetable and fish oils manipulated through partial hydrogenation or deodorization processes that are frequently utilized in the food industry. Nutritional and epidemiological studies revealed some harmful effects of these unnatural lipids for human health. However, it must be pointed out that in the chemical manipulation of oils the structures of *trans* fatty acid residues consist of geometrical and positional isomers with unshifted and shifted double bonds compared with the natural *cis* compounds. With the name "*trans* lipids" we indicate these unnatural geometrical and positional isomers. It has to be mentioned for clarity that there are a few natural lipids that exist only in the *trans* configuration, such as conjugated linoleic acid isomers, sphingolipids, and isoprene lipids.

A number of studies have confirmed that the broad spectrum of lipid mixtures plays an important role in the adaptability and flexibility of the cell membrane to environment necessities. The cell membrane is also a supporting matrix for proteins involved in many cellular processes, and therefore its physical and chemical properties can directly or indirectly affect cell metabolism. The isomerization of the *cis* double bond present in MUFA and PUFA residues of membrane phopsholipids to the corresponding more thermodynamically stable *trans* isomer is one of the lipid structural changes that has recently attracted the interest of diverse research areas. Geometric isomerism has become a topic of research involving several disciplines, such as microbiology, chemistry, biochemistry, pharmacology, nutrition, and medicine [3–5]. Recent work showing that the geometrical *cis* to *trans* lipid conversion can occur by a free radical process in a biological environment has certainly contributed to the interdisciplinary context of this subject [6,7]. The results of these fields aim at a global understanding of the occurrence of the *cis-trans* lipid conversion and its role in cell network signaling and membrane functioning.

6.2
Enzyme-Catalyzed *Cis-Trans* Isomerization of Unsaturated Fatty Acid Residues in Bacteria

In living cells, membrane properties such as permeability and "fluidity" have to be flexible to the various conditions; therefore there are different mechanisms to effect the necessary modification. Focusing on phospholipids, these mechanisms include changing the saturated/unsaturated ratio, changing the chain length or branching of the fatty acid residues, and the phospholipid polar headgroup. There is a strict correlation between the physical properties of biological membranes, going from a highly ordered to a fluid liquid crystalline phase, and the composition of lipids and fatty acid residues, hence the corresponding physical properties [8]. Saturated, *cis* and *trans* unsaturated phospholipids display very different values of phase transition temperatures (T_m) (the temperature at which the change between gel and liquid crystal phases occurs). For example, in the case of phosphatidylcholines (Scheme 6.1), where the R_1 fatty acid substituent is fixed as C16:0, and the R_2 varies through the series C18:0, 9-*trans*–C18:1 and 9-*cis*–C18:1, T_m values are 41.5 °C, 35 °C and, –3 °C, respectively [9].

It is worth underlining at this point that the molecular shapes of saturated and *cis* unsaturated lipids are quite different, since the *cis* geometry confers a kink in the lipid hydrocarbon chain, with an angle of about 30° in the acyl chain. In contrast, the molecular shape of the saturated fatty acyl chain is straight, and it is interesting to note that CTI corresponds to a "cancellation" of the bending, because *trans* isomers, like saturated compounds, are also straight (Fig. 6.1).

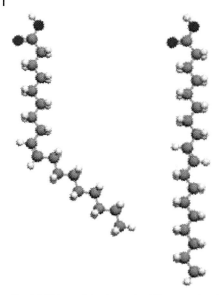

Fig. 6.1 This models shows the different molecular shape and the total volume occupied by the *cis* isomer (left) and *trans* isomer (right) of C18:1.

From the point of view of the bilayer self-organization, under physiological conditions the packing order of the lipids and the rigidity of the lipid assembly follows the order: saturated > *trans* unsaturated > *cis* unsaturated. The gross membrane property of "fluidity," as well as permeability, follows the inverted order, so that at a physiological temperature, the *cis* unsaturated residue ensures the most "fluid" state to the lipid assembly. When a stress is applied to cells, the first sign is an injury to specific membranes. Therefore, the greatest adaptive response occurs in membranes in order to keep the fluidity at a constant value, despite the changed environmental conditions. This adaptation, known as "homeoviscous adaptation" [10], is effected through the variation in the fatty acid composition of membrane lipids. In particular, the regulation of the degree of phospholipid saturation is well known in animals and plants, where the higher fluidity is obtained by the increase the double bond content [11–14]. Therefore, at low temperatures, adaptation is afforded by enrichment of polyunsaturated lipids, whereas at high temperatures the saturated fatty acid residues play the most important role for keeping the right properties of the lipid assembly.

It was in the early 1990s that the *cis* and *trans* geometries of unsaturated lipids were first discovered to be significantly involved in adaptation responses [4]. The first reports concerned the response of the psychrophilic bacterium *Vibrio* sp. strain ABE-1 [15] and of *Pseudomonas putida* P8 [16] to an increase in temperature or the presence of toxic phenol concentrations, respectively. The researchers detected in membranes *trans* lipids that were not dependent on growth, so that they did not derive from a de novo synthesis. In fact, they were also present in nongrowing cells, when lipid biosynthesis is absent and the total saturated and unsaturated fatty acid content cannot vary. Extending the analyses to other bacteria, it was shown that in obligate aerobic and anaerobic bacteria, biosynthesis provides oleic or palmitoleic acid, respectively, but formation of *trans* double bonds was not related to biosynthetic routes. The *trans* formation had an enzyme-like behavior, which increases proportionally to the chemical or environmental stress applied to the bacterial cultures. Only geometrical isomers (in particular 9-*trans*–C16:1) are found in anaerobic bacteria *Pseudomonas* sp. strain E-3, *Pseudomonas putida*, and *Vibrio* sp. strain ABE-1, as could be determined by gas chromatographic analyses. It is worth noting that microbiologists can determine the structure of the *trans* isomers present in bacterial lipids and the double bond position along the hydrocarbon chain by preparing by dimethyl disulfide addition the corresponding methylthio-derivatives, which have typical GC/MS spectra [17].

It was also observed that by adding a monounsaturated fatty acid, such as 9-*cis*–C18:1 that is not present in *P. putida* P8, it could be incorporated in the membrane, and under addition of toxic compounds, it was also converted to the corresponding *trans* isomer [18]. The total amount of monounsaturated fatty acid remained constant, but the *cis/trans* ratio changed according to the stress conditions. This was the proof of a biological path carried out by a specific enzyme, which worked without changing the position of the double bond but only its geometry. In addition, the system does not require ATP or any other cofactor such as NADP(H), glutathione, or oxygen [15,16,19]. The independence from energy providers could be explained by the higher thermodynamic stability of *trans* isomers.

The existence of a *cis-trans* isomerase was first hypothesized and then proved by cloning the *cti* gene from *Pseudomonas* strains, followed by purification and characterization of the enzyme. Based on the *cis-trans* isomerase activity test in different cell compartments, the cytoplasmic membrane was considered as the location of the enzyme, where phospholipids are also present. However, the enzyme was also purified from the periplasmic fraction and this fact was then explained because the isomerase has an N-terminal hydrophobic signal sequence (ca. 20 amino acids), which is cleaved off after targeting the enzyme to the periplasmic space [20]. The *cis-trans* isomerases are proteins of about 80 kDa molecular weight, whose sequences have been deduced from sequencing the corresponding gene [21]. Comparison of several *cis-trans* isomerase protein sequences has identified them as heme-containing proteins of the cytochrome *c*-type [20,22–24].

The optimal pH of isomerases is 7–8. The *cis-trans* isomerase in the solvent-tolerant bacterium *Pseudomonas putida* S12 mainly works for the transformation of palmitoleic acid (9-*cis*–C16:1) to its geometrical isomer 9-*trans*–C16:1. For example, in case of the addition of 3-nitrotoluene, it gives a final *cis/trans* ratio of 32:68 [25]. The *cis-trans* isomerase isolated from *Pseudomonas* sp. strain E-3 is flexible enough to convert the double bonds at positions 9, 10, or 11, but not those at positions 6 or 7, of *cis*-monounsaturated fatty acids having a chain length of 14, 15, 16, or 17 carbon atoms. CTI is 400- to 450-fold more efficient than the reverse reaction [22], and occurs on fatty acids in the free form. However, in the presence of membrane fraction, 9-*cis*–C16:1 residues esterified to phosphatidylethanolamine are isomerized, but not those esterified to phosphatidylcholine. This leads to the conclusion that in the *Pseudomonas* strain, where the levels of free fatty acids are low, it is more likely that this enzyme works directly with phospholipids and, consequently, the selectivity could stem from the recognition of the polar head.

No enzyme inhibition has been detected under anoxic conditions as well as in the presence of chelating agents, but the activity was strongly inhibited by 1 mmol L^{-1} cathecolic antioxidants, such as *a*-tocopherol and nordihydroguaretic acid. It is worth recalling that these compounds are also known as inhibitors of lipoxygenase (LOX) activity [22]. Several other stress elicitors have been tested, and it has been found that osmotic stress (caused by NaCl and sucrose), heavy metals, heat shock, and membrane-active antibiotics are all able to activate the *cis-trans* conversion [23,24].

Why do bacteria choose the *cis-trans* conversion as an adaptive response? Two main factors can be considered: (1) In the presence of high temperature, salt stress, or toxic compounds, the membrane fluidity and permeability have to decrease. This can be achieved by increasing the saturated/unsaturated ratio, which needs the activation of fatty acid biosynthesis, but this is only possible in conditions of growth. On the other hand, having a mechanism independent from the biosynthesis and directly effective on the *cis* compound allows a rapid modification of membranes to occur. In this way, the organism can rapidly cope with emerging environmental stress. After this response acting on a time-scale of minutes, other mid- and long-term mechanisms come into play, finally leading to a complete adaptation. (2) Despite the similarity of the straight molecular shape in saturated and *trans* lipids, the modifications induced in membrane properties are not completely the same. Lipid packing in the *trans* isomers has a more ordered state than in the *cis* isomers, and this has a substantial effect on the rigidity of the membrane, as in the case of saturated lipids. However, the change from *cis* to *trans* unsaturated double bonds does not have the same decreasing effect on membrane fluidity, compared with the change from *cis* to saturated fatty acids [26]. As will be described later (Section 6.3.3), model studies using liposome vesicles have been performed to elucidate the different contributions given by saturated, *cis* and *trans* fatty acid structures. Among them, it is worth noting the recent hypothesis of a basic contribution to the dimensions of the cell compartment delimited by membranes of different fatty acid composition [27]. This leads to an interpretation of the *cis* and *trans* geometries following an evolutionary point of view, which means that the *trans* geometry is compatible with the prokaryotic organisms, whereas only saturated and *cis*, and not *trans*, fatty acids have the optimal balance for the dimension and functions of eukaryotic cells.

Up to now, enzymatic isomerization has been evidenced mainly in *Pseudomonas* and *Vibrio* species. The fact that such an adaptation system was found in these organisms explains their wide occurrence in all niches of a great number of ecosystems, comprising soil, human skin, and water, also in comparison with other Gram-negative bacteria or species lacking such adaptation.

The molecular mechanism of enzymatic isomerization is far from being fully understood. In a study of *Pseudomonas* sp. strain E-3 the similarity between *cis-trans* isomerase and LOX was hypothesized on the basis of the common inhibition given by antioxidants. However, chelating agents, which do not affect isomerization, also inhibit LOX. A second mechanistic hypothesis was then formulated, consisting of the hydration–dehydration mechanism, similar to the formation of 3-*trans*-enoyl-CoA from the corresponding *cis* isomer [28]. Another recent mechanism was proposed based on analysis of carbon isotope fractionation of CTI [25]. Scheme 6.3 shows the proposed mechanism of CTI based on the enzyme–substrate complex as an intermediate, which allows the rotation of the carbon–carbon double bond to occur.

Since the involvement of the heme-binding motif was provided by site-directed mutagenesis experiments on the *cti* gene of *Pseudomonas putida* P8, which causes the loss of isomerization activity [29], the observed strong isotope fractionation provided evidence that isomerization includes a binding of the substrate to the active center of heme-containing protein of the cytochrome *c*-type. Future research on the structure of *cis-trans* isomerase should allow the mechanism to be properly defined.

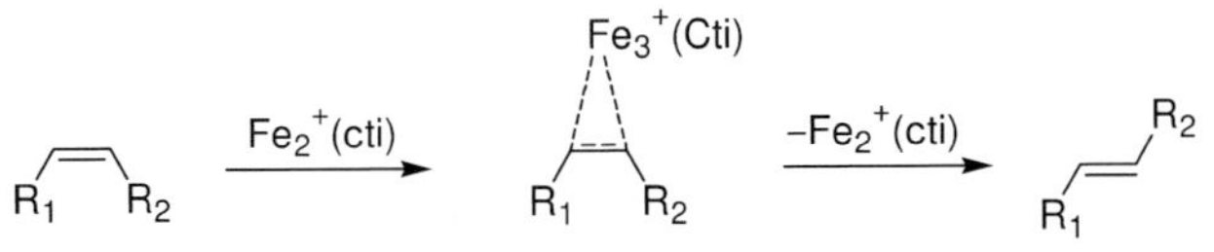

Scheme 6.3 Proposed mechanism for the enzymatic *cis-trans* isomerization of monounsaturated fatty acid by cti.

6.3
Radical-Catalyzed *Cis-Trans* Isomerization of Unsaturated Lipids and its Effect on Biological Membranes

6.3.1
Geometric Isomerization of Unsaturated Fatty Acids in Solution

Reactive species such as $RS^{\bullet}$, $RSO_2^{\bullet}$, $NO_2^{\bullet}$, or $R_3Sn^{\bullet}$ radicals and $Br^{\bullet}$ or $I^{\bullet}$ atoms have been known for a long-time to induce CTI of double bonds by addition–elimination steps [30]. Scheme 6.4 shows a reaction mechanism that consists of a reversible addition of reactive species $X^{\bullet}$ to the double bond to form the radical adduct **1**. The reconstitution of the double bond is obtained by β-elimination of $X^{\bullet}$ and the result is in favor of *trans* geometry, the most thermodynamically favorable disposition. Indeed, the energy difference between the two geometrical isomers of prototype 2-butene is 1.0 kcal mol^{-1}. It is worth noting that (1) the radical $X^{\bullet}$ acts as a catalyst for CTI, and (2) positional isomers cannot be formed as reaction products because the mechanism does not allow a double bond shift.

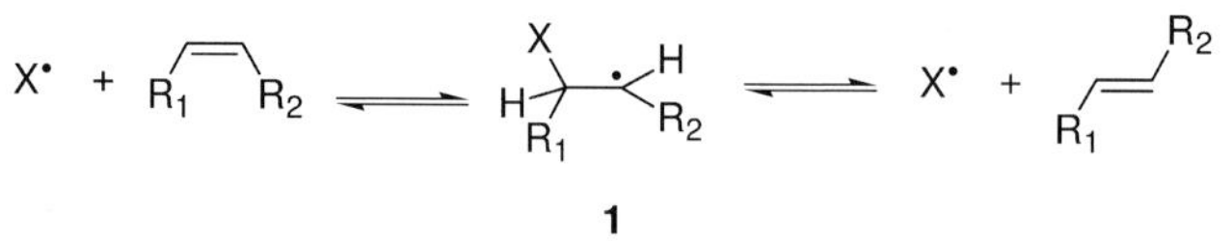

1

Scheme 6.4 Reaction mechanism for the *cis-trans* isomerization catalyzed by free radicals or atoms.

The efficiency of the isomerization process strongly depends on the characteristic of the attacking radicals. The most relevant species from a biological point of
view are the thiyl radical (RS•) and nitrogen dioxide (NO$_2$•), since both radicals
have been connected to cellular processes. The CTI by RS• is an efficient process
and detailed kinetic data are available for the reaction of methyl oleate with
HOCH$_2$CH$_2$S• radical (cf. Scheme 6.4) [31–33]. On the other hand, the available
kinetic data for CTI by NO$_2$• suggest that it cannot be very efficient as an isomerizing species, and in a biological environment this reaction should not play a role
[6]. Moreover, thiols are known to be the dominant "sink" for NO$_2$• in cell/tissues
with generation of thiyl radical [34]. Therefore, in the biological environment
thiyl radicals are likely to be the most relevant isomerizing species.

CTI reactions of methyl linoleate [35], γ-linolenate [35], and arachidonate [36]
catalyzed by thiyl radicals have been studied in some detail. Each isolated double
bond in PUFA behaves independently as discussed above. Indeed, the time profiles of methyl linoleate disappearance and formation of mono-*trans* and di-*trans*
isomers in these experiments indicated that the CTI occurs stepwise (Scheme
6.5). The number of possible geometrical isomers increases according to 2^n, where
n is the number of double bonds, and the complete analysis may be difficult and
incomplete for high unsaturation as in the case of methyl arachidonate.

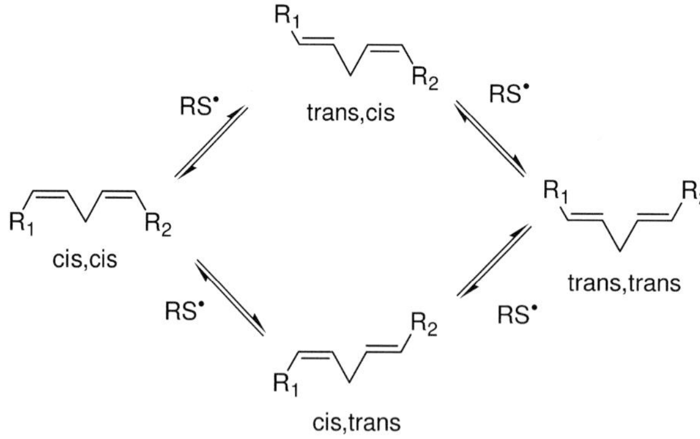

Scheme 6.5 *Cis-trans* isomerization of polyunsaturated fatty
acid residues catalyzed by thiyl radicals.

The CTI of unsaturated fatty acid residues of a variety of L-α-phosphatidyl-
cholines catalyzed by thiyl radicals were studied in alcoholic solutions. POPC (pal-
mitoyl oleoyl PC), DOPC (dioleoyl PC), and SAPC (stearoyl arachidonoyl PC), to-

gether with soybean and egg yolk lecithins, which are a mixture of phosphatidyl-cholines with different fatty acid chains, were used [31,35,37,38]. After transester-ification of the phospholipids, the fatty acid methyl esters composition was determined by gas chromatographic analysis. The results are similar to the analogous isomerization of fatty acid methyl esters (i.e. all *cis* double bonds isomerize with the same efficiency independently of their location, affording only geometrical isomers), and the step-by-step mechanism shown in Scheme 6.5 operates for PUFA residues. Figure 6.2 illustrates the experiment with egg lecithin, where the time-courses of disappearance of oleate (○), linoleate (●), and arachidonate (▲) residues are normalized to 100% for a better comparison. In the initial stage of the reaction, the arachidonate residue (having four double bonds) isomerizes twice as fast as the linoleate residue (having two double bonds) and 4 times faster than the oleate residue (having one double bond). It is also worth underlining that the two mono-*trans* isomers of linoleate residues are formed in the same amounts (see inset Fig. 6.2) as occurs with the four mono-*trans* isomers of arachidonate residues.

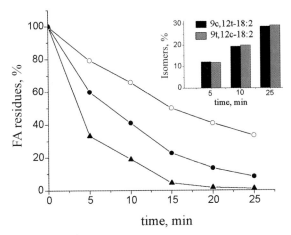

Fig. 6.2 Time profiles of disappearance of *cis* fatty acid residues obtained by photolysis of egg yolk lecithin in isopropanol. Conditions: 15 mmol L^{-1} of fatty acid contents with 7 mmol L^{-1} HOCH$_2$CH$_2$SH at 22 °C. Inset: Time-courses of the formation of two mono-*trans* isomers 9-*cis*,12-*trans*–C18:2 and 9-*trans*,12-*cis*–C18:2.

6.3.2
Isomerization of Phosphatidylcholine in Large Unilamellar Vesicles

Liposomes such as large unilamellar vesicles obtained by extrusion technique (LUVET) are generally accepted as close models of cell membranes. Figure 6.3 represents a sketch of a vesicle, where it can be seen that the system consists of two distinct compartments – the aqueous and lipid phases. The phospholipid polar heads face the aqueous internal and external phases, whereas the fatty acid chains form the hydrophobic bilayer of the model membrane.

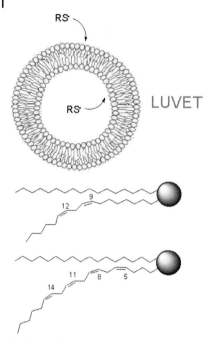

Fig. 6.3 Schematic representation of Large Unilamellar Vesicles made by Extrusion Technique (LUVET) and two potential phosphatidylcholines. Thiyl radicals (RS•) are initially generated in the aqueous compartment and may enter the lipid bilayer.

The CTI of unsaturated fatty acid residues in LUVET catalyzed by thiyl radicals was also studied in some detail [31,35,37]. Trends of the reactivity indicated the overall picture of geometric isomerization in model membranes by the action of diffusible thiyl radicals. In particular, using vesicles made of egg yolk lecithin, it was possible to demonstrate that the double bonds located closest to the membrane polar region are the most reactive towards the attack of diffusing thiyl radicals [37]. In the case of linoleic acid residues in vesicles, the double bond in position 9 was more reactive than that in position 12. Also arachidonic acid residues in vesicles were more reactive than oleic and linoleic acids, and two positions (i.e. the double bonds in 5 and 8) out of the four present in this compound were transformed preferentially. From the studies carried out so far, arachidonic acid residues in membrane phospholipids emerge as very important components to be investigated, because they allow endogenous *trans* isomers, formed by radical processes, to be distinguished from exogenous *trans* isomers derived from dietary contribution.

In particular, investigations should focus on the erythrocyte membrane phospholipids, which are the preferential storage place for arachidonic acid after biosynthesis. As shown in a previous section, nutritional investigations indicated that *trans* fatty acids are incorporated in cell membranes because the *trans* dietary precursors can be processed in vivo. In the case of arachidonic acid, as shown in the biosynthetic paths of Scheme 6.6, two double bonds (positions 11 and 14) originate from linoleic acid, the precursor taken from the diet, whereas the two other double bonds (positions 5 and 8) are formed by desaturase enzymes, which produce selectively the *cis* unsaturation. It is evident that the 5 and 8 double bonds of arachidonic acid, stored in membrane phospholipids, can only have a *cis* configuration, unless these positions have been involved in an isomerization process and converted to *trans* isomers.

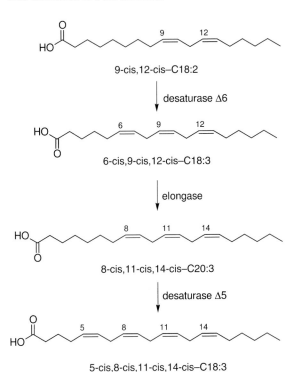

9-cis,12-cis–C18:2

desaturase Δ6

6-cis,9-cis,12-cis–C18:3

elongase

8-cis,11-cis,14-cis–C20:3

desaturase Δ5

5-cis,8-cis,11-cis,14-cis–C18:3

Scheme 6.6 Enzymatic fatty acid transformation.

Some insights on thiyl radical formation in a biological environment and its consequences to the unsaturated membrane lipids have recently been gathered during the study of a biomimetic model of liposomes in the presence of some sulfur-containing proteins. A new tandem radical damage has been proposed, which couples the generation of thiyl radicals from the protein damage and the formation of *trans* lipids in the membrane bilayer [38]. Since the early 1960s it has been known that damage caused by hydrogen atoms H• to a protein, namely ribonuclease A from bovine pancreas (RNase A), affords the release of the low-molecular-weight thiol moiety CH_3SH [39]. This knowledge has been used in the design of a biomimetic model, composed of a *cis*-unsaturated lipid vesicle (dioleoyl phosphatidylcholine, DOPC) and RNase A, to correlate the formation of thiyl radicals with the lipid damage. The protein was used at micromolar levels and lipid concentration was at millimolar level. The system underwent γ-irradiation, which induces the release of a low-molecular-weight thiol from the methionine residues. From the thiol under radical conditions, the reactive and highly diffusible thiyl radical species $CH_3S^\bullet$ could be formed. This rapidly diffuses in the lipid bilayer, causing isomerization of the double bonds (Scheme 6.7). Indeed, the formation of *trans* residues in the vesicles occurred by the catalytic cycle of the isomerization, thus amplifying the thiyl radical generation since the first instant of irradiation. Therefore, it is evident that the isomerization can be proposed as a very sensitive tool for detecting this type of protein damage at nanomolar level, which could be very difficult to test by analytical techniques. Extension of this model to other methionine-containing proteins (amyloid-β peptide [60], enkephalins) could provide a full evaluation of this damage, also in relationship with some human diseases.

It is worth noting that the radical damage to methionine-containing peptides and proteins consists of a desulfurization process, which leads to the replacement of a methionine residue with an α-aminobutyric acid in the sequence. This could be a posttranslational modification, which is linked to a postsynthetic modification of lipids by the above-reported tandem mechanism. A chemical biology approach can be proposed involving lipidomics and proteomics, in order to configure the metabolic changes related to a radical stress.

The conclusive picture emerging from the chemical studies under biomimetic conditions is that thiyl radicals are efficient catalysts for CTI of lipids in bilayers, and this process cannot be ignored when considering radical damage to biological components.

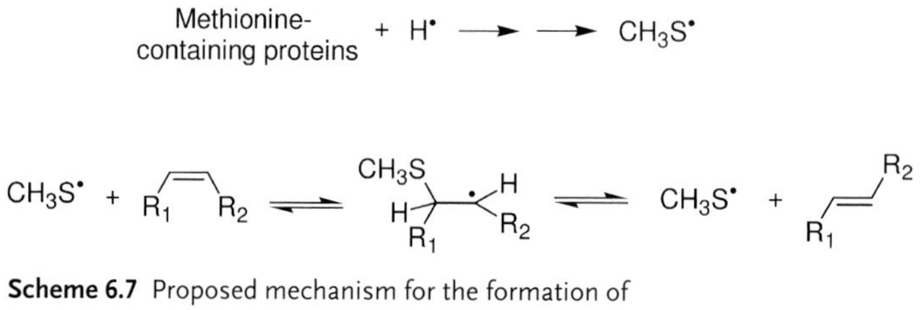

Scheme 6.7 Proposed mechanism for the formation of thiyl radicals from methionine-containing proteins under γ-irradiation.

6.3.3
Biological Consequences

As previously recalled, enzymatic CTI in eukaryotic cells is unknown and the presence of *trans* fatty acid isomers in humans has been generally attributed to exogenous sources. After a series of studies in several countries it was found that *trans* fatty acid isomers can give harmful effects on health, involving risk factors of heart attack and coronary artery disease, impairment of fetal and infant growth and inhibition of lipid metabolic pathways. Detailed information on these effects can be found in several studies, reviews, and books [3,40–42]. Recently, new regulations in the USA establishes that by 2006 the *trans* content must be indicated in the nutritional information about foods [43].

Whether an endogenous path by free radical transformation could account for *trans* isomer formation had still not been demonstrated, until recently. The working hypothesis started from the fact that several radical-based processes occur during normal cell metabolism [44], and a certain number of thiyl radicals could be derived from the intracellular sulfur-containing compounds. From the biomimetic models described in the previous section, it was clear that an isomerization process, not to be confused with a dietary contribution, could be evaluated by the detection of mono-*trans* isomers of arachidonate in membrane phospholipids. The isomer trends of the models were very useful for helping the identification of isomers. Cell cultures of human leukemia cell lines (THP-1) were incubated in the absence and presence of thiol compounds, ensuring that no *trans* compounds could come from the medium [45]. In parallel experiments, some millimolar levels of thiol compounds were added to the cell cultures during incubation, and the comparison of isomeric trends was carried out. A basic content of *trans* lipids in THP-1 cell membranes could be found during their growth without thiol, and after the addition of the amphiphilic 2-mercaptoethanol, it increased to 5.6% of the main fatty acid residues. Moreover, when a radical stress by γ-irradiation is artificially produced in the cell cultures added with thiol, a larger isomerization effect could be seen, with *trans* lipid formation up to 15.5% in membrane phospholipids. The fatty acid residues most involved in this transformation were arachidonate moieties, as expected.

This result can be considered as the first evidence for a geometrical change induced in the biological environment under radical stress, and it opens new perspectives for the role of *trans* lipids in the lipidome of eukaryotic cells.

What is the relevance of *trans* lipid geometry for biological consequences? As previously recalled, some *trans* isomers are natural, such as the conjugated isomers of linoleic acid. Because of this, the biological consequences of *trans* isomers cannot always be considered to be negative [46]. The whole scenario is not yet available. Some biological effects of *trans* lipids during cell metabolism have been examined, in particular for their interaction with lipid enzymes. In fact, they can enter the lipid cascades or be incorporated in membrane phospholipids, but give rise to different molecules, which can influence cell properties and functions [47,48].

In the case of polyunsaturated substrates, some mono-*trans* isomers have been prepared by total synthesis and examined in biological assays, showing that the mono-14-*trans* isomer of arachidonic acid can react with cytochrome P450 epoxygenase, a monooxygenase enzyme present in rat liver microsomes. The corresponding epoxide is produced, and this indicates that a natural pathway produces an unnatural compound [49]. Also the degradation of *trans*-unsaturated fatty acids by β-oxidation in peroxisomes has been tested, using the model of *Saccharomyces cerevisiae* [50]. Efficient cell growth with elaidic acid was found and the need for

β-oxidation auxiliary enzymes to metabolize the *trans* double bond in the odd-numbered position of linolelaidic acid (i.e. position 9) was also reported [51].

There are other cases of inhibition of lipid enzymatic pathways by *trans* fatty acid isomers. The above reported mono-14-*trans* isomer of arachidonic acid is inhibitor of the synthesis of thromboxane B_2 and, therefore, can prevent rat platelet aggregation [52]. The transformation of mono-*trans* isomers of linoleic acid by rat liver microsomes showed that the 9-*cis*,12-*trans* isomer is better desaturated, whereas the 9-*trans*,12-*cis* isomer (Scheme 6.1) is better elongated [53].

trans Lipid structures can be also considered as antisense analogs of natural all-*cis* PUFA molecules, and thiyl radical-catalyzed CTI represents an easy access to all-*trans* polyunsaturated compounds. A chemical biology approach can be used to address the significance of *cis* geometry in cells. A recent study of the activity of all-*trans* isomer of arachidonic acid (i.e. 5-*trans*,8-*trans*,11-*trans*,14-*trans*–C20:4) in rabbit platelet aggregation, showed that it is 10 times less active than the natural substrate, but at micromolar levels it is an efficient inhibitor of the response induced by the platelet aggregating factor (PAF) [54].

As different biological activities are displayed by each *trans* isomer, a careful determination of the type and concentration of these unnatural compounds in vivo is needed. In particular, the characterization of membrane lipids containing arachidonic residues can be important for functional lipidomics, in order to achieve a clear understanding of the contribution from endogenous or exogenous processes to the presence of *trans* isomers in vivo. Moreover, if one considers the close relationship established between free radical processes and human pathologies and aging, the functional lipidomic approach involving arachidonate geometrical isomers could provide additional useful information on the role of radical stress conditions in health and diseases [7]. This is a challenge for sophisticated analytical techniques, able to separate and identify all possible isomers, especially in the complicated case of PUFA. In fact, as previously pointed out (Section 6.3.1), the number of geometrical isomers for an unsaturated compound (equal to 2^n) can be very high (for example, for a C22:6 fatty acid, $2^6 = 64$). The presence of a *trans* lipid library achieved by radical CTI can help to simplify this issue, starting from the recognition between geometrical and positional isomers.

We have previously mentioned that lipids make part of the membrane architecture, which has the typical bilayer arrangement due to the phospholipid supramolecular organization. This leads to biological consequences – the composition of fatty acid residues with saturated and unsaturated hydrocarbon chains is crucial to regulate membrane properties, maintaining the best balance for cellular functioning and also survival. Vesicle models made of phosphatidylcholines with saturated and unsaturated fatty acid residues are useful for studies of permeability and "fluidity." Several studies have compared the effects of saturated, *cis* and *trans* unsaturated residues. An example is given in Fig. 6.4 for vesicles made up of different phospholipid compositions.

In this study, vesicles made of a saturated phosphatidylcholines (open circles), that is, dipalmitoylphosphatidylcholine (DPPC), where R_1 and R_2 substituents are both C16:0 fatty acid chains (Scheme 6.1), were compared with: (a) *cis* lipid vesi-

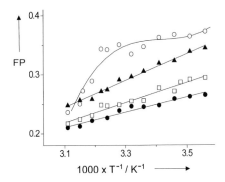

Fig. 6.4 Polarization of TMA-DPH fluorescence (FP) as a function of temperature in dipalmitoylphosphatidylcholine (DPPC) (open circles) and dioleoylphosphatidylcholine (DOPC) (solid circles) vesicles, as well as in DPPC/DOPC (1/1) (open squares) and DPPC/dielaidoylphosphatidylcholine (DEPC)/DOPC (5/4/1) (triangles) vesicles.

cles (solid circles) made of dioleoylphosphatidylcholine (DOPC), that is, R_1 and R_2 substituents are both 9-*cis*–C18:1; (b) lipid vesicles (triangles) which contain a *trans* phosphatidylcholine, dielaidoylphosphatidylcholine (DEPC), that is, R_1 and R_2 substituents are 9-*trans*–C18:1, in mixture with DPPC and DOPC to reach the ratio of 5/4/1 DPPC/DEPC/DOPC; (c) vesicles containing a 1/1 ratio of DPPC/ DOPC (open squares). It was clearly shown that vesicles with a 40% *trans* lipid content have higher polarization values than those where only *cis* residues are present, or those with a mixture of saturated and *cis* residues, demonstrating the relevant contribution of the *trans* geometry to decreasing the membrane "fluidity" [31].

Alteration of the physical properties of membranes due to the presence of *trans* fatty acid residues, can be also connected with some biological effects. Phosphatidylcholine vesicles containing geometrical isomers of PUFA residues were found to be less efficiently oxidized than the corresponding *cis* lipids [55]. Another recent investigation on model membranes containing different *trans* monounsaturated fatty acid residues (C14:1, C16:1, and C18:1) determined that the affinity for cholesterol was 40–80% higher than that in their *cis* analogs, probably due to a better interaction between the straight *trans* acyl chain and the cholesterol molecule [56]. In the same report, the behavior of rhodopsin, a prototypical member of the G protein-coupled receptor family, was evaluated as influenced by the *trans* geometry. In *trans* membrane models, the level of rhodopsin activation was diminished, in particular at lower temperatures (5 °C), where *trans* isomers are in the gel state, whereas *cis* isomers are in the fluid state.

Some recent work has also been devoted to a possible evolutionary meaning of the *trans* geometry in cells. Considering the chemical evidence that *trans* structures are the most stable and the biological fact of the natural occurrence of *cis* isomers, it is interesting to ask when and why the *cis* geometry became predominant in eukaryotic unsaturated fatty acids. Curiously, this question has been omitted from the interesting debates of the fatty acid presence during the evolution of cell membranes and humans [57–59]. The current argument that *cis* lipids are needed for optimal cell membrane balance cannot be considered exhaustive. It is not clear that *trans* isomers have been a priori excluded from the eukaryotic cell membrane composition; it could be a result of life evolution and selection. The fact that some bacteria still use *trans* lipid geometry could favor the hypothesis of an evolutionary meaning.

Critical aggregation concentration (CAC) to form spontaneously unilamellar vesicles and the resulting average dimensions were measured for *cis* and *trans* phospholipids using fluorescence and light-scattering techniques, together with electron microscopy [27]. These measurements revealed that the two geometrical lipid isomers (POPC and PEPC) aggregate at almost the same concentration, that is, in the range of 4–5 × 10^{-6} mol L^{-1}. On the other hand, the diameter of the POPC vesicle was found to be 25% larger than the corresponding *trans* fatty acid-containing vesicle. The influence of the unsaturation degree of fatty acid residues on the vesicle dimensions was therefore established, with a decrease along the series *cis* > *trans* > saturated. Further work on model vesicles will contribute to a comprehensive picture of the lipid geometry in cell evolution.

6.4
Perspectives and Future Research

The *cis-trans* isomerization of lipid structures has shown its versatility, spanning from microbiology to chemistry, including biochemistry, nutrition, and medicine. The chemical work done in vesicles can be interestingly applied to synthetic transformations in organized systems and in aqueous media, which cover both the selectivity and the environmental aspects. Also, biotechnological applications involving membrane behavior can be foreseen. Lipid and thiyl radical reactivity will be further developed in the direction of signaling pathways involved in biological processes, dealing in particular with the *trans* lipid geometry, the effect of this structural change on biological and pharmacological interactions. The combination of data achieved by different approaches is finally expected to contribute to lipidomics and the role of geometrical *trans* lipid isomers in living organisms.

References

1 D. E. Vance, J. E. Vance, Eds. *Biochemistry of Lipids, Lipoproteins and Membranes*, 4th edn. Elsevier, Amsterdam, **2002**.

2 B. G. Fox, K. S. Lyle, C. E. Rogge. Reactions of the diiron enzyme stearoyl-acyl carrier protein desaturase. *Acc. Chem. Res.* **2004**, *37*, 421–429.

3 J. L. Sébédio, W. W. Christie, Eds. *Trans Fatty Acids in Human Nutrition*. The Oily Press, Dundee, **1998**.

4 H. Keweloh, H. J. Heipieper, *Lipids* **1996**, *31*, 129–137.

5 C. Dugave, L. Demange, *Chem Rev.* **2003**, *103*, 2475–2532.

6 C. Chatgilialoglu, C. Ferreri, *Acc. Chem. Res.* **2005**, *38*, 441–448.

7 C. Ferreri, C. Chatgilialoglu, *ChemBioChem.* **2005**, *6*, 1722–1734.

8 J. E. Cronan, E. P. Gelman, *Bacteriol. Rev.* **1975**, *39*, 232–320.

9 G. Cevc, Ed. *Phospholipid Handbook*. Marcel Dekker, New York, **1993**.

10 M. Sinensky, *Proc. Natl Acad. Sci. USA* **1974**, *71*, 522–525.

11 T. Sakamoto, N. Murata, *Curr. Opin. Microbiol.* **2002**, *5*, 206–210.

12 Y. Murakami, M. Tsuyama, Y. Kobayashi, H. Kodama, K. Iba, *Science* **2000**, *287*, 476–479.

13 S. I. Allakhverdiev, Y. Nishiyama, I. Suzuki, Y. Tasaka, N. Murata, *Proc. Natl Acad. Sci. USA* **1999**, *96*, 5862–5867.

14 N. Holmberg, L. Bülow, *Trends Plant Sci.* **1998**, *3*, 61–66.

15 H. Okuyama, N. Okajima, S. Sasaki, S. Higashi, N. Murata, *Biochim. Biophys. Acta* **1991**, *1084*, 13–20.

16 H. J. Heipieper, R. Diefenbach, H. Keweloh, *Appl. Environ. Microbiol.* **1992**, *58*, 1847–1852.

17 C. Wayne Moss, M. A. Lambert-Fair, *J. Clin. Microbiol.* **1989**, *27*, 1467–1470.

18 R. Diefenbach, H. Keweloh, *Arch. Microbiol.* **1994**, *162*, 120–125.

19 R. Diefenbach, H. J. Heipieper, H. Keweloh, *Appl. Microbiol. Biotechnol.* **1992**, *38*, 382–387.

20 H. J. Heipieper, F. Meinhardt, A. Segura, *FEMS Microbiol. Lett.* **2003**, *229*, 1–7.

21 V. Pedrotta, B. Witholt, J. Bacteriol. **1999**, *181*, 3256–3261.

22 H. Okuyama, A. Ueno, D. Enari, N. Morita, T. Kusano, *Arch. Microbiol.* **1998**, *169*, 29–35.

23 G. Neumann, N. Kabelitz, H. J. Heipieper, *Eur. J. Lipid Sci. Technol.* **2003**, *105*, 585–589.

24 S. Isken, P. Santos, J. A. M. de Bont, *Appl. Microbiol. Biotechnol.* **1997**, *48*, 642–647.

25 H. J. Heipieper, G. Neumann, N. Kabelitz, M. Kastner, H. H. Richnow, *Appl. Microbiol. Biotechnol.* **2004**, *66*, 285–290.

26 C. Roach, S. E. Feller, J. A. Ward, S. Raza Shaikh, M. Zerouga, W. Stillwell, *Biochemistry* **2004**, *43*, 6344–6351.

27 C. Ferreri, S. Pierotti, A. Barbieri, L. Zambonin, L. Landi, S. Rasi, P. L. Luisi, F. Barigelletti, C. Chatgilialoglu, *Photochem. Photobiol.* **2006**, *82*, 274–280.

28 A. von Wallbrun, H. H. Richnow, G. Neumann, F. Meinhardt, H. J. Heipieper, *J. Bacteriol.* **2003**, *185*, 1730–1733.

29 R. Holtwick, H. Keweloh, F. Meinhardt, *Appl. Environ. Microbiol.* **1999**, *65*, 2644–2649.

30 P. Renaud, M. P. Sibi, Eds. *Radicals in Organic Synthesis*. Wiley-VCH, Weinheim, **2001**.

31 C. Chatgilialoglu, C. Ferreri, M. Ballestri, Q. G. Mulazzani, L. Landi, *J. Am. Chem. Soc.* **2000**, *122*, 4593–4601.

32 C. Chatgilialoglu, A. Altieri, H. Fischer, *J. Am. Chem. Soc.* **2002**, *124*, 12816–12823.

33 C. Chatgilialoglu, A. Samadi, M. Guerra, H. Fischer, *ChemPhysChem* **2005**, *6*, 286–291.

34 E. Ford, M. N. Hughes, P. Wardman, *Free Radical Biol. Med.* **2002**, *32*, 1314–1323.

35 C. Ferreri, C. Costantino, L. Perrotta, L. Landi, Q. G. Mulazzani, C. Chatgilialoglu, *J. Am. Chem. Soc.* **2001**, *123*, 4459–4468.

36 C. Ferreri, M. R. Faraone Mennella, C. Formisano, L. Landi,

C. Chatgilialoglu, *Free Radical Biol. Med.* **2002**, *33*, 1516–1526.

37 C. Ferreri, A. Samadi, F. Sassatelli, L. Landi, C. Chatgilialoglu, *J. Am. Chem. Soc.* **2004**, *126*, 1063–1072.

38 C. Ferreri, I. Manco, M. R. Faraone-Mennella, A. Torreggiani, M. Tamba, C. Chatgilialoglu, *ChemBioChem.* **2004**, *5*, 1710–1712.

39 R. Shapira, G. Stein, *Science* **1968**, *162*, 1489–1491.

40 T. L. Roberts, D. A. Wood, R. A. Riemersma, P. J. Gallagher, F. C. Lampe, *Lancet* **1995**, *345*, 278–282.

41 E. Larquè, S. Zamora, A. Gil, *Early Hum. Dev.* **2001**, *65*, S31–S41.

42 F. B. Hu, M. J. Stampfer, J. E. Manson, E. Rimm, G. A. Colditz, B. A. Rosner, C. H. Hennekens, W. C. Willett, *N. Engl. J. Med.* **1997**, *337*, 1491–1499.

43 M. Bender Brandt, L. A. LeGault, *J. Food Comp. Anal.* **2003**, *16*, 383–393.

44 C. Ferreri, S. Kratzsch, L. Landi, O. Brede, *Cell. Mol. Life Sci.* **2005**, *62*, 834–847.

45 C. Ferreri, S. Kratzsch, O. Brede, B. Marciniak, C. Chatgilialoglu, *Free Radical Biol. Med.* **2005**, *38*, 1180–1187.

46 K. W. J. Wahle, S. D. Heys, D. Rotondo, *Prog. Lipid Res.* **2004**, *43*, 553–587.

47 B. Koletzko, T. Decsi, *Clin. Nutr.* **1997**, *16*, 229–237.

48 J. L. Sébédio, S. H. F. Vermunt, J. M. Chardigny, B. Beaufrére, R. P. Mensink, R. A. Armstrong, W. W. Christie, J. Niemela, G. Hènon, R. A. Riemersma, *Eur. J. Clin. Nutr.* **2000**, *54*, 104–113.

49 U. Roy, O. Loreau, M. Balazy, *Bioorg. Med. Chem. Lett.* **2004**, *14*, 1019–1022.

50 M. Guzman, W. Klein, T. G. del Pulgar, M. J. Geelen, *Lipids* **1999**, *34*, 381–386.

51 A. Gurvitz, B. Hamilton, H. Ruis, A. Hartig, *J. Biol. Chem.* **2001**, *276*, 895–903.

52 O. Berdeaux, J. M. Chardigny, J. L. Sébédio, T. Mairot, D. Poullain, J. M. Vatèle, J. P. Noël, *J. Lipid Res.* **1996**, *37*, 2244–2250.

53 O. Berdeaux, J. P. Blond, L. Bretillon, J. M. Chardigny, T. Mairot, J. M. Vatèle, D. Poullain, J. L. Sébédio, *Mol. Cell. Biochem.* **1998**, *185*, 17–25.

54 D. Anagnostopoulos, C. Chatgilialoglu, C. Ferreri, A. Samadi, A. Siafaka-Kapadai, *Bioorg. Med. Chem. Lett.* **2005**, *15*, 2766–2770.

55 R. M. Sargis, P. V. Subbaiah, *Biochemistry* **2003**, *42*, 11533–11543.

56 S.-L. Niu, D. C. Mitchell, B. J. Litman, *Biochemistry*, **2005**, *44*, 4458–4465.

57 R. G. Ackman, *J. Food Lipids* **1997**, *4*, 295–318.

58 D. Deamer, J. P. Dworkin, S. A. Sandford, M. P. Bernstein, L. J. Allamandola, *Astrobiology* **2004**, *2*, 371–381.

59 D. Segrè, D. Ben-eli, D. W. Deamer, D. Lancet, *Origins Life Evol. B.* **2001**, *31*, 119–145.

60 V. Kadlcik, C. Sicard-Roselli, C. Houée-Levin, M. Kodicek, C. Ferreri, C. Chatgilialoglu, *Angew. Chem. Int. Ed* **2006**, *45*, 2595–2598.

7

In Silico Dynamic Studies of *Cis-Trans* Isomerization in Organic and Biological Systems*

Ute F. Röhrig, Ivano Tavernelli, and Ursula Rothlisberger

7.1
Introduction

There are a large number of computational studies on photoinduced *cis-trans* isomerization (CTI) (see, e.g., Refs [1,2] and references therein). In the present chapter, we will discuss solely photoinduced CTI, as opposed to thermally induced CTI. In addition, we will focus on studies directed towards understanding of the primary reaction of the visual cascade (i.e. the CTI of the retinal protonated Schiff base (RPSB) in rhodopsin). The accurate description of the rhodopsin photocycle poses several challenges for state-of-the-art computational methods and resources. Experimental data reveal the importance of the protein environment for ultrafast and efficient photoisomerization. However, the simulation of the dynamics of a whole protein within its native environment (e.g. aqueous solution or a cell membrane) is feasible nowadays only by using empirical force fields for the description of interatomic interactions. This approach, called "classical molecular dynamics" because it is based exclusively on Newtonian mechanics as opposed to quantum mechanics, allows for the treatment of systems up to a size of 100 000 atoms for a time of up to 1 µs, and has been applied successfully to the study of many biological processes. Since some initial structural information is necessary, applications to rhodopsin [3–8] have only emerged after publication of its crystal structure [9–12]. The accessible time-scale would allow for the exploration of the experimentally detected photointermediates [13] up to lumirhodopsin, and could possibly be extended even further with enhanced sampling techniques. However, since the electronic structure is not taken explicitly into account, chemical reactions like bond cleavage, bond formation, or electronic excitations cannot be described within this approach. This type of problem has led to the idea of combining electronic structure methods, which are able to treat chemical transformations, with classical methods in order to include environment effects. In the so-called QM/MM (quantum mechanical/molecular mechanical) approach [14–16], only one part of the system (e.g. the active site) is treated by an electronic structure method (QM re-

---

*) Please find a list of abbreviations at the end of this chapter.

cis-trans Isomerization in Biochemistry. Edited by Christophe Dugave
Copyright © 2006 WILEY-VCH Verlag GmbH & Co. KGaA, Weinheim
ISBN: 3-527-31304-4

gion) while the environment (MM region) is modeled by a classical force field. The most delicate part of a QM/MM scheme is the interface, which must ensure a physically meaningful coupling between these regions (for reviews see Refs [17,18]). To date, there has been only one study of the photoisomerization of rhodopsin, employing an excited-state QM/MM MD approach [19].

The reaction path of photochemical reactions is determined primarily by the topology of the excited singlet surfaces S_x $(x = 1, 2, \ldots)$, along with the topology of the ground state S_0 if no intersystem crossing to a triplet surface occurs. The most important characteristics of these potential energy surfaces (PESs) are the locations of minima, points of contact (e.g. conical intersections between S_0 and S_1), barriers and singlet-triplet intersection points. The shapes of the PESs are intimately tied to changes in the electronic wave function in relation to molecular geometry. Different approaches can help in the understanding of photoisomerization processes. Qualitative approaches are based on correlation diagrams expressed either in the valence bond (VB) or in the molecular orbital (MO) framework [1,2]. Although essential for the understanding of the nature and the symmetry of the electronic wave functions associated with the respective molecular geometries in the various electronic states, this approach is usually limited to the description of a reaction along one single coordinate at a time. The simultaneous analysis of two parameters already allows for an infinite number of possible paths (one-dimensional cuts) involving both variables. *Ab initio* calculations of ground-state and excited-state PESs, on the other hand, can in principle be carried out in a multidimensional parameter space. However, $3N$-6 nuclear degrees of freedom are too many for a calculation of the energies of all possible geometries even for small molecules. This problem is avoided in the *ab initio* molecular dynamics (AIMD) technique, in which the system itself chooses the relevant portion of the PES and all conformations are sampled according to their statistical weights when the system is in equilibrium. Simulations, e.g., in the microcanonical ensemble allow the efficient exploration of minima, saddlepoints and intersection regions on the PES from which the nuclear forces are computed. The energies for the other states of interest can also be calculated for all geometries sampled along the trajectory. This "on the fly" screening of excited-state surfaces can easily be extended to the canonical ensemble by coupling a thermostat to the nuclear degrees of freedom. An advantage of the MD approach is the completely unconstrained relaxation of the system on the ($3N$-6)-dimensional PES spanned by the internal degrees of freedom of the molecule. This yields the possibility of studying the trajectory following photoexcitation and identifying the most important degrees of freedom involved in the relaxation in an unbiased way. The projection of a trajectory into a one-dimensional or two-dimensional space of relevant degrees of freedom is performed here only for analysis and visualization purposes.

An elegant way of carrying out AIMD is the Car-Parrinello approach [20,21], in which the dynamics of the nuclei as well as the temporal evolution of the electronic wave function are described by Newtonian equations of motion. Using massively parallel computers, this approach allows the direct simulation of up to a few hundred atoms for 10–100 ps. Commonly, Car–Parrinello MD (CPMD) is based on density functional theory (DFT) [22,23] for the description of the electronic structure. Several theoretical frameworks for the application of DFT to excited states have been developed (e.g. the restricted open shell Kohn–Sham formalism (ROKS) [24,25] and time-dependent DFT (TDDFT) [26]). These methods will be introduced in some detail in Section 7.2.

In Section 7.3 we will provide some theoretical background for photoinduced C=C and C=N double bond isomerizations. We will consider small organic model compounds to explain different types of photochemical behavior. One of these compounds is a protonated Schiff base (PSB5, see Fig. 7.1) and serves as a model system for the RPSB involved in the vision process. In Section 7.4, we will provide results from excited-state MD simulations for the compounds mentioned in Section 7.3, before describing simulations of the CTI in rhodopsin in Section 7.5.

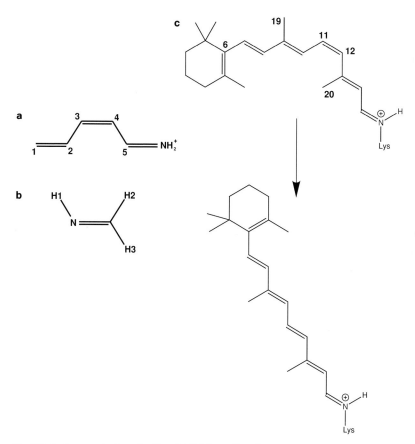

Fig. 7.1 Systems under investigation: (a) 2,4-Pentadiene-1-iminium cation (protonated Schiff base, PSB5); (b) formal-dimine; (c) retinal protonated Schiff base (RPSB, 11-*cis* to all-*trans* isomerization.

7.2
Computational Methods

7.2.1
Time-Dependent Density Functional Theory (TDDFT)

Time-dependent density functional methods (for a recent review see Ref. [27]) were developed in the early 1980s [28–30], but the rigorous theoretical background was set in 1984 by the work of Runge and Gross [26]. They proposed the equivalent of the basic DFT theorems of Hohenberg and Kohn [22] for the case of the time evolution of the electron density $\rho(r, t)$ of a many-electron system evolving under the influence of an external time-dependent potential $v_{ext}(r, t)$. The Runge–Gross theorem establishes that, starting from an initial many-electron wave function, $\Psi(r_1, \ldots, r_N, t = 0)$, at all later times $t > 0$ the density $\rho(r, t)$ determines the potential $v_{ext}(r, t)$ uniquely up to an additive purely time-dependent function. In turn, the potential $v_{ext}(r, t)$ uniquely determines the wave function $\Psi(r_1, \ldots, r_N, t)$ of the system, which therefore can be considered to be a function of the time-dependent density, $\Psi(r_1, \ldots, r_N, t) = \Psi([\rho(r, t)], t)$, where the square brackets stand for functional dependence.

A set of one-electron time-dependent Kohn–Sham (KS) equations $(m_e = 1, \hbar = 1, e = 1)$

$$i\frac{\partial}{\partial t}\phi_j(r, t) = \left(-\frac{1}{2}\nabla^2 + v_{KS}(r, t)\right)\phi_j(r, t), \qquad j = 1, \ldots, N_e \tag{1}$$

can be derived from the the variational equation $\delta A[\rho, t]/\delta\rho(r, t) = 0$ applied to the quantum mechanical action

$$A(\Psi) = \int_{t_0}^{t_1} dt \langle \Psi(t)|i\frac{\partial}{\partial t} - \hat{H}(t)|\Psi(t)\rangle \tag{2}$$

whose exact formulation in terms of electron density $\rho(r, t)$ was given by van Leeuwen [31]. The KS orbitals in Eq. (1) evolve under the influence of the single particle potential $v_s(r, t)$, which consists of the external potential $v_{ext}(r, t)$, the Hartree potential $v_H(r, t)$ and the exchange-correlation (xc) potential $v_{xc}(r, t)$. This last is defined by the functional derivative

$$v_{xc}(r, t) = \frac{\delta A_{xc}[\rho]}{\delta\rho(r, t)} \tag{3}$$

and its general functional form is unknown. The simplest approximation is the so-called adiabatic approximation, which assumes locality in time for v_{xc}

$$v_{xc}([\rho], r, t) = \frac{\delta A_{xc}[\rho]}{\delta\rho(r, t)} \cong \frac{\delta E_{xc}[\rho]|_t}{\delta\rho(r)|_t} = v_{xc}([\rho(r)|_t], r) \tag{4}$$

and gives the familiar exchange-correlation potential of time-independent DFT [23] evaluated with the density obtained at time t. Throughout this work we will make use of this approximation. We will refer to the TDDFT scheme based on the propagation of the KS orbitals ϕ_j in Eq. (1) as P-TDDFT. Numerically, one solves the set of Eq. (1) by approximating the time evolution operator, $\mathcal{U}(t, t_0)$, and propagates the states as

$$\phi_j(t) = \mathcal{U}(t, t_0)\phi_j(t_0) \tag{5}$$

where

$$\mathcal{U}(t, t_0) = \hat{T}\exp\left(-i\int_{t_0}^{t}\mathcal{H}_{KS}(\tau)\,d\tau\right) \tag{6}$$

and $\hat{T}$ is the time ordering operator. A first approach we adopt here is based on the iterative scheme developed by Baer et al. [32] combined with a two-step Runge–Kutta scheme to maintain order Δt^3 accuracy. A first guess for the potential at time $t_0 + \Delta t/2$, $v_{eff}(\mathbf{r}, t_0 + \Delta t/2)$ is obtained by evolving the KS states using the effective potential at time t_0. The full-time evolution is then achieved by evolving the wave functions for the full time step Δt, using the approximated potential computed from the half step. For a given effective potential $v_{eff}(\mathbf{r}, t)$, the solution of the time-dependent Schrödinger-like equations, for both half and full steps, is accomplished by iterating until convergence the set of integral equations

$$\phi_j^{(n)}(t_0 + \Delta t) = \phi_j^{(0)}(t_0 + \Delta t) - i\int_{t_0}^{t_0+\Delta t} d\tau\, \mathcal{H}_{KS}\left(\{\phi^{(n-1)}(\tau)\}, \tau\right)\phi_j^{(n-1)}(\tau) \tag{7}$$

The integrals are computed by Chebyshev interpolation in the time domain.

A second and more widely used approach for the computation of excitation energies within DFT is based on the linear-response formulation of the time-dependent perturbation of the electronic density. The basic quantity in linear response TDDFT (LR-TDDFT) is the time-dependent density–density response function [33]

$$\chi(r, t, r', t') = \frac{\delta\rho(r, t)}{\delta v_s(r', t')}\Big|_{v_{KS}} \tag{8}$$

which relates the first-order density response $\rho'(r, t)$ to the applied perturbation, $v_{ext}(r, t)$

$$\rho'(r, t) = \int d^3r'dt'\, \chi(r, t, r', t')\, v_{ext}(r', t') \tag{9}$$

where v_{KS} is the ground state KS potential and $v_s([\rho], r, t) = v_{KS}([\rho|_t]) + v_{ext}(r, t)$. The response function for the physical system of interacting electrons, $\chi(r, t, r', t')$

is related to the computationally more advantageous single particle KS response, $\chi_s(r, t, r', t')$,

$$\chi(r, t, r', t') = \chi_s(r, t, r', t') + \int dr_1 dt_1 \int dr_2 dt_2 \, \chi_s(r, t, r_1, t_1)$$

$$\times \frac{\delta v_{Hxc}(r_1, t_1)}{\delta \rho(r_2, t_2)} \chi(r_2, t_2, r', t') \tag{10}$$

and the problem of finding excitation energies of the interacting system reduces to the search of the poles of the response function.

In terms of the set of KS orbitals $\{\phi_i(r), \phi_a(r)\}$ (we use indices i and j for occupied orbitals (occupation $f = 1$) and indices a and b for virtual orbitals (occupation $f = 0$), respectively), the matrix elements for the single particle density response function induced by a perturbation with frequency ω become

$$\chi_{i,a}^s(\omega) = \frac{1}{\omega - (\varepsilon_i - \varepsilon_a)} \tag{11}$$

while the full response matrix obeys (σ, τ being spin variables)

$$\sum_{jb\tau} \left[S_{ia\sigma,jb\tau} - K_{ia\sigma,jb\tau}(\omega) \right] \delta P_{jb\tau}(\omega) = v_{ia\sigma}^{ext}(\omega) \tag{12}$$

where, formally, $\chi^{-1} = [S - K]$, $\delta\rho_\sigma(r, \omega) = \sum_{ia} \phi_{i\sigma}(r) \delta P_{ia\sigma}(\omega) \phi_{a\sigma}^*(r)$, and

$$S_{ia\sigma,jb\tau} = \delta_{\sigma,\tau} \delta_{i,j} \delta_{a,b}(\omega - (\varepsilon_j - \varepsilon_b)) \tag{13}$$

$$K_{ia\sigma,jb\tau}(\omega) = \int dr dr' \frac{\phi_{i\sigma}^*(r)\phi_{a\sigma}^*(r) \; \phi_{j\tau}(r')\phi_{b\tau}(r')}{|r - r'|} +$$

$$\int d(t - t') \; e^{i\omega(t-t')}$$

$$\int dr dr' \; \phi_{i\sigma}^*(r)\phi_{a\sigma}(r) \frac{\delta^2 A_x c[\rho]}{\delta\rho_\sigma(r, t) \, \delta\rho_{tau}(r', t')} \phi_{j\tau}^*(r')\phi_{b\tau}^*(r') \tag{14}$$

The excitation energies are obtained from the zeros of the expression in the square brackets of Eq. (12), which is equivalent to finding the square root of the eigenvalues of the following matrix [33]

$$\Omega_{ia\sigma,jb\tau} = \delta_{\sigma,\tau} \delta_{i,j} \delta_{a,b}(\varepsilon_{b\tau} - \varepsilon_{j\tau})^2 + 2\sqrt{\varepsilon_{a\sigma} - \varepsilon_{i\sigma}} K_{ia\sigma,jb\tau} \sqrt{\varepsilon_{b\tau} - \varepsilon_{j\tau}} \tag{15}$$

Additional steps are required for an efficient implementation in a plane wave code [34]. The adiabatic local density approximation (ALDA) of the TDDFT kernel, $f_{xc}^{\sigma\tau}(r, r', t, t') = \dfrac{\delta v_{xc}^\sigma(r, t)}{\delta\rho_\tau(r', t')}$, is obtained in the time-independent and specially local limit, $f_{xc}^{\sigma\tau}(r, r') = \delta(r - r') \frac{\delta v_{xc}^{\sigma}}{\delta \rho_{\tau}(r)}|_{\rho_{\tau}(r')}$, and is generally computed for the most commonly used time-independent xc-functionals.

In order to perform *ab initio* molecular dynamics in excited states, the forces on the atoms are computed on the fly. There are two different implementations of force calculations for the two TDDFT schemes, P-TDDFT and LR-TDDFT. In the first case, the excited state is obtained with the promotion of one electron from the highest occupied molecular orbital (HOMO) to a selected unoccupied molecular orbital (the lowest one, LUMO, in our case). The corresponding KS excited single-determinant configuration is taken for the computation of the electronic density, which is then used to compute the forces on the atoms according to the Hellmann–Feynman electrostatic theorem [35]

$$F_I = -\nabla_I E = -\int d^3 r \, n(r) \nabla_I v_{ions} - \nabla_I \left(\frac{1}{2} \sum_{I \neq J} \frac{Z_I Z_J}{|R_I - R_J|} \right) \quad (14)$$

where v_{ions} is the component of the KS potential, v_{KS}, induced by the electrostatic interaction with the nuclei.

In case of LR-TDDFT, the forces on the nuclei are derived within the Tamm–Dancoff approximation [36,37] from nuclear derivatives of the excited-state energies using the extended Lagrangian formalism introduced by Hutter [34]. In general, LR-TDDFT MD simulations are about 70–90 times faster than P-TDDFT MD simulations. The LR-TDDFT scheme has also been combined with our QM/MM approach [38,39] in order to enable the calculation of excitation spectra [40–42] and excited-state dynamics in condensed-phase systems.

The LR-TDDFT MD simulations presented here are purely adiabatic and proceed in the following way. First a simulation in S_0 is carried out in order to obtain an equilibrated system at the target temperature of 300 K using a Nosé thermostat [43,44]. A random configuration (nuclear coordinates and velocities) is taken from this simulation, and the system is vertically excited into a selected singlet state. The positions of the nuclei are updated according to the forces computed as the gradients of the excited-state energy. In a second iteration the ground state density and KS orbitals for the latest geometry are computed, and the linear response calculation provides the new excited-state energy and the new forces on the nuclei. This approach allows to adiabatically follow the instantaneous electronic excited-state PES in the Born–Oppenheimer MD scheme. Different approaches have been used to control the nuclear temperature during the relaxation along the excited-state PES. One consists in a rescaling of the nuclear velocities, once the canonical ensemble kinetic energy of the classical degrees of freedom

$$\frac{3}{2} N k_B T = \langle \sum_{i=1}^{N} \frac{p_i^2}{2m_i} \rangle \quad (15)$$

exceeds the tolerance from a target value. This scheme is conceived to mimic intermolecular collision in dilute gas phase, where hot molecules exchange kinetic energy with cold molecules by infrequent collisions.

7.2.2
Restricted Open-Shell Kohn–Sham Theory (ROKS)

ROKS is based on the sum method by Ziegler, Rauk, and Baerends [45], and allows for MD simulations in the first excited singlet state (S_1) [24]. The theoretical framework has been generalized to arbitrary spin states [25], and a new algorithm for solving the self-consistent field equations has been introduced recently [46].

The transition of one electron from the HOMO to the LUMO in a closed-shell system leads to four different excited wave functions (Fig. 7.2a). While two states $|t_1\rangle$ and $|t_2\rangle$ correspond to energetically degenerate triplets, the mixed states $|m_1\rangle$ and $|m_2\rangle$ are not eigenfunctions of the total-spin operator. They can be combined to form another triplet state $|t_3\rangle$ and the singlet state $|s\rangle$ (Fig. 7.2b). The total energy of the S_1 state is then given by

$$E_s = 2E_m^{KS} - E_t^{KS} \tag{18}$$

and the wave function can be expressed as

$$|s[\{\phi_i\}]\rangle = \sqrt{2}|m[\{\phi_i\}]\rangle - |t[\{\phi_i\}]\rangle \tag{19}$$

where $\{\phi_i\}$ denotes the complete set of orbitals. Using the exchange-correlation potentials

$$v_{xc}^\alpha = \frac{\delta E_{xc}[\rho^\alpha, \rho^\beta]}{\delta \rho^\alpha} \quad \text{and} \tag{20}$$

$$v_{xc}^\beta = \frac{\delta E_{xc}[\rho^\alpha, \rho^\beta]}{\delta \rho^\beta} \tag{21}$$

for α and β spin, two sets of Kohn–Sham equations are obtained. One set applies to the doubly occupied orbitals

$$\begin{aligned}\Big\{&-\frac{1}{2}\nabla + V_H + v_{ext}(\mathbf{r}) \\ &+ v_{xc}^\alpha[\rho_m^\alpha(\mathbf{r}), \rho_m^\beta(\mathbf{r})] + v_{xc}^\beta[\rho_m^\alpha(\mathbf{r}), \rho_m^\beta(\mathbf{r})] \\ &-\frac{1}{2}v_{xc}^\alpha[\rho_t^\alpha(\mathbf{r}), \rho_t^\beta(\mathbf{r})] - \frac{1}{2}v_{xc}^\beta[\rho_t^\alpha(\mathbf{r}), \rho_t^\beta(\mathbf{r})]\Big\}\phi_i(\mathbf{r}) = \sum_{j=1}^{n+1} \Lambda_{ij}\phi_j(\mathbf{r})\end{aligned} \tag{22}$$

and one to the two singly occupied orbitals a

$$\left\{\frac{1}{2}\left[-\frac{1}{2}\nabla + V_H + +v_{ext}(\mathbf{r})\right]\right.$$
$$\left.+v_{xc}^{a}[\rho_m^{a}(\mathbf{r}), \rho_m^{\beta}(\mathbf{r})] - \frac{1}{2}v_{xc}^{a}[\rho_t^{a}(\mathbf{r}), \rho_t^{\beta}(\mathbf{r})]\right\}\phi_a(\mathbf{r}) = \sum_{j=1}^{n+1}\Lambda_{aj}\phi_j(\mathbf{r})$$

(23)

and b (Fig. 7.2a).

$$\left\{\frac{1}{2}\left[-\frac{1}{2}\nabla + V_H + +v_{ext}(r)\right]\right.$$
$$\left.+v_{xc}^{a}[\rho_m^{a}(\mathbf{r}), \rho_m^{\beta}(\mathbf{r})] - \frac{1}{2}v_{xc}^{a}[\rho_t^{a}(\mathbf{r}), \rho_t^{\beta}(\mathbf{r})]\right\}\phi_b(\mathbf{r}) = \sum_{j=1}^{n+1}\Lambda_{bj}\phi_j(\mathbf{r})$$

(24)

These equations can be solved through minimization, using an algorithm for orbital-dependent functionals [46,47].

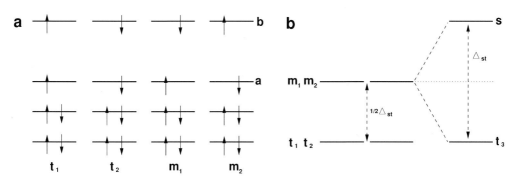

Fig. 7.2 Restricted open shell Kohn–Sham theory (ROKS). (a) Four determinants resulting from HOMO–LUMO transition of one electron; (b) the mixed states $|m_1\rangle$ and $|m_2\rangle$ can be combined to form a triplet state $|t_3\rangle$ and the singlet state $|s\rangle$. The singlet–triplet splitting Δ_{st} corresponds to twice the splitting between triplets and mixed states.

ROKS has been applied to the study of CTI in gas phase [24,48–52]. It has also been combined with a CPMD–QM/MM approach, and thus permits the simulation of the photoisomerization of the RPSB in rhodopsin (Section 7.5), taking into account the protein environment. The computational cost of a ROKS MD simulation is roughly twice as high as a ground-state simulation. It represents therefore the most efficient approach for excited-state MD simulations.

7.3
Theoretical Aspects of CTI

The qualitative minimal description of the electronic aspects of a double bond photoisomerization requires wave functions representing two electrons delocalized over two interacting atomic p-orbitals, which form a π-bond in the optimized S_0 structure. In the following we will make use of qualitative descriptions in the molecular orbital (MO) and valence bond (VB) frameworks [1,2] in order to derive a general understanding of the topologies of the different PESs describing the electronic states of the system as a function of the twist angle θ around the central double bond. We will restrict the discussion to singlet states.

We first briefly consider the photoexcitation of ethylene as a prototype of a $\pi \rightarrow \pi^*$ isomerization reaction. In the following discussion, we will focus on analogies and differences with ethylene. For a minimal basis set description of the π-system of ethylene, only the two nonorthogonal atomic orbitals (AO) p_{C_A} and p_{C_B}, localized on the two carbon atoms, need to be taken into account. All possible MO configurations (three singlets and three triplets) can be constructed as Slater determinants of the two AOs*

$$|H\rangle = \frac{1}{\sqrt{2 + 2S_{AB}}}(|p_{C_A}\rangle + |p_{C_B}\rangle) \qquad (25)$$

$|H\rangle$

$$|L\rangle = \frac{1}{\sqrt{2 - 2S_{AB}}}(|p_{C_A}\rangle - |p_{C_B}\rangle) \qquad (26)$$

$|L\rangle$

where H stands for HOMO, L for LUMO, and S_{AB} is the overlap of AOs $|p_{C_A}\rangle$ and $|p_{C_B}\rangle$. Alternatively, the three VB electronic singlet states ($^1|p_{C_A}^2\rangle$, $^1|p_{C_B}^2\rangle$, $^1|p_{C_A}p_{C_B}\rangle$) can be generated directly starting from the same localized AOs. The photoisomerization occurs after $\pi \rightarrow \pi^*$ excitation of one electron and triggers the twisting of the carbon–carbon bond on the S_1 PES (Fig. 7.3).

In the planar ground state structure, S_0 can be described by the configuration $^1|H^2\rangle$, S_1 by the configuration $^1|HL\rangle$, and S_2 by the configuration $^1|L^2\rangle$. As the molecule is twisted towards 90°, the configuration $^1|H^2\rangle$ rises in energy, while $^1|L^2\rangle$ drops. The two MOs become degenerate at 90°, where p_{C_A} and p_{C_B} are or-

---

*) The three singlets, $^1|H^2\rangle = \frac{1}{\sqrt{2}}|H\bar{H}|$, $^1|HL\rangle$ $= \frac{1}{2}(|H\bar{L}| - |\bar{H}L|)$, $^1|L^2\rangle = \frac{1}{\sqrt{2}}|L\bar{L}|$, and the three triplets $^{3(1)}|HL\rangle = \frac{1}{\sqrt{2}}|HL|$, $^{3(0)}|HL\rangle = \frac{1}{2}(|H\bar{L}| + |\bar{H}L|)$, $^{3(-1)}|HL\rangle = \frac{1}{\sqrt{2}}|\bar{H}\bar{L}|$, are

given as linear combination of Slater determinants, $|ab|$, where electrons of β spin are indicated by a bar.

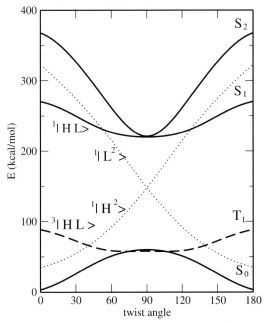

Fig. 7.3 Qualitative state diagram for ethylene as a function of the twist angle, adapted from J. Michl, V. Bonačić-Koutecký, *Electronic Aspects of Organic Photochemistry*, Wiley, 1990. Reproduced with permission from Mol. Phys. 2005, 103, 963–981. Copyright Taylor & Francis Ltd. http://www.tandf.co.uk/journals.

thogonal, and therefore geometries near the 90° twist are biradicaloids. In a more accurate description, configuration interaction introduces a mixing of the two MOs $^1|H^2\rangle$ and $^1|L^2\rangle$, and the corresponding states S_0 and S_2, split up and develop at 90° a maximum and a minimum, respectively. The first excited singlet state, S_1, has a zwitterionic nature (hole-pair character, clearly expressed by the VB description, $1/\sqrt{2}(|p_A^2\rangle - |p_B^2\rangle)$) and therefore possesses a minimum at the 90° twist, where the electron repulsion is the smallest.

7.3.1
Protonated Schiff Bases

In the case of protonated Schiff bases (e.g. PSB5, see Fig. 7.1a), the topologies of the S_0, S_1, and S_2 PES as a function of the isomerizing angle θ are qualitatively similar to the case of ethylene. However, due to the presence of a positive charge in the delocalized π-system, the twisting around the central C=C bond must be viewed as the breaking of a charged π bond. Because of the higher electronegativity of the Schiff base end compared with the polyenic end, state S_1 (characterized by a hole-pair like electronic structure) is lowered in energy, and the S_0–S_1 separation becomes significantly smaller than in ethylene, even forming a conical intersection in some molecules. On the other hand the S_1–S_2 separation becomes larger, and the two states do not touch in contrast to the case of ethylene [53]. This means that S_1 keeps its hole-pair ionic character throughout the isomerization reaction.

7.3.2
Formaldimine

Formaldimine (Fig. 7.1b) is the prototype of a Schiff base and can be viewed as a strongly perturbed ethylene. The main difference is the sp^2-type electron lone-pair localized on the nitrogen atom. The qualitative analysis of the electronic structure (at MO or VB level) of formaldimine is therefore based on a three-orbitals (p_C, p_N, sp_N^2), four-electrons model. In addition to the three singlet states described in the case of ethylene, there are two configurations of the type $n\pi^*$, in which sp_N^2 is occupied by a single electron, and the remaining three electrons occupy the orbitals with π symmetry. In the singlet ground state, S_0, the molecule has two electrons in the sp_N^2 lone pair on the nitrogen and two in the bonding π orbital of the C=N bond. Photoexcitation leads to a $n \rightarrow \pi^*$ transition, while twisting of the double bond towards 90° produces a so-called tritopic biradicaloid. Upon twisting, the electronic configuration representing the ground state remains of covalent nature, because p_C can now interact with sp_N^2. However, this interaction is not as strong as the p_C–p_N interaction in the planar structure, and therefore the energy increases. The covalent nature of S_0 is an important difference compared to the biradical character of S_0 in twisted ethylene. In contrast to ethylene, S_1 never acquires a zwitterionic character for all geometries between the planar and the 90° twisted structure.

We will come back to this theoretical considerations when assessing the achievements and failures of TDDFT in different cases.

7.4
CTI in PSB5 and Formaldimine

7.4.1
Protonated Schiff Base (PSB5)

In the visual pigment rhodopsin, an incoming photon leads to the ultrafast and highly efficient 11-*cis* to all-*trans* isomerization of the RPSB, thus initiating the first step of the visual cascade. Since even the isolated chromophore (Fig. 7.1c) is too large for many high-level ab initio calculations, numerous studies have concentrated on smaller model systems, especially on the 2,4-pentadiene-1-iminium cation (PSB5, Fig. 7.1a). However, it has to be remembered that the photochemistry of this molecule in gas phase will differ significantly from the photochemistry of the RPSB in the protein environment, which provides, for example, a counterion close to the positively charged nitrogen.

CASSCF/CASPT2 calculations [53,54] predict that in vacuo the spectroscopic state of PSB5 is S_1, an ionic single HOMO-LUMO transition, while S_2, a covalent doubly excited state, is not involved. In S_1, the initial motion along the minimum energy path is dominated by stretching modes, leading to a planar stationary structure with decreased bond length alternation (elongated central double bond, see Fig. 7.4 and Table 7.1 for the CASPT2 S_1 planar optimized structure). After the "turning point" (d_{C3-C4} = 1.53 Å, $\phi_{C2-C3-C4-C5}$ = 25°), the twisting motion becomes dominant, and the molecule undergoes a barrierless relaxation towards a conical intersection with S_0 at a twist angle $\phi_{C2-C3-C4-C5}$ of about 80°. During this process, a shift of electron density from the carbon end towards the Schiff base end is observed, which reaches its maximum at the conical intersection.

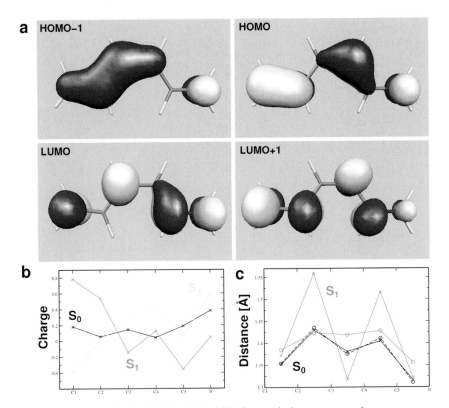

Fig. 7.4 (a) Frontier orbitals of PSB5. (b) Hirshfeld charges (hydrogens summed into heavy atoms). S_1 and S_2 calculated with LR-TDDFT. (c) Bond lengths in planar optimized S_0 (black) and S_1 (dotted) structures. LR-TDDFT values shown with crosses, CASPT2 values from Ref. [80] shown with circles.

Table 7.1 PSB5: Average bond lengths (in Å) from MD simulations in S_1.

	ROKS	LR-TDDFT	P-TDDFT
C1–C2	1.434	1.391	1.414
C2–C3	1.386	1.520	1.396
C3–C4	1.444	1.357	1.441
C4–C5	1.405	1.476	1.406
C5–N6	1.349	1.353	1.350

The optimized planar S_1 geometries from CASPT2 and LR-TDDFT calculations are very different [55,56]. In CASPT2, the bond length alternation is diminished, while in LR-TDDFT it is largely increased, the central double bond being as short as 1.318 Å (Table 7.1). It is clear that in this state no double bond isomerization can be observed. A full optimization of the S_1 state without the restriction to planarity yields a minimum structure with a dihedral angle $\phi_{C3-C4-C5-N} = 106°$ and a pyramidalization angle of the nitrogen atom (defined as the angle between the H–N–H plane and the C=N bond) of 24° (Fig. 7.5a).

Multiple LR-TDDFT MD simulations in S_1 were carried out . For comparison, one simulation was carried out with ROKS. Figure 7.5a shows the time evolution of the S_1 and S_0 energies along one S_1 trajectory (temperature control 300 ± 100 K). A large decrease in the S_1 energy by 2 eV is observed at the same time as an increase of the S_0 energy of the same amount, but the two surfaces do not intersect. The S_1 relaxation is characterized by a large increase in the single bond lengths (Fig. 7.5b) and a rotation around these bonds ($\phi_{C1-C2-C3-C4} = 100°$). The final structure after 100 fs looks similar to the optimized S_1 structure (inset in Fig. 7.5a).

We also performed MD simulations using the P-TDDFT scheme. Here, we initially excited one electron from the HOMO to the LUMO. The nature of this excitation is confirmed by a LR-TDDFT calculation which predicts a contribution of about 83% of the HOMO–LUMO transition to the S_1 density. We find a remarkable difference between the LR-TDDFT and the P-TDDFT schemes: while LR-TDDFT decreases the double bond lengths, P-TDDFT increases them in agreement with CASPT2 and ROKS calculations (Fig. 7.6). The central double bond adopts an average length of 1.441 Å (Table 7.1) in good agreement with the ROKS value. Fig. 7.6 (upper panel) shows the time evolution of the dihedral angles. $\phi_{C2-C3-C4-C5}$ reaches $\approx 40°$, but no isomerization event is observed during the simulated time of 200 fs. Since P-TDDFT simulations are computationally very demanding, simulations limited to this timescale.

The S_1 geometry optimized with ROKS is in good agreement with the CASPT2 geometry and shows the correct reduction of the bond length alternation. ROKS dynamics in the S_1 state show flucuations of the dihedral angle $\phi_{C2-C3-C4-C5}$ up

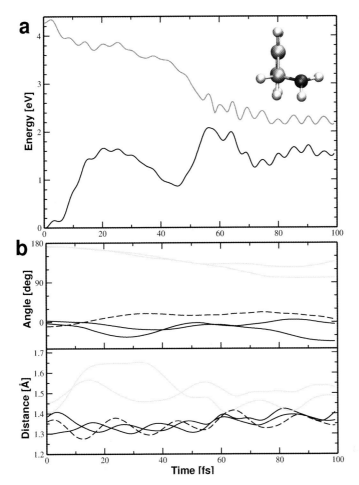

Fig. 7.5 LR-TDDFT simulations of PSB5. (a) Time evolution of the S_1 energy (dotted) and the S_0 energy (black) along one S_1 simulation (temperature control 300 ± 100 K). Inset: optimized S_1 structure, view along the C3–C4 bond. (b) Dihedral angles (black: around double bonds, light grey: around single bonds) and bond lengths (black: double bonds, light grey: single bonds) from the same simulation. The central dihedral angle $\theta_{C2-C3-C4-C5}$ and the central double bond d_{C3-C4} are shown as dashed lines.

to 41° at room temperature. At elevated temperature (630 ± 100 K) $\phi_{C2-C3-C4-C5}$ fluctuates further to 56°, while the terminal C1–C2 double bond isomerizes completely (rotation of 360° in 88 fs). It can therefore be assumed that the barrier towards isomerization of a double bond is small in ROKS.

In summary, LR-TDDFT results for PSB5 are in disagreement with the experimental detection of photoinduced double bond isomerizations in similar compounds. The shape of the S_1 PES is different from the one computed with CASSCF/CASPT2, P-TDDFT, or ROKS. This seems to concern mainly the bond

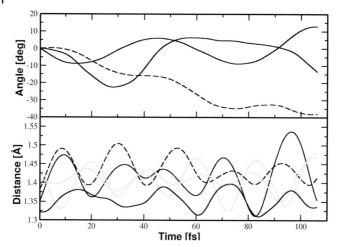

Fig. 7.6 P-TDDFT S_1 simulation of PSB5. Upper panel: time evolution of the dihedral angles around the double bonds (d_{C3-C4} shown as dashed line). The dihedral angles around the single bonds do not deviate much from 180° and are omitted for clarity. Lower panel: time evolution of the single bond lengths (light grey) and of the double bond lengths (black). Reproduced with permission from Mol. Phys. 2005, 103, 963–981. Copyright Taylor & Francis Ltd. http://www.tandf.co.uk/journals.

length alternation, since it has been shown that LR-TDDFT yields reasonable energies along the twisting coordinate [52,55,57]. However, once the molecule is trapped in the local minimum characterized by short double bonds, the barrier towards double bond isomerization is very high and cannot be overcome by kinetic energy.

Finding a valid explanation for the different capabilities of the two TDDFT schemes to reproduce the correct bond alternation is not straightforward. We believe that the major reason for the different description of the photoisomerization is the linear-response approximation. In particular, in P-TDDFT no approximation is introduced for the TDDFT response kernel, $f_{xc}^{\sigma\tau}(r, r', t, t')$, which is only required in the LR-TDDFT formulation. This is especially the case for excited configurations with ionic character, which are electronically very different from the ground state (see Section 7.3). In this case the correlation between the KS orbital energy differences and the TDDFT excitation energies (especially in case of the Tamm–Dancoff approximation) is weak and the validity of LR-TDDFT becomes questionable. An alternative explanation could be related to the difficulties of TDDFT in describing charge transfer states [52,55] and distorted geometries (stretched bonds or angles). The structures obtained during the excited-state dynamics can indeed be a source of error at the stage of the ground-state wavefunction optimization, which is needed before each LR-TDDFT MD step.

Since LR-TDDFT breaks down for the description of the photodynamics of PSB5 and P-TDDFT is computationally too demanding for the description of the full RPSB, we employ ROKS for the QM/MM description of the photoreaction in rhodopsin (see Section 7.5).

7.4.2
Formaldimine

Formaldimine (methyleneimine) is very reactive and decomposes by polymerization, oxidation, or hydrolysis. It plays an important role in chemistry as the smallest member of the large class of imines (Schiff bases) and in addition it is of astrophysical interest, having been detected in dark interstellar dust clouds [58]. In the laboratory, H_2CNH can only be observed transiently by pyrolysis of amines or different azido compounds, and its electronic absorption spectrum has only been recorded very recently [59]. The spectrum shows a broad and structureless peak with a maximum at 250 nm (4.96 eV) in close agreement to our calculated vertical excitation energy of 4.92 eV (LR-TDDFT). After radiationless decay to the ground state the molecule possesses enough internal energy for fragmentation.

The molecular structure of formaldimine in the ground state obtained by microwave spectroscopy [60] agrees very well with our calculated structure (Table 7.2, Fig. 7.7). Theoretical investigations of the photoisomerization in formaldimine have found the twisting pathway to be preferred over the in-plane mechanism [24,50,61–64]. Formaldimine is a five atomic molecule and therefore has nine internal degrees of freedom that are all free to relax in excited-state MD simulations. For the analysis of the generated trajectories we will focus on the valence angle α_{CNH}, the dihedral angle ϕ_{H3CNH} (for atom numbering see Fig. 7.1), and the pyramidalization angle of the carbon atom ω_{Pyr}, defined as the angle between the H–C–H plane and the C=N bond.

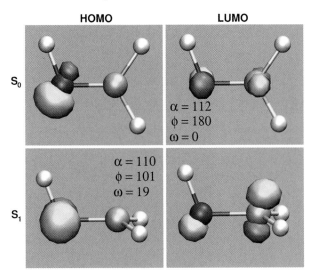

Fig. 7.7 Frontier orbitals of formaldimine in the S_0 (top) and in the S_1 (bottom) optimized structures. Geometry optimization in S_1 was carried out with LR-TDDFT. Angles are given in degree.

Table 7.2 Optimized structures of formaldimine.

	$S_0{}^b$	S_1	
		LR-TDDFT	ROKS[c]
d_{CN} (Å)	1.274 (1.273)	1.378	1.392
d_{NH} (Å)	1.036 (1.021)	1.045	1.038
d_{CH2} (Å)	1.106 (1.09)	1.098	1.093
d_{CH3} (Å)	1.101 (1.09)	1.098	1.093
α_{CNH} (deg)	110.8 (110.4)	110.4	112.0
β_{NCH2} (deg)	125.1 (125.1)	119.0	118.0
γ_{NCH3} (deg)	118.5 (117.0)	119.0	118.0
χ_{HNCH2} (deg)	0.0	−100.8	−105.0
ϕ_{HNCH3} (deg)	180.0	100.8	105.0
ω_{Pyr} (deg)	0.0	18.6	–
Adiabatic energy (eV)	0.0	3.2	2.8

a ω_{Pyr} is the pyramidalization angle of the carbon atom (angle
between the H–C–H plane and the C=N bond). For numbering
see Fig. 7.1b.
b Experimental data for S_0 from Ref. [60] in parentheses.
c ROKS/BLYP data from Ref. [24].

In the S_1 simulations without rescaling of the nuclear velocities, an intersection between S_1 and S_0 is reached several times (see Fig. 7.8a for an example). The crossing point is characterized by the coordinates $\phi_{H3CNH} \approx 100°$, $\alpha_{HNC} \approx 100°$, $\omega_{Pyr} \approx 12°$. This point is not located at the minimum of the PES ($\phi = 101°$, $\alpha = 110°$, $\omega = 19°$, see Fig. 7.7) and differs especially in the angle α. All surface crossings observed in our simulations are characterized by large values for the derivative in time of the energy gap between the two adiabatic states which translate, according to Landau–Zener theory [65,66], into a high probability for a jump between the surfaces. For all crossings described in this section, the probabilities for the jump to S_0 are larger than 99%, leading to a radiationless transition. Different possibilities exist for the course of such non-adiabatic trajectories (Fig. 7.8a): in the first trajectory (green), the system jumps from state S_1 to state S_0 at the first crossing point. Once in the ground state, it relaxes rapidly towards the basin of attraction on the reactant side (Fig. 7.8b). In case the system remains on the excited-state surface until the second surface crossing (blue curve), the relaxation on S_0 leads instead to the isomerized product. The time spent in S_1 between the two crossings (36 fs) corresponds to a full angle bending period and brings the system back to the surface crossing conformation. The additional two curves

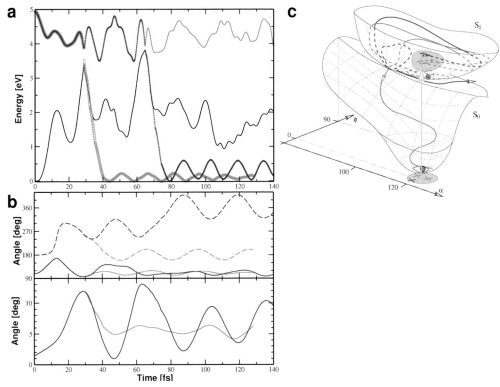

Fig. 7.8 LR-TDDFT simulations of formaldimine without temperature control: (a) Time evolution of the S_0 (black) and S_1 (red) energies. Two different trajectories are shown in blue and in green. (b) Relaxation of the angles α_{CNH} (upper panel, full lines), ϕ_{H3CNH} (upper panel, dashed lines) and ω_{Pyr} (lower panel) along the green and the blue trajectories from (a). (c) Qualitative PES diagram. The colors correspond to the trajectories shown in (a). Reproduced with permission from Mol. Phys. 2005, 103, 963–981. Copyright Taylor & Francis Ltd. http://www.tandf.co.uk/journals.

(black and red) of Fig. 7.8a correspond to the energy time-series for the excited and ground states respectively, when the forces driving the dynamics are computed from the S_1 surface for the full trajectory. In this case, after the first 100 fs of relaxation, the system thermalizes to the minimum on the excited-state PES, and no further S_1/S_0 intersection is observed.

If a temperature control scheme is applied in order to mimick energy redistribution by intermolecular collisions, the fluctuations of the S_1 energy after initial relaxation are very small, and the system does not have enough kinetic energy to reach the point of intersection with S_0, which is characterized by a small angle α_{HNC} (data not shown). In addition, the high dimensionality of the phase space prevents, also for such a relatively small molecule, an efficient sampling of the phase space leading to a decreased probability to meet the surface crossing region.

In summary, LR-TDDFT simulations of the excited-state dynamics of formaldimine show that the S_1 PES minimum does not coincide with the crossing between the S_1 and the S_0 surfaces. This finding is in agreement with one-dimensional ab initio CI calculations of twisted formaldimine, which describe the S_0 and the S_1 energy as a function of the angle a_{HNC} [62]. The details of the relaxation in S_1 can be shown to depend crucially on the applied thermostatting scheme. Although our simulations show that isomerization occurs through a twisting mechanism, it is evident that initially a strong gradient leads towards a stretching of the angle a_{HNC}.

7.5
CTI in Rhodopsin

7.5.1
Introduction

Light absorption by the G protein-coupled receptor rhodopsin leads to vision via a complex signal transduction pathway that is initiated by the photoisomerization of the RPSB. Within the protein, the 11-*cis* to all-*trans* isomerization of the RPSB is ultrafast (200 fs) [67] and very efficient (quantum yield 0.65) [68], in contrast to the same photoreaction in solution. This fact is puzzling in view of the steric confinement of the RPSB to a small binding pocket that should hamper the large movements required to adopt an all-*trans* conformation (Fig. 7.1c). Two models have been proposed to explain a more volume-conserving CTI: the hula twist (HT) [69] and the bicycle pedal (BP) [70] mechanisms. While in the HT a rotation of a single bond adjacent to the isomerizing double bond is assumed, in the BP a second double bond isomerization is proposed. Both models are at variance with the finding of an all-*trans* RPSB in bathorhodopsin and structural characteristics of the following photointermediates, unless a further isomerization or single bond rotation is assumed.

Much work has been devoted to understanding the molecular mechanism of this photoreaction [13]. Early semi-empirical studies, modeling the protein cavity by an effective steric potential, showed that the protein cavity restricts possible isomerization pathways and leads to the formation of a strained intermediate [71]. Experimental evidence reveals that bathorhodopsin, the first thermally equilibrated intermediate in the signaling cascade, exhibits a strained all-*trans* structure of the RPSB and stores 32 ± 1 kcal mol^{-1} of the photon energy [72]. Two different energy storage mechanisms have been discussed: electrostatic energy storage by charge separation between the protonated Schiff base and its counterion Glu113, and mechanical energy storage in form of strain energy. From the crystal structures of bovine rhodopsin [9–12] it is known that the RPSB is twisted in the C11–C12 region. Recent theoretical studies establish the rotational direction of the twist as uniquely negative and identify the major steric influences [3,4,12,73]. Experimental data show that removal of the C20 methyl group and thus lowering of the steric strain around the C11–C12 bond slows down the photoreaction and decreases the quantum yield [74], in agreement with quantum chemical calculations [75] that predict a faster isomerization for pretwisted bonds.

7.5.2
Classical and QM/MM Studies of the CTI in Rhodopsin

Our rhodopsin model system is based on the crystal structures of bovine rhodopsin [10] embedded in a membrane mimetic environment (Fig. 7.3) [3]. The BLYP [76,77] functional is used for the QM subsystem, which is evolved according to the Car–Parrinello algorithm [20]. For the description of the electronically excited S_1 state, ROKS [24,25] is employed.

As a starting point, a ground-state QM/MM MD simulation was carried out for 5 ps [78]. Taking different snapshots from this simulation, 23 excited-state QM/MM trajectories of about 100 fs each were simulated. The excited-state geometry of the RPSB is characterized by the well-known inversion of the bond length pat-

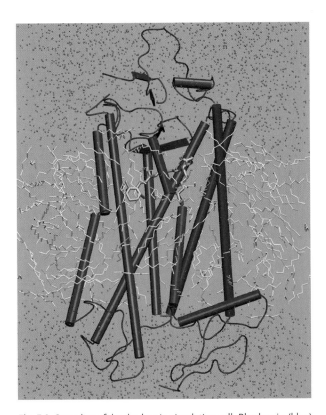

Fig. 7.9 Snapshot of the rhodopsin simulation cell. Rhodopsin (blue) is represented with the extracellular side on top. The retinal chromophore (turquoise) is accommodated within the seven transmembrane helices. The membrane is mimicked by a layer of n-octane (white) surrounded by an aqueous phase (red).

tern (Fig. 7.10, lower panel). In S_1, especially the bonds C9-C10 (1.44 Å), C11–C12 (1.43 Å), and C13–C14 (1.43 Å) are elongated, thus lowering the barrier towards isomerization. Whereas the electronic structure shows no selectivity for rotation around any of these double bonds, the protein environment favors C11–C12 bond isomerization by imposing steric constraints. In fact, the dihedral angles from C7 to C11 and from C12 to N deviate in S_1, similar to S_0, only by at most 15° from a perfect *trans* conformation (Fig. 7.10, upper panel). In contrast, the pretwisted dihedral angle $\phi_{C10-C11-C12-C13}$ rotates towards more negative values, with fluctuations up to −72° and an average of −35° (Fig. 7.10, middle panel). An angle appropriate for isomerization ($\approx$ −90°) [75] is not reached, indicating the presence of a residual barrier, which is probably due to the limited accuracy of our ROKS QM/MM approach.

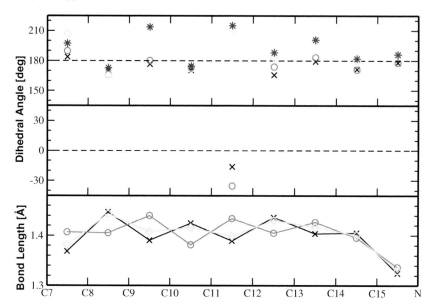

Fig. 7.10 QM/MM simulations of rhodopsin. Dihedral angles (upper and middle panel) and bond lengths (lower panel) along the conjugated carbon chain of the RPSB in the dark state (black), in S_1 (dotted), and in the all-*trans* ground state (light grey). The dihedral angles of the all-*trans* ground state obtained from classical MD are shown in black stars. Reproduced with permission from *J. Am. Chem. Soc.* **2004**, *126*, 15328–15329. Copyright 2004 American Chemical Society.

The barrier is very small, as we can show by performing an excited-state MD simulation in which the initial nuclear velocities of the RPBS are increased, so that the local temperature corresponds to 690 K (Fig. 7.11a, solid blue line, ROKSht). This approach allows small energy barriers to be crossed without imposing an a priori chosen reaction path and has been used previously [79].

This enhanced kinetic energy is sufficient to allow the barrier crossing in a very short time: $\phi_{C10-C11-C12-C13}$ rotates within 50 fs to −103° and fluctuates then

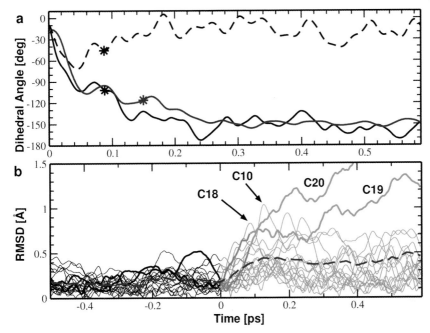

Fig. 7.11 Photoisomerization in rhodopsin. (a) Time evolution of the dihedral angle $\phi_{C10-C11-C12-C13}$. One ROKS trajectory (dashed black line), one trajectory (same initial conformation) at higher temperature (solid black line, ROKSht), and the average of the 23 isomerizing classical simulations (blue) are shown. The S_1–S_0 transitions are indicated by stars. (b) Displacement of each heavy atom of the RPSB from its average position in the dark state (black), in S_1 (red), and in the all-*trans* ground state configuration (green) in ROKSht. The average value is indicated by a blue dashed line, and the two methyl groups C19 and C20 are highlighted by bold lines. Reproduced with permission from *J. Am. Chem. Soc.* **2004**, *126*, 15328–15329. Copyright 2004 American Chemical Society.

around an average value of –100°. No recrossing to the –35° configuration is observed, indicating that the –100° configuration is at least a local minimum on the ROKS S_1 energy surface. When released to the ground state after 90 fs, the RPSB evolves towards a highly twisted all-*trans* structure, without isomerization or rotation of any other dihedral angle. Fig. 7.11b shows the displacement of each RPSB heavy atom from its average position in the dark state (black), in S_1 (red), and in the all-*trans* S_0 configuration (green). Remarkably, in the excited state, no atom moves more than 0.8 Å, that is only 0.3 Å more than the maximal thermal displacement in the dark state. In the following 500 fs of ground-state relaxation, only the methyl groups C19 and C20 move further away from their position in the dark state, and the average root mean square deviation (RMSD) is 0.4 Å. The strain is propagated through the carbon chain, as it can be seen from the deviation of the dihedral angles up to almost 40° (Fig. 7.10, upper panel). The surprisingly small difference between the primary photoproduct and the dark state structure is

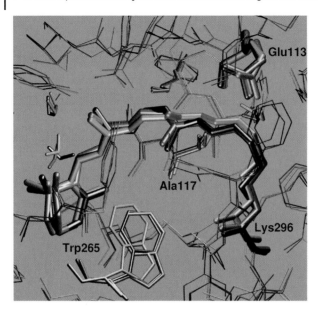

Fig. 7.12 Photoisomerization in rhodopsin. Superposition of the chromophore structure in the protein binding pocket in the dark state (black), at the S_1–S_0 transition (red), and after 500 fs of relaxation in the isomerized state (green). Reproduced with permission from *J. Am. Chem. Soc.* **2004**, *126*, 15328–15329. Copyright 2004 American Chemical Society.

also evident in Fig. 7.12, where the dark state structure (0 fs, black), the transition structure (90 fs, red), and the all-*trans* ground-state structure (590 fs, green) are superimposed.

A second approach for the study of the small residual isomerization barrier was applied, namely a restrained excited state dynamics in which the dihedral angle $\phi_{C10-C11-C12-C13}$ was varied stepwise from –65° to –100°. Also in this way, we obtain the same highly twisted all-*trans* ground-state structure, suggesting that the isomerization pathway is sterically tightly restricted.

In order to estimate the energy stored in the system after isomerization, we optimized the chromophore structure both in the 11-*cis* and in the all-*trans* configuration in rhodopsin, starting from the initial dark state conformation of ROKSht and fixing all atoms except for the RPSB. The completely frozen protein environment favors a slightly different chromophore conformation than the one observed during the MD simulations, but the twisted overall structure is the same. We obtain a total energy difference between the *cis* and the *trans* isomer of 28 kcal mol^{-1} in good agreement with the experimentally determined energy storage in bathorhodopsin (32 kcal mol^{-1}) [72]. In the all-*trans* conformer, the internal energy of the RPSB (+18 kcal mol^{-1}) and the van der Waals (steric) interaction energy between the RPSB and the protein (+10 kcal mol^{-1}) increase substantially, while the electrostatic interaction energy remains unaffected (less than

0.1 kcal mol⁻¹ difference). New close contacts (distance below 2.5 Å) are formed with residues from helices III (Ala117, Thr118), V (Met209), VI (Tyr268), VII (Ala292), and extracellular loop E II (Gly188), while some close contacts with Glu113 (H III), Cys187 (E II), and Trp265 (H VII) are released. Electrostatic energy storage is negligible, consistent with the finding that the saltbridge between the protonated Schiff base and the counterion Glu113 remains stable. This result indicates that the photon energy is stored in the internal strain of the chromophore and in steric interactions with the protein.

Since the QM/MM simulations suggest that the photoreaction is limited to a single pathway which is essentially determined by steric constraints, classical MD simulations (CLASSinv) have been used in order to gain a statistical picture of the isomerization at a low computational cost. Here, the empirical torsional potential around the C11–C12 bond, which is represented by a sinusoid with minima at 0 and 180°, is switched for 150 fs to a sinusoid of the same height, but with one single minimum at 0°. After 150 fs, the potential is switched back to its original form. Comparison of the average all-*trans* structures of the first 500 fs after classical isomerization shows a remarkable similarity between the structures generated by multiple classical MD trajectories and the DFT results (RMSD = 0.2 Å), once again emphasizing the predominating sterical influence of the protein. Even the strain propagation along the conjugated carbon chain can be described properly by the classical model, as it is evident from the average deviation of the torsional angles from a planar conformation (Fig. 7.10) that parallels the DFT results. Some differences can only be found in the dihedral angle between the ionone ring and the carbon chain ($\phi_{C5-C6-C7-C8}$ = –46° in CLASSinv, –27° in ROKSht) due to the neglect of conjugation effects in the classical force field.

7.6
Summary and Conclusions

In summary, ab initio and hybrid QM/MM molecular dynamics studies provide a valuable tool for the investigation of photoinduced CTI. Excited-state MD simulations of two small organic molecules, PSB5 and formaldimine, were carried out in order to assess the performance of different excited-state methods for the description of *cis-trans* isomerization reactions. In contrast to the majority of previous studies of the same model systems, our approach has the advantage to allow a completely unconstrained and unbiased relaxation of the excited molecule on its (3N-6)-dimensional PES.

In the case of the protonated Schiff base model, the forces computed from the LR-TDDFT S_1 PES lead to a single bond rotation instead of double bond isomerization. This failure might be related to the local approximation of the exchange-correlation functional or, as suggested by the significantly different results obtained with P-TDDFT, to a breakdown of the linear response approximation. Further investigations in this respect are needed. P-TDDFT and ROKS correctly predict an inversion of the bond length pattern and a twisting around double bonds.

In formaldimine, LR-TDDFT simulations in S_1 lead to double bond isomerization, and the S_1 PES seems to be correctly reproduced. Molecular dynamics simulations demonstrate that the S_1 PES minimum does not coincide with a S_1/S_0 crossing. While the isomerization occurs via a twisting mechanism, it is shown that on the S_1 PES a strong gradient leads initially to an increase of the bonding angle a_{CNH}.

We showed that TDDFT excited-state MD simulations can provide interesting insights into photoreactions, but the correct description of the excited-state PES has to be carefully tested.

Hybrid QM/MM simulations of the photoreaction in rhodopsin based on ROKS yield insights into the atomic details of the reaction mechanism. MD simulations show that the protein binding pocket selects and accelerates the isomerization exclusively around the C11–C12 bond via pre-formation of a twisted structure, in agreement with static ab initio calculations and experimental studies.

It has been shown that the 11-*cis* to all-*trans* isomerization is possible within the protein binding pocket with minor atomic rearrangements, producing a highly strained chromophore. No simultaneous single or double bond isomerization, as assumed by the HT and the BP mechanisms, is observed. Instead, all dihedral angles, especially those around the double bonds, are twisted up to an average value of 40°. In other words, the two-dimensional picture of the 11-*cis* to all-*trans* isomerization of the RPSB, which would require a large volume change (Fig. 7.1c), has to be replaced by a three-dimensional picture, in which a twisted all-*trans* RPSB can occupy a similar volume as the 11-*cis* RPSB (Fig. 7.12). Calculation of different energy contributions shows that the photon energy is stored mostly in steric energy in bathorhodopsin. Hence, the initial step can be viewed as the compression of a molecular spring, which can then in the following steps release its strain by altering the protein environment in a highly specific manner.

List of Abbreviations

AIMD	ab initio molecular dynamics
AO	atomic orbital
BLYP	Becke–Lee–Yang–Parr exchange and correlation functional
BP	bicycle pedal
CASSCF	complete active space self-consistent field
CI	configuration interaction
CPMD	Car–Parrinello molecular dynamics
DFT	density functional theory
eV	electron volt
HOMO	highest occupied molecular orbital
HT	hula twist
KS	Kohn–Sham

LDA   local density approximation
LUMO   lowest unoccupied molecular orbital
MD   molecular dynamics
MO   molecular orbital
QM/MM   hybrid quantum/classical (quantum mechanical/molecular mechanical)
PES   potential energy surface
ROKS   restricted open shell Kohn-Sham
RPSB   retinal protonated Schiff base
S_0 singlet ground state
S_1 first excited singlet state
TDDFT   time-dependent density functional theory
VB   valence bond theory

References

1 J. Michl, V. Bonačić–Koutecký, *Electronic Aspect of Organic Photochemistry*, Wiley–Interscience, New York, **1990**.

2 M. Klessinger, J. Michl, *Excited States and Photochemistry of Organic Molecules*, VCH, New York, **1995**.

3 U. F. Röhrig, L. Guidoni, U. Rothlisberger, *Biochemistry* **2002**, *41*, 10799–10809.

4 J. Saam, E. Tajkhorshid, S. Hayashi, K. Schulten, *Biophys. J.* **2002**, *83*, 3097–3112.

5 P. S. Crozier, M. J. Stevens, L. R. Forrest, T. B. Woolf, *J. Mol. Biol.* **2003**, *333*, 493–514.

6 T. Huber, A. V. Botelho, K. Beyer, M. F. Brown, *Biophys. J.* **2004**, *86*, 2078–2100.

7 M. C. Pitman, A. Grossfield, F. Suits, S. E. Feller, *J. Am. Chem. Soc.* **2005**, *127*, 4576–4577.

8 V. Lemaître, P. Yeagle, A. Watts, *Biochemistry* **2005**, *44*, 12667–12680.

9 K. Palczewski, T. Kumasaka, T. Hori, C. A. Behnke, H. Motoshima, B. A. Fox, I. L. Trong, D. C. Teller, T. Okada, R. E. Stenkamp, M. Yamamoto, M. Miyano, *Science* **2000**, *289*, 739–745.

10 D. C. Teller, T. Okada, C. A. Behnke, K. Paleczewski, R. E. Stenkamp, *Biochemistry* **2001**, *40*, 7761–7772.

11 T. Okada, Y. Fujiyoshi, M. Silow, J. Navarro, E. M. Landau, Y. Shichida,

Proc. Natl Acad. Sci. USA **2002**, *99*, 5982–5987.

12 T. Okada, M. Sugihara, A. N. Bondar, M. Elstner, P. Entel, V. Buss, *J. Mol. Biol.* **2004**, *342*, 571–583.

13 R. A. Mathies, J. Lugtenburg, *The primary photoreaction of rhodopsin*, in Handbook of Biological Physics, D. G. Stavenga, W. J. DeGrip, E. N. Jr. Pugh, Eds, volume 3, Elsevier, Amsterdam, **2000**, 55–90.

14 A. Warshel, M. Levitt, *J. Mol. Biol.* **1976**, 103, 227.

15 U. C. Singh, P. A. Kollmann, *J. Comput. Chem.* **1986**, *7*, 718.

16 M. J. Field, P. A. Bash, M. Karplus, *J. Comput. Chem.* **1990**, *11*, 700.

17 J. Gao, M. A. Thompson, Eds. *Combined Quantum Mechanical and Molecular Mechanical Methods*, Oxford University Press, Oxford, **1998**.

18 P. Sherwood, *Hybrid quantum mechanics/molecular mechanics approaches*, in Modern Methods and Algorithms of Quantum Chemistry, J. Grotendorst, Ed. volume 3 of NIC Series, Forschungszentrum Jülich, **2000**, 285–305.

19 U. F. Röhrig, L. Guidoni, A. Laio, I. Frank, U. Rothlisberger, *J. Am. Chem. Soc.* **2004**, *126*, 15328–15329.

20 R. Car, M. Parrinello, *Phys. Rev. Lett.* **1985**, *55*, 2471.

21 D. Marx, J. Hutter, *Ab initio molecular dynamics: theory and implementation*, in Modern Methods and Algorithms of Quantum Chemistry, J. Grotendorst, Ed., volume 1 of NIC Series, Forschungszentrum Jülich, **2000**, 301.

22 P. Hohenberg, W. Kohn, *Phys. Rev. B* **1964**, *136*, 864.

23 W. Kohn, L.J. Sham, *Phys. Rev. A* **1965**, *140*, 1133.

24 I. Frank, J. Hutter, D. Marx, M. Parrinello, *J. Chem. Phys.* **1998**, *108*, 4060–4069.

25 M. Filatov, S. Shaik, *Chem. Phys. Lett.* **1998**, *288*, 689–697.

26 E. Runge, E. K. U. Gross, *Phys. Rev. Lett.* **1984**, *52*, 997–1000.

27 K. Burke, E. K. U. Gross, *J. Chem. Phys.* **2005**, *123*, 062206.

28 A. Zangwill, P. Soven, *Phys. Rev.* **1980**, *21*, 1561.

29 M. J. Stott, E. Zaremba, *Phys. Rev.* **1980**, *21*, 12.

30 G. D. Mahan, *Phys. Rev. A* **1980**, *22*, 1780.

31 R. van Leeuwen, *Phys. Rev. Lett.* **1998**, *80*, 1280.

32 R. Baer, R. Gould, *J. Chem. Phys.* **2001**, *114*, 3385.

33 M. E. Casida, *Time-dependent density functional response theory of molecular systems: theory, computational methods, and functionals*, in Recent Developments and Applications of Modern Density Functional Theory, J. M. Seminario, Ed. Elsevier, Amsterdam, **1996**, 391.

34 J. Hutter, *J. Chem. Phys.* **2003**, *118*, 3928–3934.

35 R. Feynman, *Phys. Rev.* **1937**, *56*, 340.

36 A. L. Fetter, J. D. Walecka, *Quantum Theory of Many-Particle Systems*. McGraw-Hill, San Francisco, **1971**.

37 S. Hirata, M. Head-Gordon, *Chem. Phys. Lett.* **1999**, *314*, 291–299.

38 A. Laio, J. VandeVondele, U. Rothlisberger, *J. Chem. Phys.* **2002**, *116*, 6941–6947.

39 A. Laio, J. VandeVondele, U. Rothlisberger, *J. Phys. Chem. B* **2002**, *106*, 7300–7307.

40 U. F. Röhrig, I. Frank, J. Hutter, A. Laio, J. VandeVondele, U. Rothlisberger, *ChemPhysChem* **2003**, *4*, 1177–1182.

41 M. Sulpizi, P. Carloni, J. Hutter, U. Rothlisberger, *Phys. Chem. Chem. Phys.* **2003**, *5*, 4798–4805.

42 M. Sulpizi, U. F. Röhrig, J. Hutter, U. Rothlisberger, *Int. J. Quantum Chem.* **2005**, *101*, 671–682.

43 S. Nosé, *J. Chem. Phys.* **1984**, *81*, 511–519.

44 S. Nosé, *Mol. Phys.* **1984**, *52*, 255–268.

45 T. Ziegler, A. Rauk, E. J. Baerends, *Theor. Chim. Acta* **1977**, *43*, 261–271.

46 S. Grimm, C. Nonnenberg, I. Frank, *J. Chem. Phys.* **2003**, *119*, 11574–11584.

47 S. Goedecker, C. J. Umrigar, *Phys. Rev. A* **1997**, *55*, 1765.

48 C. Molteni, I. Frank, M. Parrinello, *J. Am. Chem. Soc.* **1999**, *121*, 12177–12183.

49 C. Molteni, I. Frank, M. Parrinello, *Comp. Mat. Sci.* **2001**, *20*, 311–317.

50 N. L. Doltsinis, D. Mary, *Phys. Rev. Lett* **2002**, *88*, 166402.

51 C. Nonnenberg, S. Grimm, I. Frank, *J. Chem. Phys.* **2003**, *119*, 11585–11590.

52 F. Schautz, F. Buda, C. Filippi, *J. Chem. Phys.* **2004**, *121*, 5836–5844.

53 R. González-Luque, M. Garavelli, F. Bernardi, M. Merchán, M. A. Robb, *Proc. Natl Acad. Sci. USA* **2000**, *97*, 9379–9384.

54 M. Garavelli, P. Celani, F. Bernardi, M. A. Robb, M. Olivucci, *J. Am. Chem. Soc.* **1997**, *119*, 6891–6901.

55 M. Wanko, M. Garavelli, F. Bernadi, T. A. Niehaus, T. Frauenheim, M. Elstner, *J. Chem. Phys.* **2004**, *120*, 1674–1692.

56 I. Tavernelli, U. Röhrig, U. Rothlisberger, *Mol. Phys.* **2005**, *103*, 963–981.

57 S. Fantacci, A. Migani, M. Olivucci, *J. Phys. Chem. A* **2004**, *108*, 1208–1213.

58 P. D. Godfrey, R. D. Brown, B. J. Robinson, M. W. Sinclair, *Astrophys. Lett.* **1973**, *13*, 119–121.

59 A. Teslja, B. Nizamov, P. J. Dagdigian, *J. Phys. Chem. A* **2004**, *108*, 4433–4439.

60 R. Pearson Jr., F. J. Lovas, *J. Chem. Phys.* **1977**, *66*, 4149–4155.

61 V. Bonačić–Koutecký, M. Persico, *J. Am. Chem. Soc.* **1983**, *105*, 3388–3395.

62 V. Bonačić–Koutecký, J. Michl, *Theor. Chim. Acta* **1985**, *68*, 45–55.

63 P. J. Bruna, V. Krumbach,
S. D. Peyerimhoff, *Can. J. Chem.* **1985**,
63, 1594–1608.

64 R. Sumathi, *J. Mol. Struct. (THEO-
CHEM)* **1996**, *364*, 97–106.

65 L. D. Landau, *Phys. Z. Sowjetunion* **1932**,
2, 46.

66 C. Zener, *Proc. R. Soc. London, Ser. A*
1932, *137*, 696.

67 R. W. Schoenlein, L. A. Peteanu,
R. A. Mathies, C. V. Shank, *Science*
1991, *254*, 412–415.

68 J. E. Kim, M. J. Taubner, R. A. Mathies,
Biochemistry **2001**, *40*, 13774–13778.

69 R. S. H. Liu, A. E. Asato, *Proc. Natl
Acad. Sci. USA* **1985**, *82*, 259.

70 A. Warshel, *Nature* **1976**, *260*, 679–683.

71 A. Warshel, N. Barboy, *J. Am. Chem.
Soc.* **1982**, *104*, 1469–1476.

72 R. R. Birge, B. W. Vought, *Methods Enzymol.* **2000**, *315*, 143–163.

73 M. Sugihara, V. Buss, P. Entel,
M. Elstner, T. Frauenheim, *Biochemistry*
2002, *41*, 15259–15266.

74 G. G. Kochendoerfer, P. J. E. Verdegem,
I. van der Hoef, J. Lugtenburg,
R. A. Mathies, *Biochemistry* **1996**, *35*,
16230–16240.

75 A. Sinicropi, A. Migani, L. De Vico,
M. Olivucci, *Photochem. Photobiol. Sci.*
2003, *2*, 1250–1255.

76 A. D. Becke, *Phys. Rev. A* **1988**, *38*, 3098.

77 C. Lee, W. Yang, R. G. Parr, *Phys. Rev. B*
1988, *37*, 785.

78 U. F. Röhrig, L. Guidoni,
U. Rothlisberger, *ChemPhysChem* **2005**,
6, 1836–1847.

79 M. Garavelli, F. Bernardi, M. Olivucci,
M. J. Bearpark, S. Klein, M. A. Robb,
J. Phys. Chem. A **2001**, *105*, 11496–
11504.

80 C. S. Page, M. Olivucci, *J. Comput.
Chem.* **2003**, *24*, 298–309.

8

Chemical Aspects of the Restricted Rotation of Esters, Amides, and Related Compounds

Christophe Dugave

8.1
Thermodynamic and Kinetic Aspects of *Cis-Trans* Isomerization

Carbon combined with multivalent electron-rich elements such as nitrogen, oxygen, or sulfur gives rise to a large number of small functional groups such as esters, thioesters, amides, carbamates, ureas, etc. which are found in synthetic compounds as well as in certain proteins, peptides, and other biomolecules. These groups may exist as a mixture of two or more Z/E isomers due to the deconjugation of heteroatoms and the restriction of free rotation about the σ-bond. The energy difference between Z and E isomers is usually below 5 kcal mol^{-1} with a maximum for esters, but many factors can influence the $Z:E$ ratio, such as steric hindrance, aromatic stabilization, dipole–dipole interactions and lone pair-σ interaction as well as intra- and intermolecular H-bonds [1] (Fig. 8.1). As anticipated, the less crowded Z isomer (analogous to the *trans* isomer of olefins) is the more abundant for nonbulky substituents, whereas larger groups tend to decrease the difference between the two isomers.

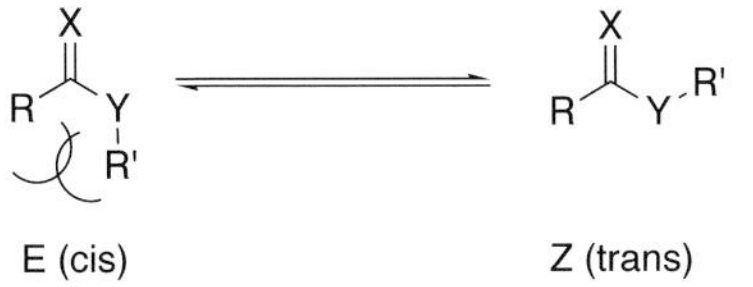

E (cis)   Z (trans)

Fig. 8.1 Schematic representation of E (*cis*) and Z (*trans*) isomers of esters (X = O, Y = O), thioesters (X = O, Y = S), amides (X = O, Y = NH) and thioxoamides (X = S, Y = NH).

Both isomers may be observed by several techniques, in particular dynamic NMR spectroscopy, UV spectroscopy and UV resonance Raman ground and excited state studies [2], which provide valuable thermodynamic data. Unfortunately, except for amides, which have been thoroughly studied due to their biological role in peptides and proteins, the lack of large systematic studies makes the establishment of general rules difficult. Ab initio studies have been mainly per-

cis-trans Isomerization in Biochemistry. Edited by Christophe Dugave
Copyright © 2006 WILEY-VCH Verlag GmbH & Co. KGaA, Weinheim
ISBN: 3-527-31304-4

formed in vacuo and, though they indicate a tendency, they have to be considered with prudence since they do not usually consider solvent effects that strongly influence the $Z:E$ equilibrium. For example, the calculated free energy difference for methyl formate is reduced from 5.2 kcal mol^{-1} in the gas phase to 1.6 kcal mol^{-1} in acetonitrile.

8.1.1
Esters and Thioesters

In the formate esters, the free energy difference decreases from esters to thioesters and amides for the methyl and *tert*-butyl groups, and *E*-methyl formate is only observed in very small amounts ($\leq 0.3\%$) at low temperature (190 K), as shown by ^{13}C NMR spectroscopy [3]. The energy is significantly lower for the bulky *tert*-butyl moiety relative to the methyl group and this seems to be a function of steric hindrance since ethyl formate and *iso*-propyl formate have intermediary values [4,5]. In the *tert*-butyl formate, a strong electronic repulsion accounts for the relatively high proportion of E isomer (10%).

E and Z isomers of esters can display significantly different properties such as acidity, as observed in the case of methyl acetate: the E isomer has been calculated to be more acidic than the Z form [6], and this theoretical result has been related to the unusually low pK_a of Meldrum's acid ($pK_a = 7.3$ to be compared to dimethyl malonate $pK_a = 15.9$) [7,8]. In contrast, the $Z:E$ ratio of methyl thionoformate (HCOSMe) is 97:3 in acetone ($\Delta G° = 1.29$ kcal mol^{-1}) in the 177–192 K range, whereas the ratio is 77:29 for cyclopropyl thionoformate in the same conditions ($\Delta G° = 0.31$ kcal mol^{-1}). The energy difference is even close to zero ($\Delta G° = 0.13$ kcal mol^{-1}) in the particular case of phenyl thioformate [9].

Steric interactions dramatically increase the free energy difference for acetates and thioacetates. For example, $\Delta G°$ is respectively 9.3 and 4.9 kcal mol^{-1} for methyl acetate and methyl thioacetate, compared to $\Delta G° = 2.1$ and 1.3 kcal mol^{-1} for methyl formate and methyl thioformate respectively. In turn, the barrier to rotation remains the same order of magnitude with $\Delta G^{\ddagger}_{E\rightarrow Z} = 12.4$ (methyl acetate) and 10.1 kcal mol^{-1} (methyl thioacetate).

Thioesters are obligatory intermediates in the in vivo formation of esters, in particular those found in complex lipids and also play a part in the formation of peptides, fatty acids, sterols, terpenes, and porphyrins. They may be found in protein either as transitory constituents or as stable motifs, and all involve a cysteine residue of the protein. The most famous example is the thiol-acryloyl moiety of photoactive yellow protein, which undergoes a thioester *cis-trans* isomerization (CTI) directed by photoisomerization of the acrylate and intramolecular restraints (see Chapter 5). A thioester linkage is also found as a high-energy intermediate in stable proteins such as the complement components C3 and C4 and the protease inhibitor a_2-macroglobulin, in which it is implicated in the entrapment of proteases [10]. Thioesters are also found in their invertebrate equivalents, thioester-containing proteins (TEPs), which play an important role in innate immunity in insects [11]. In the recently published structure of the complement component

C3, the Cys988–Gln991 Z-thioester linkage may react with nucleophilic groups (mainly OH and to a lesser extent NH) at cell surfaces and proteins in order to increase internalization [12]. Thioesters also reversibly connect cysteine residues to palmitate [13] and other fatty acids [14] in order to facilitate protein–membrane interactions and protein trafficking.

8.1.2
Amides and Thioxoamides

Amide *cis-trans* isomerism is well documented [15–17], in particular for peptides and proteins since it plays a central role in the structure and biological activity of these molecules. This point will be detailed further in Chapter 9 and peptides will be only tackled here for the purposes of comparison with other chemically related motifs.

Due to the pronounced amide resonance which results from the ability of the nitrogen atom to delocalize its lone electron pair over the whole amide moiety, amide has a planar shape with two minimum energy structures for $\omega = 0°$ (*cis* rotamer) and 180° (*trans* rotamer). Although the *trans* isomer is the more stable for steric reasons, the *cis*:*trans* ratio is higher relative to esters with an enthalpic difference between 0.5 and 6 kcal mol^{-1}, with a ratio for methyl formamide of about 10:90 in water [18]. As already observed with esters, the transition from formyl derivatives to acetyl compounds results in a dramatic rise in free energy difference ($\Delta\Delta G° = +3.58$ kcal mol^{-1} for esters in acetonitrile), which disfavors the formation of a *cis* isomer (≪0.1% for methyl acetate and 1.4% for methyl acetamide in water). The same trends are observed for peptides and proteins that are made of a regular succession of secondary and tertiary amide bonds. Small flexible peptides usually have an all-*trans* configuration, which is far more stable, though inter- and intramolecular interactions, solvation, and protonation can modulate this situation.

The pseudo-double bond character of amides is much more pronounced than for esters due to the conjugation of the H–N–C=O moiety and is correlated to the ability of distorted amides to be hydrolyzed to bases [19]. For this reason, the barrier to interconversion is significantly higher that for the ester series, with $\Delta G^{\ddagger}$ typically ranging from 16 to 22 kcal mol^{-1} [17]. However, the rotational barrier is not solely due to conjugation and also partly arises from the orientation of the nitrogen lone pair which is perpendicular to the amide plane [20]. Therefore, the rates of isomerization are considerably slower than for esters. This means that both isomers can be observed by simple techniques, for example at room temperature by ^{1}H and ^{13}C NMR spectrometry and UV spectrophotometry [21].

Amide CTI is a first-order reversible reaction (unimolecular process) characterized by a kinetic constant $k_{obs} = k_{t\rightarrow c} + k_{c\rightarrow t}$. In secondary amides, k_{obs} is entirely determined by $k_{c\rightarrow t}$, suggesting that a stabilized *cis* isomer corresponds to a decelerated *cis* → *trans* isomerization rather than an accelerated *trans* → *cis* reaction, whereas both rate constants usually contribute in a similar way for tertiary amides. The kinetic constant $k_{c\rightarrow t}$ was determined for a set of Gly- and Ala-containing dipeptides to be in the 0.3–0.7 s^{-1} range, depending on the ionization state of the peptide, a situation basically different from CTI in larger peptides which is not sensitive to the ionization state of the side-chains. The CTI was also investigated by UV resonance Raman spectrophotometry and the activation energy barriers were estimated to be 13.8 ± 0.8 kcal mol^{-1} ($k_{c \to t}$ = 2.3 ± 0.3 s^{-1}) and 11.0 mol^{-1} ($k_{c \to t}$ = 14 ± 2 s^{-1}) for N-methyl acetamide (NMA) and the Gly-Gly dipeptide, respectively, with energy differences of 2.6 kcal mol^{-1} for NMA and 3.1 for Gly-Gly [2]. A highly negative $\Delta S^{\ddagger}$ value was calculated for the Gly-Ala dipeptide, suggesting a major reorganization of the solvent molecules surrounding the peptide when it is approaching the transition state. This is also true for simple amide derivatives such as formamide since it was calculated that resonance accounts for only one half of the rotational barrier (7.3 vs. 15.6 kcal mol^{-1}) and for two-thirds of thioformamide (13.7 vs. 21.0 kcal mol^{-1}) [20]. Consequently, CTI of thioxoamides is slower than for the corresponding amides. Thermodynamic and kinetic data were also obtained by ^{1}H and ^{13}C NMR spectroscopy, magnetization transfer experiments and line-shape analysis for poly-Ala oligopeptides containing a Tyr residue.

Tertiary amides have significantly lower energy differences than the corresponding secondary amides. For example, $\Delta G°$ is 5.0 kcal mol^{-1} for acetyl-N-methyl glycine and only 0.3 kcal mol^{-1} for acetyl glycine [22]. On the other hand, the energy barrier to rotation remains about 18–22 kcal mol^{-1} and this explains why unsymmetrical N-disubstituted amides are often observed as a mixture of interconverting *cis-trans* isomers at room temperature.

Tertiary thioxoamides have *cis:trans* ratios roughly similar to those of the corresponding amides (Table 8.1). As observed with secondary thioxoamides, the energy barrier is greatly increased by the C=O to C=S substitution as calculated from ^{1}H and ^{13}C dynamic NMR experiments. As an example, dimethyl-[d_3]acetamide and dimethyl-[d_3]thioacetamide have $\Delta G^{\ddagger}$ values of about 21.4 and 23.5 kcal mol^{-1}, respectively, at 298 K in [d^6]DMSO, and the same tendency has been observed in a large set of thioxoamides. An excellent linear correlation was obtained with $\Delta G^{\ddagger}_{\text{thioxoamide}}$ = 1.13/1.11 $\Delta G^{\ddagger}_{\text{amide}}$ [23]. This was also observed with a set of thioxylated peptides whose CTI was 25- to 125-fold slower than that of all-amide peptides (Table 8.1) [24,25]. As observed with nonpeptide thioxoamides, $\Delta G^{\ddagger}$ is relatively insensitive to solvents [23], implying a bulky transition state of rotation with minor solvent reorganization. The enhanced acidity of the thioxoamide N–H proton promotes the formation of structures such as the C_7 γ-turn and C_{10} β-turn which are not usually found in all-amide peptides due to strong CS–N–H$\cdots$O=C hydrogen bonding [26].

Table 8.1 Kinetic and thermodynamic constants of Ala-Xaa-Pro-Phe-pNA/Ala-Xaa-ψ[CS-NH]Pro-Phe-pNA at pH 7.8, 10 °C.

Xaa	*cis* content (%)[a]	$k_{c \to t}$ (10^{-5} s^{-1})	$\Delta H^{\ddagger}$ (kcal mol^{-1})[c]	$\Delta S^{\ddagger}$ (cal mol^{-1})[d]
Gly	33/33	610[b]/25	17.9/15.7	5.6/19.4
Ala	9.8/8.5	690[b]/7	19.1/19.1	0.6/10.0
Abu	9.2/9.6	650[b]/5	19.5/19.3	0.8/9.9
Leu	8.1/6.5	620[b]/8	21.2/15.4	6.0/22.7
Phe	12.6/10.5	360[b]/11	1.91/19.3	5.5/27.3

a SD ≤ 1%.
b SD ≤ 30 × 10^{-5} s^{-1}.
c SD ≤ 0.25 kcal mol^{-1}.
d SD ≤ 1.3 cal mol^{-1}.

Amide surrogates such as *N*-aminoamides and *N*-hydroxyamides have been thoroughly investigated due to their potential as protease-resistant pseudopeptides, but little is known about the CTI of these molecules. It seems that *N*-amination and *N*-hydroxylation weakly perturb the local conformation of the peptide bond, though they dramatically affect the hydrogen-bonding network since CO–NH–NH$_2$ is a weak proton donor and CO–NH–OH is a strong proton donor. In particular, *N*-hydroxyamides tends to form 8- and 11-membered rings in peptides in dichloromethane instead of the usual 10-membered ring of the corresponding amides.

8.1.3
Oxalamides and Hydrazides

Several peptidomimetic motifs such as oxalamides and hydrazides have motifs that seem to be very similar to amides, but these flexible templates display several preferential conformations due to the existence of multiple dihedral angles and hence have been used to tune peptide conformations in bioorganic and medicinal chemistry (Fig. 8.2) [27,28].

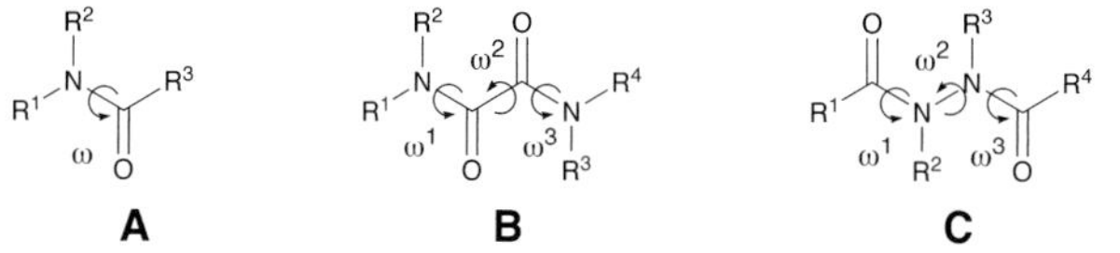

Fig. 8.2 Dihedral angles of amides (A), oxalamides (B) and hydrazides (C).

Theoretical study of the conformational preferences, energy differences, and charge distribution of NdMO (B with R^1 = R^4 = CH$_3$ and R^2 = R^3 = H) showed that the *trans-trans-trans* form is the most stable conformation in both gas phase and water due to the absence of unfavorable interactions between the oxygen atoms and the methyl groups. NdAH (B with $R^1 = R^4 = CH_3$ and $R^2 = R^3 = H$) exists as four low-energy isomers with energy differences below 2.4 kcal mol^{-1}: a *trans-gauche-trans* isomer in the gas phase and a *cis-gauche-trans* isomer in aqueous solution, both of which are stabilized by an intramolecular H-bond. Water has an important effect by changing the molecular dipoles of both NdMO and NdAH by about 30%. NdMO has some similarities with NMA (A with $R^1 = R^3 = CH_3$ and $R^2 = H$); in particular there is a resemblance between the amide resonance and charge distribution of NdMO. Conversely, NdAH presents pyramidality of the nitrogen atoms and hence a reduced resonance of the two amide groups. Moreover, the N–N bond behaves as a single bond that can rotate almost freely [29]. Therefore, hydrazino- and azapeptides have found numerous applications in peptide chemistry and biology since the increase in dihedral angles and potential H-bond donors and acceptors gives access to a wide variety of stable secondary structures [30,31].

Recently, Marshall and coworkers have reported on the particular structural behavior of azaproline (azPro) derivative Ac-azPro-NHMe (Fig. 8.3) which tends to form an eight-member hydrogen bond ($i \rightarrow i + 1$) in a typical β-turn conformation without introducing any steric bulk. It stabilizes a *cis*-amide bond whereas proline favors a seven-member H-bond ($i + 2 \rightarrow i$) with a *trans*-amide conformation (γ-turn) (Fig. 8.3) [27]. Additional *N*-methylation decreases the energy difference between *cis*- and *trans*-azPro conformers and hence severely perturbs the β-turn conformation [28]. In nonproline derivatives, permethylation inverted the tendency since calculation at the B3LYP/6–31G* level showed that the *cis*-Ac-azAla-NHMe was 1.36 kcal mol^{-1} more stable than the *trans*-amide conformer, whereas *trans*-Ac-azAla-NHMe was favored by 1.24 kcal mol^{-1} [32].

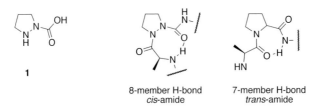

1

8-member H-bond
cis-amide

7-member H-bond
trans-amide

Fig. 8.3 Structure of azaproline (azPro) **1** and *cis*-Ala-azPro 8-membered ring compared with the *trans*-Ala-Pro 7-membered ring.

8.1.4
Carbamates and Ureas

The carbamate *anti* rotamer is usually favored by 1.0–1.5 kcal mol^{-1} for steric and electrostatic reasons [33]. Energy difference may be even close to zero in the case of certain carbamates which may be roughly found as a 50:50 mixture of *syn* and *anti* isomers. This is the case with many amino acid Boc derivatives, which appear as two well-defined upfield singlets at 1.23 and 1.28 ppm in ^{1}H NMR spectroscopy at room temperature instead of the lone singlet at 1.40–1.45 ppm. Rotational barriers are high, albeit lower than for the corresponding amides, a result which may be explained by the multiplicity of potential H-bonds capable of stabilizing a tetrahedral nitrogen transition state. This is supported by the low solvent effect on carbamate CTI relative to the corresponding amide compounds. For example, $\Delta G^{\ddagger}$ is 16.7 and 15.5 kcal mol^{-1} in carbon tetrachloride and 19.3 and 15.5 kcal mol^{-1} in D$_2$O:CD$_3$OD 75:25 for amide **2** and carbamate **3**, respectively (Fig. 8.4) [34]. It is noteworthy that all carbamates studied by Cox and Lectka displayed unprecedented negative $\Delta S^{\ddagger}$ values ranging from −21 cal mol^{-1} (compound **3a**) to −6 cal mol^{-1} (compounds **3b** and **3d**) in D$_2$O:CD$_3$OD 25:75 with no significant influence of the R substituent on $\Delta G^{\ddagger}$ (15.4–15.6 kcal mol^{-1}). Moreover, multiple H-bonds are well known to stabilize a given isomer over intermolecular association [33,35].

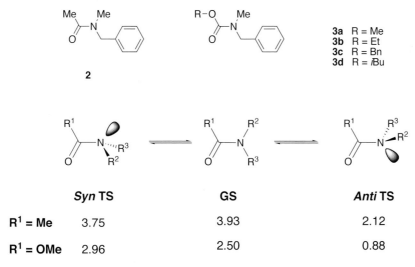

		Syn TS	GS	Anti TS
R^1 = Me		3.75	3.93	2.12
R^1 = OMe		2.96	2.50	0.88

Fig. 8.4 (A) Structures of amide **2** and carbamate **3a–d**. Dipole moments of ground state (GS) and two possible *syn* and *anti* transition states calculated at the 6-31G**//6-31G* level for R^2 = R^3 = Me [34].

trans-trans-Ureas are the most stable, although intra- and intermolecular H-bond-mediated stabilizations of certain isomers in urea-containing compounds are much more frequently found than in amide-containing molecules and participate in the selection of particular structures that display high stabilities [36]. Oligourea peptide are γ-peptide analogs while carbamates may mimic β-peptides. For this reason, intramolecular stabilization requires a minimum chain length as demonstrated for N,N′-linked oligourea peptide analogs since optimal stabilization was obtained for heptamers and longer oligoureas [37]. As a consequence,

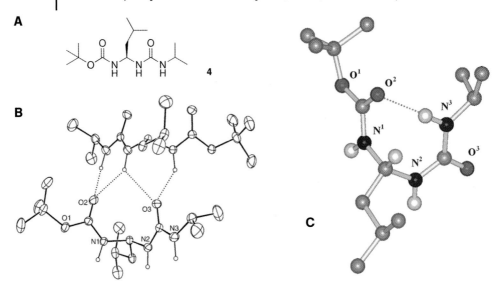

Fig. 8.5 Molecular structures of a urea peptide (A) in the crystal state (B, ORTEP representation with ellipsoids shown at the 40% probability level) and in a CDCl$_3$ solution (MM2 energy-minimized conformation) [38].

not all possible rotamers are represented and there may be some considerable differences between the solid-state and solution structures as observed with the model urea peptide **4** (Fig. 8.5) [38].

Due to competitive conjugation, ureas have barriers to rotation lower than the corresponding amides. Urea isomerization proceeds either via a classical CTI ($\Delta H^{\ddagger} = 18.5 \pm 1.6$ kcal mol^{-1}) or a whole-body flip ($\Delta H^{\ddagger} = 15.5 \pm 1.2$ kcal mol^{-1}), i.e. from *trans-trans* to *cis-cis*.

8.2
Influence of the Environment on CTI

8.2.1
Solvent and Concentration

Theoretical calculations by statistical mechanical simulation, in agreement with experimental data, have shown that polar solvents affect the energy of *E* isomers of esters more than that of the *Z* rotamer. The energy difference was reduced from 5.2 kcal mol^{-1} in the gas phase ($\varepsilon = 1$) to 1.6 kcal mol^{-1} in the acetonitrile ($\varepsilon = 35.9$) and from 8.5 to 5.2 kcal mol^{-1} for methyl formate and methyl acetate, respectively. This may be explained by the increased polar character of the carbonyl group in the more polar solvent [39]. This was also observed in DMSO ($\varepsilon = 47.2$) where the proportion of the minor *E* isomer tends to increase [1]. The keto-enol tautomerization of formamide, investigated at the Hartree-Fock level using the 6-31G** basis set, suggested that various types of nondissociative proton transfer processes (direct, dimeric, and solvent-assisted) might account for solvent and concentration effects [40].

Proton NMR studies of *N*-methyl formamide (NMF) and NMA at high dilution in deuterated solvents have shown that the level of *cis* isomer of NMF is 8% in water, 10.3% in chloroform, 8.8% in benzene, and 9.2% in cyclohexane, while the level of *cis*-NMA (a model for the secondary peptide bond) is 1.5% in water and does not change very much in nonpolar solvents [18]. Ab initio molecular calculations suggest that the small difference in dipole moments in *cis* and *trans* forms explain the relative insensitivity of amides to solvent change, unlike esters [22,41]. This may be explained by nearly identical free energies of solvation for the two isomers [18]. The energy difference between *cis* and *trans* isomers in aqueous solution (ΔG° = 2.5 kcal mol^{-1}) accounts for the preferential *trans* conformation adopted by most peptide bonds. Similar results were obtained with nonproline tertiary amides [22].

CTI of the Xaa-Pro peptide bond directly depends on the solvent's ability to donate a hydrogen bond which stabilizes both isomers and restricts isomerization, whereas it is quite independent of solvent dielectric constant and solvent polarity. From inversion transfer ^{13}C NMR spectroscopy experiments, a barrier energy difference $\Delta G^{\ddagger}_{aprotic} - \Delta G^{\ddagger}_{protic} = 1.3 \pm 0.2$ kcal mol^{-1} was determined for the Ac-Gly-Pro-Ome peptide [42]. This result suggested that a nonpolar environment accelerates CTI of the Xaa-Pro moiety and strongly supported the hypothesis of a deconjugated transition state [43]. This was confirmed by Cox and Lectka who investigated the solvent effect on secondary and tertiary amides including proline [34].

In contrast to amides, the barrier to CTI of carbamate was insensitive to solvent. However, this situation dissimulates two distinct effects that annihilate each other: (1) a decrease in $\Delta H^{\ddagger}$ (with a minimum for 75% D$_2$O/CD$_3$OD) which reflects a charge separation stabilizing the transition state; (2) an increase in negative $\Delta S^{\ddagger}$ indicating a stronger solvation of the most polar transition state relative to the less polar ground state [34,44]. The effect of H-bonding on the *anti:syn* ratio (K) was particularly pronounced with carbamates [45]. For example, esterification of Boc-Ala-OH resulted in a jump from 8% *anti*-Boc-Ala-OH to 92% *anti*-Boc-Ala-OMe in CDCl$_3$ at −54.3 °C, and this was also observed with intermolecular interactions since addition of acetic acid produced a stronger effect on Boc-Ala-OH relative to the corresponding methyl ester [35]. This influence of H-bonds was further confirmed with dual H-bond donor/acceptor molecules [33]. Concentration has a similar effect on Boc-Ala-OH, since a CDCl$_3$ increase from 7.9 mmol L^{-1} to 161.9 mmol L^{-1} caused a jump from 8% to 43% *syn* isomer, respectively.

8.2.2
pH and Salts

The pH dependence of amide CTI is much more documented than for esters. Although the carbonyl oxygen is the preferred site of protonation by strong Brønsted acids, amide CTI is usually not sensitive to pH unless ionizable groups surround the isomerizing peptide bond. In dipeptides, *cis:trans* ratios and CTI rate constants remain relatively stable between 4 and 8, though the pK_a of each *cis* and *trans* isomer may be significantly different [22]. For lower and higher pH values, however, the changes in ionic environment modify both the *cis:trans* ratio as well as kinetics of isomerization. This suggests that ionization states strongly influence the electronic distribution in the amide moiety which controls both geometric preferences and electronic transition during CTI.

Usually, metal ions as Lewis acids do not influence the *cis:trans* ratio of secondary amides. Conversely, the isomer preference of the amino acyl-prolyl bond is strongly shifted to the *cis*-conformation (from 10 to 70%) by addition of high concentrations (0.2–0.47 mol L^{-1}) of Li^+ ions in anhydrous solvents such as trifluoroethanol or tetrahydrofuran [46]. The Li^+ effect is still misunderstood, and it seems reasonable to imagine that the Lewis acid cation is complexed by the basic oxygen of the amide carbonyl and hence perturbs the H-bond network which stabilizes the *trans* conformer [46]. However, this effect is limited to certain amino acids preceding proline. Nonproline tertiary amides such as the MeLeu^9-MeLeu^{10} moiety in the cyclic undecapeptide cyclosporine are dramatically affected by addition of Lewis acids and shift from 100% *cis* in tetrahydrofuran to 100% *trans* upon addition of LiCl. Li^+ as well as Ag^+ disrupt amide resonance by coordination to the amide nitrogen and seem to be able to lower $\Delta G^{\ddagger}$ for CTI of *N,N*-dimethylacetamide [16]. An early computational study has predicted a lowering of $\Delta G^{\ddagger}$ upon coordination of Li^+ ions [47] which might cause an increase of $k_{t \to c}$. Pt^{II}–H bonding is also suspected to stabilize the *cis* isomer of acetamide. This interaction, which is closely related to the H-bond, is predominant in nonoxygenated solvents since oxygen is able to compete with platinum and thus favors the *trans* isomer [48].

Surprising effects of several detergents on CTI have been reported. Bairaktari et al. described an exclusive *cis*-Ile^1-Lys^2 conformation of the 17-mer peptide bombolitin dissolved in aqueous sodium dodecylsulfate which suggested a dramatic shift of $\Delta G°$ caused by the micellar environment [49]. Fischer and coworkers also demonstrated the effects of detergents on Xaa-Pro CTI in a series of tetrapeptides Suc-Ala-Xaa-Pro-Phe-pNA including the thioxopeptide Suc-Ala-Leuψ[CS-N]Pro-Phe-pNA. The *trans* isomer was favored by an energy difference of $\Delta\Delta G°$ averaging 2.8 ± 0.2 kcal mol^{-1}. An up to 23-fold rate enhancement was observed and the micelle-induced decrease of $\Delta G^{\ddagger}$ was entirely enthalpy-driven. In saturating conditions, the charge of the detergent did not influence greatly the rates of isomerization, suggesting that the hydrophobic peptide was deeply inserted inside the lipophilic environment of the micelle. Such an effect, observed in hydrophobic solvents that lower the energy barrier to isomerization, was also the basis of the FK506 binding protein (FKBP) catalysis of CTI [50].

8.2.3
Temperature

As anticipated, temperature modulates the *Z:E* ratio and increases the rates of CTI. In most cases, esters and thioesters cannot be differentiated at room temperature whereas both rotamers of the corresponding amides, carbamates (Table 8.2), and ureas are observed in these conditions, but are not affected to the same extent.

Table 8.2 Percentage of *anti* isomer, free energy difference, exchange rate and energy barrier for Boc-Ala-OH [51].

T (K)	% *anti* isomer	$\Delta G°$ (kcal mol^{-1})[a]	$k_{a \to s}$ (s^{-1})	$\Delta G^{\ddagger}$ (kcal mol^{-1})[b]
219	8	1.06	–	–
240	13	0.91	–	–
261	37	0.28	–	–
283	55	−0.10	5.5	15.5
295	64	−0.33	16	15.6
308	69	−0.49	52	15.6
321	76	−0.74	105	15.8
335	81	−0.99	300	15.9

a $\Delta H° = 4.8 \pm 0.3$ kcal mol^{-1}; $\Delta S° = 17 \pm 1$ cal mol^{-1} K^{-1}.
b ± 0.2 kcal mol^{-1}.

In particular, for the secondary amides of nonproline dipeptides, $\Delta S^{\ddagger}$ may vary depending on the protonation state [21] and hence influence the temperature-dependent energy barrier in a way described by the equation $\Delta G^{\ddagger} = \Delta H^{\ddagger} - T\Delta S^{\ddagger}$, though amide and carbamate CTI are usually essentially enthalpy-driven. The solvent dielectric constant ε also varies depending on temperature (i.e. 47.6 at −50 °C and 32.1 at 50 °C for acetonitrile) and strongly influences the charge distribution on the pseudo double bond.

8.3
The Study of CTI of Amides and other Conjugated π-Systems

8.3.1
Spectroscopic Techniques

8.3.1.1 NMR Spectroscopy

NMR spectroscopic techniques have proved to be extremely useful for the determination of *cis:trans* ratios as well as for the monitoring of CTI and the analysis of thermodynamic parameters. Most research efforts have been directed towards amides that are implicated in important biological processes and, to a lesser extent, towards carbamates because they allowed interesting comparative studies with amides, in particular in terms of CTI mechanisms. Many nuclei are potential probes in NMR spectroscopy for determining *cis:trans* ratios as well as interconversion kinetics such as ^{1}H, ^{13}C, and ^{18}F, since CTI affects both coupling and δ as well as intra- and intermolecular interactions (Fig. 8.6). Usually, *cis* and *trans* isomers are identified by nuclear Overhauser effect (NOE) experiments due to proximity effects generated by a given isomer. In the case of multi-isomer equilibrium, the situation may be complicated for little populated conformers that cannot be identified unambiguously. When possible, the monitoring of individual signals after a sudden modification of the environment gives access to microscopic rate constants separately [52].

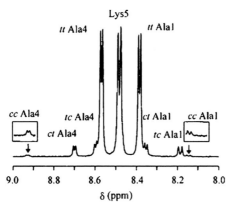

Fig. 8.6 NH-region of a ^{1}H NMR spectrum of Ac-Ala-Pro-Pro-Ala-Lys-NH$_2$ in a 20 mmol L^{-1} phosphate buffer pH 5 at 4 °C. Insets magnify signals of the *cc* isomer (1%) [52].

Line-shape analysis, magnetization transfer experiments, 2D NOESY etc. can be used for kinetic studies of CTI, but they require relatively high peptide concentrations (Fig. 8.7) [53]. Mainly used for the investigation of model systems, they can also be utilized in complex cases for the determination of individual rate constants and for the monitoring of protein folding.

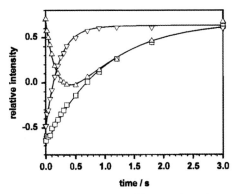

Fig. 8.7 Relative intensities of the time-dependent magnetization of CH$_3$ of *cis*-Ala-Tyr in H$_2$O:D$_2$O 9:1 pH 5.9 in a T_1 experiment (□) and the magnetization-transfer experiments with the *cis* signal parallel (△) and antiparallel to the stationary magnetic field (▽) at 316 K. Solid lines represent the correspondingly fitted biexponential decays with $T_1 = 0.984$ s, $k_{c \to t} = 3.7$ s^{-1} and a *cis* content of 0.61% [53].

Quantitative studies of the solvent and temperature effects coupled to molecular mechanics studies give access to a wide variety of parameters [44]. The monitoring of CTI using fluorine as an internal probe (i.e. 4-fluorophenylalanine) is also particularly attractive due to the high sensitivity of [18]F [54].

8.3.1.2 Spectrometric and Fluorimetric Assays

Spectrophotometric and fluorimetric assays have been used for more than 15 years for the kinetic study of CTI in peptides. Fluorescence studies were first used to monitor the refolding of protein: the assay was based on the change in fluorescence intensity of a tryptophan residue at 320–350 nm upon irradiation at 260–290 nm when its environment was modified by protein refolding (usually, an increase of fluorescence resulting from the burying of the indole moiety inside a more hydrophobic environment). However, such assays evaluate the overall structure of the protein, that is the sum of multiple additive processes, and, in view of CTI studies, they have been limited to simple models such as RNase T1 [55]. Several UV/visible and fluorescence assays have since been developed to determine the kinetic constants of amide CTI. All use tetrapeptides derived from the chromogenic substrate for the protease α-chymotrypsin: Suc-Ala-Ala-Pro-Phe-pNA. In particular, the first assay for monitoring CTI was based on the cleavage of the *para*-nitroaniline moiety ($\lambda_{max} = 390$ nm) by a set of proteases including α-chymotrypsin, trypsin, subtilysin, pepsin, and dipeptidyl-peptidase [17] which selectively hydrolyze tetrapeptides with a *trans*-Xaa-Pro sequence, thus stopping reverse isomerization. Sensitivity was greatly improved by operating at low temperature (4–10 °C) and dissolving the substrate in a 0.47 mol L^{-1} LiCl in trifluoroethanol, which inverses the *cis:trans* ratio (from 20:80 to 80:20) of the peptide relative to aqueous buffers. Therefore, sudden addition of the peptide solution to an aqueous

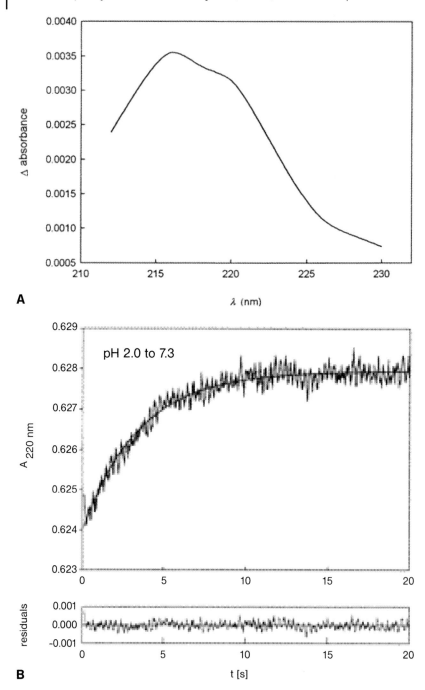

Fig. 8.8 Absorbance difference of Gly-Gly *cis* and *trans* isomers. (A) Time-course of CTI of Gly-Gly recorded at 220 nm in a 50 mmol L⁻¹ phosphate buffer pH 7.3 following a pH jump from a 50 mmol L⁻¹ phosphate buffer pH 2.0 at 25 °C (B) [21].

buffer (containing a nonlimiting concentration of protease) started the CTI process, which resulted in a two-slope curve: (1) a fast hydrolysis of the preexisting *trans* isomer; (2) a slow (CTI-dependent) hydrolysis of the newly formed *trans* isomer.

Kinetic constants are obtained from the second part of the two-slope curve: treatment of the data as a pseudo first-order reaction gives $k_{c \to t}$, while interpolation of the slope the ratio $K_{cis/trans}$. Although it may be marred by additive errors, $k_{c \to t}$ can be deduced from the equation: $K_{cis/trans} = k_{c \to t}/k_{c \to t}$.

A protease-free assay using Suc-Ala-Ala-Pro-Phe-DFA (DFA is 2,4-difluoroaniline) was also developed, but its sensitivity at 246 nm is low (<0.006 absorbance unity) for a typical peptide concentration of 21 µmol L^{-1} [56]. The same problem limits the applicability of the assay reported by Fischer and coworkers based on the pH sensitivity of the secondary amide bonds. A pH jump (i.e. from 2.0 to 7.3) led to a slight variation of absorbance at 220 nm which reflects the pH-dependent CTI (Fig. 8.8). However, this assay cannot be used routinely with oligopeptides, although it gives kinetic data that cannot be easily obtained by other methods [21].

The fluorescence of the *ortho*-aminobenzoate motif was utilized to elaborate a continuous fluorimetric assay based on intramolecular fluorescence quenching through isomer-differential collision of the fluorophore and a quencher, i.e. nitrophenylalanine or a *para*-nitrophenyl moiety at the C-terminus [57]. Relative fluorescence recorded at 410 nm (λ_{ex} = 317 nm) is cleared from the participation of the preexisting *trans* isomer, but this assay is limited to peptides bearing fluorophores and properly positioned fluorescence quencher moieties.

8.3.1.3 Other Spectroscopic Techniques

Several other techniques may be employed to investigate CTI dynamics. UV resonance Raman ground and excited state studies have given access to the thermodynamic parameters of the *trans* to *cis* photochemical isomerization of secondary amides upon laser excitation at 206.5 nm at various temperatures [2].

Infrared (IR) spectrometry [58], as well as circular dichroism [59], provide interesting information about the tendency of peptides and pseudopeptides to be structured in various solvents through intramolecular interactions (in particular H-bonds) that may be favored by only one conformation [26,60]. Transient 2D-IR spectroscopy is an attractive technique for studying ultrafast conformational transitions in peptides [61], but, to my knowledge, has only been used to investigate rotation about ϕ and ψ dihedral angles. Thermal Fourier transform IR microspectroscopy was used to investigate the reversible CTI isomerization of captopril in the solid state [62], a process that may be ascribed to changes in the H-bond pattern [63].

Circular dichroism (CD) spectra transition between *cis* and *trans* conformers was successfully applied to the monitoring of light-induced CTI of thioxoamide bonds to determine rate constants and energy barriers (Fig. 8.9) [25].

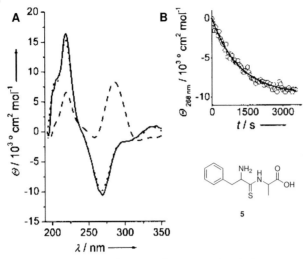

Fig. 8.9 (A) CD spectra of compound 5 (1.4×10^{-4} mol L^{-1}) in a 50 mmol L^{-1} sodium phosphate buffer pH 7.0. Thermally equilibrated peptide (———), peptide after irradiation (– – –) and re-equilibrated peptide (. . .). (B) First-order kinetics of the molar ellipticity of compound 5 at 268 nm [25].

After irradiation, the CD response of compound **5** reversibly changed to the opposite sign at 268 nm, suggesting that geometry around the chromophore N–C=S bond was changed. The rate constant obtained was comparable to those calculated from UV-visible spectrophotometry.

CTI may exert kinetic control of amide bond cleavage and, therefore, may sign the formation of specific ions and rearrangement products in mass spectrometry. High-resolution ion mobility/time-of-flight techniques have shown the presence of multiple conformers that arise from populations of *cis-* and *trans*-Xaa-Pro motifs [64] and may be differentiated depending on their relative ability to be protonated. CTI also influences the amide cleavage pathway under low-energy collision in tandem mass spectrometry that leads to the formation of the cyclic dipeptide diketopiperazine [65,66].

8.3.2
Separation of Z and E Isomers

Cis-trans isomers of peptides can be separated by several techniques including thin-layer chromatography (TLC), high-performance liquid chromatography (HPLC), and capillary electrophoresis.

In organic solvents, protected peptides containing clustered proline residues often display multiplicity of spots which correspond to slowly interconverting Xaa-Pro isomers, however, the resolution strongly depends on the mixture composition as well as peptide sequence. As an example, Boc-Pro-Pro-OMe appears as

two well-separated spots in chloroform:methanol 95:5 (Rf = 0.74 and 0.65) on silica gel plates but as a single spot in a 80:20 mixture (Rf = 0.72) [67].

Low-temperature HPLC separation of peptides allows the separation of *cis* and
trans isomers of constrained peptides such as the intramolecular disulfide of
Cys-Leu-Pro-Arg-Glu-Pro-Gly-Leu-Cys which gave two peaks at 0 °C coalescing at
higher temperature (Fig. 8.10). The resolution was also strongly dependent on the
solvent of injection since HPLC chromatograms obtained from a concentrated
LiCl/TFE solution of peptides showed significant modifications [68]. When the
rate of interconversion is sufficiently slow relative to experimental time, both isomers of smaller and nonconstrained peptides can be separated with excellent resolution at 0 °C [69,70].

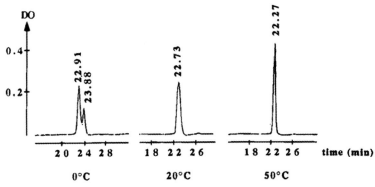

Fig. 8.10 HPLC profile at 215 nm of peptide Cys-Leu-Pro-Arg-
Glu-Pro-Gly-Leu-Cys recorded at 0, 20, and 50 °C (C18 column
Vydac 218TP, 3 × 250 mm, 0.7 mL min⁻¹, linear gradient
of A:B from 100:0 to 0:100 in 30 min (A: 5% acetonitrile +
0.05% TFA; B: 60% acetonitrile + 0.05% TFA) [68].

Capillary electrophoresis is also able to distinguish both *cis* and *trans* isomers of
linear peptides [67] and thioxopeptides irradiated at 254 nm [25,71]. Although
capillary electrophoresis is anticipated to separate products by charge, and charge
itself would not be affected by conformational changes, this may be explained by
the distinct solvation of conformers which therefore have different abilities to be
ionized in solution.

8.3.3
Models and Mimics for the Study of Amide CTI: Towards Multiple CTI Pathways

The mechanism of CTI in pseudo soluble bonds, especially peptides, has been
thoroughly investigated. Among the possible ways of amide CTI, many are plausi-
ble as biologically relevant and have highlighted new trends in the comprehension
of complex processes such as protein folding and denaturation, peptides/protein
conformational transitions etc, and their biological effects such as CTI-modulated
activity as well as macroscopic consequences (i.e. neurodegeneration). There

seem to be several distinct ways of isomerizing amides and related compounds, including catalysis by Brønsted acids, H-bonds, nucleophilic catalysis, metal cations, π-bases, light, etc. These different aspects were summarized in the excellent review by Cox and Lectka [16] who have done most of the work described in this section. Since then, several other possible CTI mechanisms have been proposed and will be discussed herein. Despite their differences, all of them result in a partial or complete deconjugation of the amide bond.

8.3.3.1 Acid/H-Bond-Catalyzed CTI

As mentioned previously, the oxygen carbonyl is the preferred site of protonation of amides by strong Brønsted acids, but this does not exclude CTI through the occurrence of a small but kinetically significant quantity of a *N*-protonated intermediate, as indicated by the 130-fold increase in the rate of CTI of dimethyl acetamide when the pH of the solution jumps from 7.0 to 1.8. An approximately 2500-fold acceleration of CTI ($\Delta\Delta G^{\ddagger} = 5$ kcal mol^{-1}) was also observed with the model amide **12b**, which isomerized through a ketone-amine-like transition state as suggested by the amide C=O IR stretch shift from 1637 to 1684 cm^{-1} (Table 8.3). The X-ray structure of compound **12b** revealed a bridging hydrogen which links both amide and amine nitrogen atoms via a "moderately strong" H-bond [72,73].

The existence of an intramolecular H-bond [60] was also postulated in the case of proline derivatives **13a,b**, with a 260-fold acceleration of CTI rates for compound **13b**, whereas this was not observed with the corresponding esters [74,75]. Significant acceleration of the CTI was also observed with oxazolidine carboxylate and 3- and 4-thiazolidine carboxylate derivatives, a result which may be explained by a puckering of proline due to stereoelectronic effects. Ab initio calculations as well as experimental evidence have highlighted the participation of an H-bond between the amide N–H and the properly positioned proline nitrogen which lowers the energy barrier to isomerization [15,17]. A similar, though less pronounced phenomenon was observed with a 4-mer peptide containing a (2*S*,4*S*)-4-fluoroproline residue [17,76] as well as acyclic peptides containing a His-Pro moiety [77] and several cyclic peptides [78–80]. Finally, structural evidence for a O=C–N–H—N intramolecular H-bond was provided by X-ray crystallographic study of the locked proline mimic **14** which displays an unambiguous signal in IR spectroscopy wherein a stretch at 3430 cm^{-1} was observed [75]. Although some of these models are basically different from the Xaa-Xaa (Xaa is nonproline amino acid) and Xaa-Pro sequences, they reproduce particular situations (environment, conformation, catalytic activity) that accelerate the amide CTI in vivo. Significant acceleration factors were observed for short peptides containing the His-Pro sequence, suggesting that an intramolecular H-bond with either a proximate amino acid residue or a properly positioned side-chain (in particular arginine guanidinium) was relevant to the noncatalyzed CTI of certain biologically active peptides such as angiotensin [77] and certain proteins such as dehydrofolate reductase [81].

8.3.3.2 Nucleophilic/Basic Catalysis of CTI

Nucleophilic catalysis of amide CTI was among the first mechanisms proposed for spontaneous and enzyme-catalyzed isomerization since it involves a tetrahedral (protease-like) transition state that disrupts amide resonance. Although this hypothesis was lately proposed with the peptidyl-prolyl isomerase Pin-1 [17], such a transition state was physically observed by X-ray crystallographic study of the potassium salt of naphthalene derivative **15** (Table 8.3) whose existence was suggested by IR and NMR data [73]. It is noteworthy that other X-ray structures of tetrahedral intermediates have been described previously [82,83]. Cox and Lectka also reported preliminary work on the base-catalyzed CTI of cyanomalonamide **17**: addition of sodium methanolate resulted in a fall of 4.1 kcal mol^{-1} in the CTI energy barrier, while it was reduced by 2.2 kcal mol^{-1} with 1.1 equivalents of a proton sponge, which all suggests an enolate transition state [16]. Recently, the first example of a π-base-catalyzed CTI was reported with simple esters and amide compounds on the one hand and rhenium complexes on the other, and the structure of the adduct resolved by X-ray crystallography clearly showed a decrease in the resonance of the amide via insertion inside the carbonyl [84]. Although this reaction might not be relevant to biochemical processes, it opens the discussion about this novel isomerization pathway.

8.3.3.3 Cation-Catalyzed CTI

Perturbation of the *cis:trans* ratio by Lewis acids [85] suggested that metal cations might catalyze the CTI. Because of their cyclic structure and their conformation in solution, prolines in peptides seems to be natural sites for cation binding involving both the pyrrolidine nitrogen and the proline carbonyl. Cu^{2+} complexation by proline derivative **13c** efficiently lowered the barrier to amide rotation by about 2 kcal mol^{-1}. Free enthalpy of activation shifted from 18.1 kcal mol^{-1} at 25 °C in THF to 13.8 kcal mol^{-1} at −25 °C when the acetyl group was replaced with a benzyl carbamate [51]. The influence of the azaphilic tendency of Lewis acids on CTI was also investigated with the "rigged" bis-pyridylamide system **19**. Tight coordination of Cu^{2+} operated through a marked quaternarization of the amide nitrogen with an associated reduction of 6 kcal mol^{-1} in $\Delta G^{\ddagger}$ with only 5 mol% Cu(OTf)$_2$, representing a 25 000-fold rate enhancement [51].

8.3.3.4 Light-Induced CTI

Excitation of the dominantly *trans*-secondary amide by a laser pulse in the 206–208 nm range was reported to cause a photochemical *trans* to *cis* isomerization which was monitored by resonance Raman spectroscopy. Activation barriers of 13.8 ± 0.8 kcal mol^{-1} and 11.0 ± 0.7 kcal mol^{-1} were obtained for NMA and Gly-Gly **21**, respectively. Temperature dependence experiments determined a Gibbs free energy gap between *cis* and *trans* conformers of 2.6 ± 0.4 and 3.1 ± 0.5 kcal mol^{-1} for NMA and **21**, respectively [2,86].

Thioxopeptides display a strong ($n \rightarrow \pi^*$) light absorption with a maximum around 270 nm for the *cis* isomer and a maximum near 300 nm for the *trans* isomer. Therefore, the equilibrium of the *cis* and *trans* conformers of a thioxopeptide **21a** is dual-directionally photoswitchable [71]. The kinetic and thermodynamic parameters of thioxopeptides were determined by Schiene-Fischer and coworkers, who showed that replacement of the canonic amide group with the thioxoamide bond causes a large rise in energy barrier and a subsequent slowing down of CTI on the macroscopic scale [25]. In contrast, the CTI of thioxylated amides on the molecular scale is an ultrafast process which occurs on the subnanosecond timescale [87].

Ab initio analyses of the CTI of pseudo-double bonds – mainly amides and carbamates – are in agreement with experimental results since they address the intrinsic thermodynamic properties of amides (gas phase) as well as solvent effects and intermolecular interactions. Since the pioneering work by Maigret and coworkers in 1970, important comprehensive ab initio studies of amide CTI have been carried out using quantum mechanical computations at various theoretical levels. Several basic models such as formamide and acetamide derivatives [88] and evolved models such as formyl glycinamide [89], N-acetyl-proline-N'-methylamide [90] have been validated for investigating amide CTI. Recently, N-acetyl-proline-N'N'-dimethylamide [91] and formyl-prolyl-prolinamide [92] have highlighted the particular case of the poly-proline moiety. As a general trend, the Hartree-Fock HF/6-31+G(d) level with the conductor-like polarizable continuum model (CPCM) method seems to be the most appropriate method in describing CTI of secondary and tertiary amide bonds in solution.

Table 8.3 Listing of synthetic models used for modeling amide CTI.

Compound		Ground State	Method	Putative Transition State	Ref.
12a	R = Ph		HOTf		[72]
12b	R = 2-FPh				
13a	R = 2-FPh, R' = nHx, X = NH		-		[74]
13b	R = 4-(CO$_2$Me)Ph R' = Ac, X = NH		-		[75]
13c	R = Ac, R' = 2-FBn X = N–Bn		Cu(OTf)$_2$		[51]
14	R = 4-BrPh		-		[74,75]
15	R = Me		KHMDS		[72]
16	R = Me		KIm		[73]
17			NaOMe		[16]
18a	R = Me, X = OEt		-		[84]
18b	R = Me, X = Pyr				
18c	R—CO—X = N-MeSuc				
19			Cu(OTf)$_2$		[51]
20			hv 206-208 nm		[2]
21a	R = Bn, R' = Me		hv 254 nm		[71]
21b	R = Me, R' = Bn				
21c	R = Me, R' = Me				

Abbreviations: Ac: acetyl; Bn: benzyl; Cu(OTf)$_2$: copper triflate; Et: ethyl; HOTf: trifluoromethanesulfonic acid; KHMDS: potassium hexamethyldisilazane; KIm: potassium imidazole; nHx: nHexyl; Me: methyl; Ph: phenyl; Pyr: pyrrole; Suc: succinimide

References

1 M. E. Jung, J. Gervay, *Tetrahedron Lett.* **1990**, *31*, 4685–4688.

2 P. Li, X. G. Chen, E. Shulin, S. A. Asher, *J. Am. Chem. Soc.* **1997**, *119*, 1116–1120.

3 T. B. Grindley, *Tetrahedron Lett.* **1982**, *23*, 1757–1760.

4 M. Oki, H. Nakanishi, *Bull. Chem. Soc. Japn.* **1970**, *43*, 2558–2566.

5 H. Nakanishi, H. Fujita, O. Yamamoto, *Bull. Chem. Soc. Jap.* **1978**, *51*, 214–218.

6 K. Byun, Y. Mo, J. Gao, *J. Am. Chem. Soc.* **2001**, *123*, 3974–3579.

7 E. M. Arnett, J. A. Harrelson Jr. *J. Am. Chem. Soc.* **1987**, *109*, 809.

8 K. B. Wiberg, K. E. Laidig, *J. Am. Chem. Soc.* **1988**, *110*, 1872–1874.

9 D. M Pawar, A. A. Khalil, D. R. Hooks, K. Collins, T. Elliot, J. Stafford, L. Smith, E. A. Noe, *J. Am. Chem. Soc.* **1998**, *120*, 2108–2112.

10 A. W. Dodds, S. K. A. Law, *Immunol. Rev.* **1998**, *166*, 15–26.

11 S. Blandin, E. A. Ievashina, *Mol. Immunol.* **2004**, *40*, 903–908.

12 B. J. Janssen, E. G. Huizinga, H. C.. A. Raaijmakers, A. Roos, M. R. Daha, K. Nilsson-Ekdahl, B. Nilsson, P. Gross, *Nature* **2005**, *437*, 505–5111.

13 J. E. Smotrys, M. E. Linder, *Annu. Rev. Biochem.* **2004**, *73*, 559–587.

14 E. M. Van Cott, L. Muszbek, M. Laposata, *Prostaglandins Leukot Essent Fatty Acids* **1997**, *57*, 33–37.

15 G. Fischer, *Chem. Soc. Rev.* **2000**, *29*, 119–127.

16 C. Cox, T. Lectka, *Acc. Chem. Res.* **2000**, *33*, 849–858.

17 C. Dugave, L. Demange, *Chem. Rev.* **2003**, *103*, 2475–2532.

18 A. Radzicka, L. Pedersen, R. Wolfenden, *Biochemistry* **1988**, *27*, 4538–4541.

19 A. J. Bennet, K. P. Wang, H. Slebocka-Tilk, V. Somayaji, R. S. Brown, *J. Am. Chem. Soc.* **1990**, *112*, 6383–6385.

20 D. Lauvergnat, P. C. Hiberty, *J. Am. Chem. Soc.* **1997**, *119*, 9478–9482.

21 C. Schiene-Fischer, G. Fischer, *J. Am. Chem. Soc.* **2001**, *123*, 6227–6231.

22 T. Ozawa, Y. Isoda, H. Watanabe, T. Yuzuri, H. Suezawa, K. Sakakibara, M. Hirota, *Magn. Reson. Chem.* **1997**, *35*, 323–332.

23 C. Piccinni-Leopardi, O. Fabre, D. Zimmermann, J. Reisse, F. Cornea, C. Fulea, *Can. J. Chem.* **1977**, *55*, 2649–2655.

24 M. Schutkowski, S. Wöllner, G. Fischer, *Biochemistry* **1995**, *34*, 13016–13026.

25 J. Zhao, J-C. Micheau, C. Vargas, C. Schiene-Fischer, *Chem. Eur. J.* **2004**, *10*, 6093–6101.

26 R. A. Shaw, E. Kollat, M. Hollosi, H. H. Mantsch, *Spectrochim. Acta A* **1995**, *51*, 1399–1412.

27 W-J. Zhang, A. Berglund, J. L-F. Kao, J-P. Couty, M. C. Gershengorn, G. R. Marshall, *J. Am. Chem. Soc.* **2003**, *125*, 1221–1235.

28 Y. Che, G. R. Marshall, *J. Org. Chem.* **2004**, *69*, 9030–9042.

29 C. Alemán, J. Puiggali, *J. Org. Chem.* **1999**, *64*, 351–368.

30 L. Guy, J. Vidal, A. Collet, A. Amour, M. Reboud-Ravaux, *J. Med. Chem.* **1998**, *41*, 4833–4843.

31 R. Günther, H-J. Hofmann, *J. Am. Chem. Soc.* **2001**, *123*, 247–255.

32 H. J. Lee, J. W. Song, Y. S. Choi, H. M. Park, K. B. Lee, *J. Am. Chem Soc.* **2002**, *124*, 11881–11893.

33 A. L. Moraczewski, L. A. Banaszynski, A. M. From, C. E. White, B. D. Smith, *J. Org. Chem.* **1998**, *63*, 7258–7262.

34 C. Cox, T. Lectka, *J. Org. Chem.* **1998**, *63*, 2426–2427.

35 D. Marcovici-Mizrahi, H. E. Gottlieb, V. Marks, A. Nudelman, *J. Org. Chem.* **1996**, *61*, 8402–8406.

36 F. S. Schoonbeeck, J. H. van Esch, R. Hulst, R. M. Kellogg, B. L. Feringa, *Chem. Eur. J.* **2000**, *6*, 2633–2643.

37 A. Violette, M-C. Averlant-Petit, V. Semetey, C. Hemmerlin, R. Casimir, R. Graff, M. Marraud, J-P. Briand, D. Rognan, G. Guichard, *J. Am. Chem. Soc.* **2005**, *127*, 2156–2164.

38 V. Semetey, C. Hemmerlin, C. Didierjean, A-P. Schaffner, A. Gimenez Giner, A. Aubry, J-P. Briand, M. Marraud, G. Guichard, *Org. Lett.* **2001**, *3*, 3843–3846.

39 K. B. Wiberg, M. W. Wong, *J. Am. Chem. Soc.* **1993**, *115*, 1078–1084.

40 V. Barone, C. Adamo, C. Minichino, *J. Mol. Struct. (Theochem)* **1995**, *330*, 325–333.

41 W. L. Jorgensen, J. Gao, *J. Am. Chem. Soc.* **1988**, *110*, 4212–4216.

42 E. S. Eberhardt, S. N. Loh, A. P. Hinck, R. T. Raines, *J. Am. Chem. Soc.* **1992**, *114*, 5437–5439.

43 D. K. Sukumaran, M. Prorok, D. S. Lawrence, *J. Am. Chem. Soc.* **1991**, *113*, 706–707.

44 P. R. Rablen, *J. Org. Chem.* **2000**, *65*, 7930–7937.

45 M. J. Deetz, C. C. Forbes, M. Jonas, J. P. Malerich, B. D. Smith, O. Wiest, *J. Org. Chem.* **2002**, *67*, 3949–3952.

46 J. L. Kofron, P. Kuzmic, V. Kishore, E. Colon-Bonilla, D.H. Rich *Biochemistry* **1991**, *30*, 6127–6134.

47 A. M. Armbruster, A. Pullman, *FEBS Lett.* **1974**, *49*, 18–21.

48 R. Cini, F. P. Fanizzi, F. P. Intini, G. Natile, C. Pacifico, *Inorg. Chim. Acta* **1996**, *251*, 111–118.

49 E. Bairaktari, D. F. Mierke, S. Mammi, E. Peggion, *J. Am. Chem. Soc.* **1990**, *112*, 5383–5383.

50 M. L. Kramer, G. Fischer, *Biopolymers* **1997**, *42*, 49–60.

51 C. Cox, D. Ferraris, N. N. Murthy, T. Lectka, *J. Am. Chem. Soc.* **1996**, *118*, 5332–5333.

52 U. Reimer, G. Scherer, M. Drewello, S. Kruber, M. Schutkowski, G. Fischer, *J. Mol. Biol.* **1998**, *279*, 449–460.

53 G. Scherer, M. L. Kramer, M. Schutkowski, U. Reimer, G. Fischer, *J. Am. Chem. Soc.* **1998**, *120*, 5568–5574.

54 J. E. Vance, D. A. Leblanc, R. E. London, *Biochemistry* **1997**, *36*, 13232–13240.

55 T. Kiefhaber, F. X. Schmid, *J. Mol. Biol.* **1992**, *224*, 231–240.

56 B. Janowski, S. Wöllner, M. Schutkowski, G. Fischer, *Anal. Biochem.* **1997**, *252*, 299–307.

57 C. Garcia-Echeverria, J. L. Kofron, P. Kuzmic, D. H. Rich, *J. Am. Chem. Soc.* **1992**, *114*, 2758–2759.

58 T. Jordan, I. Mukerji, Y. Wang, T. G. Spiro, *J. Mol. Struct.* **1996**, *379*, 51–64.

59 J. E. Baldwin, T. D. W. Claridge, C. Hulme, A. Rodger, C. J. Schofield, *Int. J. Peptide Protein Res.* **1994**, *43*, 180–183.

60 B. Ishimoto, K. Tonan, S-I. Ikawa, *Spectrochim. Acta Part A* **1999**, *56*, 201–209.

61 J. Bredenbeck, J. Helbing, R. Behrendt, C. Renner, L. Moroder, J. Wachtveitl, P. Hamm, *J. Phys. Chem. B* **2003**, *107*, 8654–8660.

62 S-L. Wang, S-Y. Lin, T-F. Chen, C-H. Chuang, *J. Pharm. Sci.* **2001**, *90*, 1034–1039.

63 E. S. Eberhardt, R. T. Raines, *J. Am. Chem. Soc.* **1994**, *116*, 2149–2150.

64 A. E. Counterman, D. E. Clemmer, *Anal. Chem.* **2002**, *74*, 1946–1951.

65 B. Paizs, S. Suhai, A. G. Harrison, *J. Am. Soc. Mass Spectrom.* **2003**, *14*, 1454–1469.

66 B. Paizs, S. Suhai, *J. Am. Soc. Mass Spectrom.* **2004**, *15*, 103–113.

67 R. D. Husain, J. McCandless, P. J. Stevenson, T. Large, D. J. S. Guthrie, B. Walker, *J. Chromatogr. Sci.* **2002**, *40*, 1–6.

68 C. Francart, J-M. Wieruszeski, A. Tartar, G. Lippens, *J. Am. Chem. Soc.* **1996**, *118*, 7019–7027.

69 A. Kálmán, F. Thunecke, R. Schmidt, P. W. Schiller, C. Horváth, *J. Chromatogr. A* **1996**, *729*, 155–171.

70 S. Bouabdallah, H. Trabelsi, T. Ben Dhia, S.. Sabbah, K. Bouzouita, R. Khaddar, *J. Pharm. Biomed. Anal.* **2003**, *31*, 731–741.

71 J Zhao, D. Wildemann, M. Jakob, C. Vargas, C. Schiene-Fischer, *Chem. Commun.* **2003**, *21*, 2810–2811.

72 C. Cox, T. Lectka, *Org. Lett.* **1999**, *1*, 749–752.

73 C. Cox, H. Wack, T. Lectka, *J. Am. Chem. Soc.* **1999**, *121*, 7963–7964.

74 C. Cox, V. G. Young, T. Lectka, *J. Am. Chem. Soc.* **1997**, *119*, 2307–2308.

75 C. Cox, T. Lectka, *J. Am. Chem. Soc.* **1998**, *120*, 10660–10668.

76 L. Demange, PhD Dissertation, Université Paris-Sud Orsay, France, **2001**.

77 U. Reimer, N. El Mokdad, M. Schutkowski, G. Fischer, *Biochemistry* **1997**, *36*, 13802–13808.

78 P. K. Pallaghy, W. He, E. C. Jimenez, B. M. Olivera, R. S. Norton, *Biochemistry* **2000**, *39*, 12845–12852.

79 D. L. Rabenstein, T. Shi, S. Spain, *J. Am. Chem. Soc.* **2000**, *122*, 2401–2402.

80 T. Shi, S. M. Spain, D. L. Rabenstein, *J. Am. Chem. Soc.* **2004**, *126*, 790–796.

81 F. L. Texter, B. D. Spencer, R. Rosenstein, C. R. Matthews, *Biochemistry* **1992**, *31*, 5687–5691.

82 A. Kirby, I. V. Komarov, P. D. Wothers, Feeder, N. *Angew. Chem. Int. Ed. Engl.* **1998**, *37*, 785–786.

83 M. Adler, M. Marsch, N. S. Nudelman, G. Bosche, *Angew. Chem. Int. Ed. Engl.* 38, 1261–1263 (1999).

84 S. H. Meiere, F. Ding, L. A. Friedman, M. Sabat, W. D. Harman, *J. Am. Chem. Soc.* **2002**, *124*, 13506–13512.

85 E. Gaggelli, N. D'Amelio, N. Gaggelli, G. Valensin, *ChemBioChem.* **2001**, *2*, 524–529.

86 P. Tarabek, M. Bonifacic, S. Naumov, D. Beckert, *J. Phys. Chem. A* **2004**, *108*, 929–935.

87 H. Satzger, C. Root, W. Zinth, D. Wildemann, G. Fischer *J. Phys. Chem. B* **2005**, *109*, 4770–4775 (2005).

88 Y. K. Kang, H. S. Park, *J. Mol. Struct. (Theochem)* **2004**, *676*, 171–176.

89 H. A. Baldoni, G. N. Zamarbide, R. D. Enriz, E. A. Jauregui, Ö. Farkas, A. Perczel, S. J. Salpietro, I. G. Csizmadia, *J. Mol. Struct. (Theochem)* **2000**, *500*, 97–111.

90 Y. K. Kang, *J. Mol. Struct. (Theochem)* **2002**, *585*, 209–221.

91 Y. K. Kang, H. S. Park, *Biophys. Chem.* **2005**, *113*, 93–101.

92 I. Hudaky, A. Perczel, *J. Mol. Struct. (Theochem)* **2003**, *630*, 135–140.

9

Amide *Cis-Trans* Isomerization in Peptides and Proteins

Stephan Wawra and Gunter Fischer

9.1
Imidic and Secondary Amide Peptide Bond Conformation

9.1.1
Simple Amides

Imidic and secondary amide peptide bonds, echoing motifs of the backbone of oligopeptides and proteins, are among the most widespread functional groups of bioactive molecules. Nineteen out of the 20 gene-coded amino acids form secondary amide peptide bonds, whereas the peptide bond preceding proline is imidic in its chemical nature. Planarity and strong conformational preferences are general features of peptide bonds. Nevertheless, slight deviation from planarity is common in folded polypeptide chains [1–3]. The *cis-trans* isomers of simple secondary and substituted tertiary amides, such as *N*-methyl acetamide and *N,N*-dimethyl acetamide respectively, because of the predictive capability of their molecular parameters, have often been used for the analysis of thermodynamics and kinetics of *cis-trans* isomerization (CTI) in polypeptide chains. This field is covered by many reviews [4–8]. Thus, restrictions in the number of energy minima in peptide bond torsion (dihedral angle ω) as well as in the ease in surmounting the energy barrier separating the isomers is in keeping with the classic view of amide bond resonance, and is best illustrated by assuming a partial double bond character of the C–N bond of the peptide unit (Fig. 9.1) [9]. For simple amides, geometric isomers with the angle $\omega \approx 0°$ (*cis*) and $\omega \approx 180°$ (*trans*) coexist in solution as demonstrated in the mid 1950s, together with the slow C–N bond rotation by ^{1}H NMR spectroscopy for dimethyl formamide. The barriers of rotation can be calculated from the rate constants of CTI yielding $\Delta G^\ddagger$ values for *N*-methyl acetamide of 79 kJ mol^{-1} for the *cis* to *trans* transition and $\Delta G^\ddagger = 89$ kJ mol^{-1} for the reverse reaction (30 °C, water) [10]. Detailed knowledge on the thermodynamic origin of CTI remains nonetheless limited, and predictions of ratios of isomers and barrier heights across different chemical structures are difficult. In the transition state of peptide bond rotation, twisted C (carbonyl) as well as N pyramidal configurations

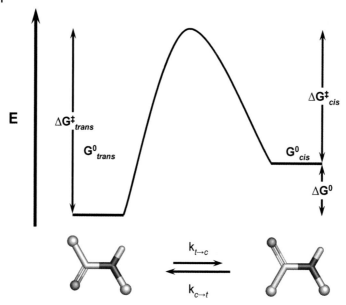

Fig. 9.1 Reaction profile for peptide bond *cis-trans* isomerization. The kinetic constants for the *trans* to *cis* ($k_{t \to c}$) and *cis* to *trans* ($k_{c \to t}$) isomerization are dependent on the Gibbs free energy of activation ($\Delta G^{\ddagger}$). The ground state energy difference ΔG° ($G^{\circ}_{cis} - G^{\circ}_{trans}$) determines the population of the *cis* and *trans* isomers in the equilibrium state.

have to be considered. Indeed, ground-state nonplanarity causes reduction of the rotational barriers in a series of constrained amides [11].

Clearly, CTI leads to a periodic backbone contraction/expansion of the polypeptide chain involved, as could be inferred from the isomer-specific distances of the C_{α} atoms directly attached to the isomerizing peptide bond. For prolyl bonds in native proteins this distance is about 0.8 Å shorter in the *cis* isomer when compared to the respective *trans* isomer [12]. This atomic translation produces a mechanical moment that was hypothesized to be involved in the functional cycle of motor proteins [13].

Provided that stabilizing tertiary interactions or steric constraints are absent in the respective molecules, secondary amides show a preference for the *trans* isomer in solution. It is obvious that the *cis* form is energetically less stable due to steric repulsion of the neighboring C_{α} atoms. The steric advantage of the *trans* conformation is profoundly diminished in imidic peptide bonds, leading to comparable populations of the isomers in solution. The free enthalpy differences between the two isomers range between 2 and 16 kJ mol^{-1} for linear molecules with tertiary and secondary amide bonds respectively. Despite the existence of many probes for CTI, the strong thermodynamic advantage of one isomer renders the estimation of the isomer ratio difficult [14]. The three-dimensional fold of a protein is able to force both types of proteinaceous peptide bonds, the imidic and the secondary amid peptide bond, into *cis* conformation. Due to the potentially charged termini adjacent to the isomerizing bond, dipeptides are less suitable for modeling CTI of polypeptides. The influence of the R substituent (R-CO-NH-R′) on the s-*cis*/s-*trans* isomer ratio was also studied for esters using methyl formate (R = H) and acetate (R = CH$_3$) in acetonitrile. The free energy differences between both conformers are ~7 kJ mol^{-1} for R = H and ~22 kJ mol^{-1} for R = CH$_3$ respectively [15]. Sterically demanding residues in the R′ position tend to increase the *cis* isomer in esters, thiol esters, and amides as characterized by the free energy differences of formate derivatives. The rotational barrier is mainly determined by the resonance stabilization of the planar ground-state as has been inferred from ab initio valence bond calculations of formamide [16]. The orientation of the free electron pair of nitrogen, which is arranged perpendicularly to the plane in the transition state of rotation, appears to be another contributing factor. Recently, it has been demonstrated that stereoelectronic effects are as important as steric effects traditionally used to explain the energetic preference for the *trans* conformation. Hyperconjugative delocalization of a nonbonding pair of electrons at peptide carbonyl oxygen was thought to additionally stabilize the *trans* conformation by interacting with the subsequent carbonyl carbon. The proposed $n \rightarrow \pi^*$ interaction of the O_{i-1} with the C_i contributes to conformational stabilization by about 3 kJ mol^{-1}, as was determined for *N*-formyl-L-proline methyl ester [17]. Multiple peptide bonds of a molecule will ultimately lead to the formation of 2^n peptide bond isomers. The isomer ratios are characterized by statistical distribution unless structural constraints (such as found in folded proteins) strongly stabilize one isomer relative to others. Ring-substituted acetanilides illustrate the influence of polar substituent effects on the barrier to rotation, and thus the resonance interaction between the NH and the carbonyl group. A linear relationship exists between the barrier height and the amide carbonyl stretching mode, supporting the view of a dominating role of resonance for CTI [18].

9.1.2
Secondary Amide Peptide Bonds

Due to the very small population of *cis* isomers of secondary amide peptide bonds, often termed nonprolyl peptide bonds, few data exist on the effect of CTI on the chemical and biological properties of polypeptides. In the mid 1970s Ramachandran and Mitra calculated a probable appearance for a *cis*-Ala-Ala peptide bond of 0.1% in the middle position of a tetrapeptide [19]. Experimentally, nonprolyl peptide bonds were found to have a *cis* content of about 0.5% in dipeptides and of ~0.15% in longer oligopeptides that often serve as references for CTI of unfolded proteins [20,21]. In ^{1}H NMR studies on linear oligopeptides containing alanine residues flanking aromatic amino acids, large isomer-specific differences for the chemical shift of alanyl methyl side-chain have been demonstrated [21]. An upfield shift occurred in the range of 0.3 and 0.5 ppm for the *cis* isomer of the peptides that allows the calculation of the *cis*/*trans* ratio from the integral signal intensities. The free energy difference between the *cis* and *trans* isomer is about 11.4 kJ mol^{-1} for the zwitterionic form of the dipeptide Gly-Gly [20,21]. This magnitude is similar to the enthalpic difference measured for the *N*-methyl acetamide conformers (10.7 kJ mol^{-1}, 1.4% *cis*, D$_2$O) over 30 years ago [10].

Despite this low thermodynamic stability the permanent existence of a single secondary amide peptide bond in *cis* conformation per 1000 amino acid residues is the minimal population that has to be considered for unfolded polypeptide chains. This *cis* peptide bond fluctuates across the polypeptide chain in relation to the sequence-specific propensity of a secondary amide peptide bond to adopt the *cis* conformation. As could be found in folded proteins, nonprolyl *cis* peptides are frequently located in the *β*-region of a φ/ψ plot [22]. It was hypothesized that *cis* peptide bonds represent high-energy structures able to store potential energy for increasing chemical reactivity [23]. Interconversion rates for the reversible CTI of secondary amide peptide bonds typically lead to half times of about 1 s for dipeptides, which decreases about 4-fold when the peptide bond is positioned in the middle of a longer peptide chain.

Despite the similarity of *cis* to *trans* interconversion rates, secondary amide peptide bonds and imidic peptide bonds differ greatly in their capacity to form the *cis* isomeric state. This means that the decreased *cis* population of secondary amide peptide bonds results from higher rates for CTIs. In the case of the Ala-Tyr dipeptide a 250-fold difference for the interconversion rate constants was observed ($k_{c \to t}$ = 0.6 s^{-1}, $k_{t \to c}$ = 2.4 × 10^{-3} s^{-1}, 298 K) [21]. Generally, chain elongation causes destabilization of the *cis* conformation when located apart from charged termini, and thus leads to a decreased *cis* population as well as a lowered barrier to rotation in *cis* to *trans* direction.

For example, the Ala-Tyr dipeptide exhibits a $\Delta G^{\ddagger}$ value of about 76.7 kJ mol^{-1} (0.48% *cis*) whereas the same peptide bond in the pentapeptide Ala-Ala-Tyr-Ala-Ala shows a $\Delta G^{\ddagger}$ of 64.4 kJ mol^{-1} (0.11% *cis*) [21]. The peptide bonds Tyr38-Pro39 of native RNase T1 and Tyr92-Pro93 of native RNase A retained in their *cis* state when Pro is mutated to Ala. These observations made it possible to compare data for nonprolyl peptide bond isomerization between partially folded protein chains and oligopeptides. Rather similar values were found for the Tyr-Ala CTI in RNase A ($k_{c \to t}$ = 0.9–1.4 s^{-1}, 298 K) and RNase T1 ($k_{c \to t}$ = 0.702 s^{-1}, $k_{t \to c}$ = 2.1 × 10^{-3} s^{-1}, 288 K) [24,25], and these parameters match rates of Tyr-Ala isomerization in Ala-Ala-Tyr-Ala-Ala ($k_{c \to t}$ = 1.77 s^{-1}, $k_{t \to c}$ 2.0 × 10^{-3} s^{-1} (298 K) [21]. The *cis* Glu166-Thr167 bond in the *cis* Pro167Thr TEM-1 *β*-lactamase variant is also characterized by a rate constant between 1 × 10^{-3} s^{-1} and 4 × 10^{-3} s^{-1} for the *trans* to *cis* interconversion [26]. Therefore the *trans* to *cis* isomerization can be rate-limiting in protein folding under native-like conditions, as was shown for a proline-free variant of the *α*-amylase inhibitor tendamistat [27]. This seems to be proven in the discovery of a novel protein class, the secondary amide peptide *cis/trans* isomerases (APIases), which can accelerate interconversion of these peptide bonds conformers [28].

Most of the secondary amide *cis* peptide bonds found in protein structures occur in functionally important regions (e.g. close to the active site of proteins). In the proximity of secondary amide *cis* peptide bonds the two neighboring C$_\alpha$ (*i* and *i*–1) atoms approach one another, enabling *π*-electron systems of aromatic side-chains to interact with the respective C$_{i\beta}$ atom. In this case the hydrogen atom of the $C_{i\beta}$ atom always points to the center of the aromatic ring. This attractive C–H $\cdots \pi$ interaction involves electrostatic dipole–quadrupole polarizabilities. Therefore, in the majority of cases the intimate side-chain/side-chain interactions involve aromatic amino acids [22]. Experimental results with a panel of proline-containing oligopeptides were consistent with a *cis* isomer stabilizing effect of an aromatic side-chain located N-terminally to the isomerizing bond [29,30], and a similar interaction might be effective for secondary amide peptide bonds too. It was noted that many of the proteins containing nonproline *cis* peptide bonds were found to be carbohydrate binding or processing proteins [22].

At a nearly unchanged *cis/trans* ratio, secondary thioxoamide peptide bonds show isomerization rates up to 1000-fold slower when compared with the oxopeptide congeners. A thioxoamide is an isosteric replacement of the normal peptide bond with only a slight change in the electron distribution of the ground-state. It allows for the photo-switching of *cis/trans* equilibria of polypeptides under mild photochemical conditions. Irradiation of thioxopeptides in the UV/Vis range leads to a markedly increased *cis:trans* ratio in the photostationary state [31,32].

9.1.3
Imidic Peptide Bonds

As the only example of an imidic peptide bond-forming proteinaceous amino acid, proline has long been recognized to play a unique role in the conformational dynamics and reactivity of native and unfolded polypeptide chains, as well as being a structural determinant [33,34].

There is a stabilizing effect in the *cis* imide bond preceding proline that can also be found for proline surrogates, which can be introduced by posttranslational modifications of the proline ring and by nonribosomal synthesis of peptide chains [35–40].

Typically, a single N-acylated proline moiety located in an unordered peptide chain leads to conformational heterogeneity, distributing the isomers into populations that contain from 5% up to 40% *cis* isomers dependent on the sequence flanking proline. The free energy barrier for the *cis/trans* interconversion ($\Delta G^{\ddagger}$) of unstructured peptides is between 60 and 100 kJ mol^{-1}. The exchange rate between the *cis-trans* isomers arising from the interconversion barrier is often slower than those of biological reactions coupled to the isomerizing polypeptide segment. Relaxation times of ten to hundreds of seconds can be observed for prolyl bond-based CTI of peptides and proteins (Fig. 9.2). For example, CTI occurs during protein folding uncoupled from much faster events such as polypeptide chain collapse, secondary structure formation, and segment rearrangements. Therefore, a kinetic trace is often monitored as a discrete step following faster events based on CTI of critical proline residues in the refolding of denatured proteins [41]. The experimental $\Delta S^{\ddagger}$ value of 0 ± 5 J mol^{-1} K^{-1} is quite small, confirming the enthalpically driven mechanism of interconversion [42]. Molecular dynamic calculations indicate that the CTI of a Ser-Pro moiety is favored by approximately 12.5 kJ mol^{-1} for an asymmetric movement ($-180° \leftrightarrow 0°$ = anticlockwise) rather than the clock-

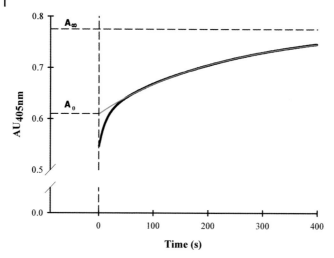

Fig. 9.2 Typical progress curve obtained in the isomer-specific proteolysis (ISP) assay developed by G. Fischer in 1984 [34]. Extrapolation of the slow kinetic phase (black solid line) allows the calculation of the zero time (t_0) absorbance value A_0 and the infinite reaction time absorbance A_∞. The difference gives an estimation of the cis:trans ratio if it is set relative to the amplitude of absorbance of the rapid reaction phase (0.04 mg mL^{-1} succinyl-Ala-Phe-Pro-Phe-4-nitroanilide, 35 mmol L^{-1} HEPES, pH 7.8, 10 °C, 0.3 mg mL^{-1} chymotrypsin, kinetic values: $k_{c \to t} = 4.3 \times 10^{-3}$ s^{-1}, ~15.5% cis at t_0).

wise transition (180° ↔ 0°). Additionally, these calculations show that the activation barrier for the isomerization process is strongly dependent on the ψ angle of proline. If the ψ value of proline in the a-helical region is about −30°, the barrier height for the anticlockwise *trans* to *cis* isomerization was calculated to be 60 kJ mol^{-1} and for the *cis* to *trans* about 40 kJ mol^{-1}.

The clockwise transition with a ψ value of −30° has to surmount a barrier of 75 kJ mol^{-1} [43,44]. Despite describing mechanistic details correctly, the calculated barriers to rotation were found to be smaller than the barriers found experimentally [10].

The proline φ angle is highly constrained (~ −63°) by the 5-membered pyrrolidine ring and confers rigidity to proline-containing peptides [45]. Generally, the conformational space around proline is restricted differently with regard to CTI, leading to a most severe reduction in the conformational space for *cis* isomers. Thus, loop formation involving a *cis* prolyl bond was found to be very rapid since the end-to-end contact in a four residue loop was in the range of 6 ns. It is an entropic advantage that led to a 5-fold decrease for the end-to-end contact formation of the *trans* isomer [46]. A common topographical feature seen in type VI β-turns is the *cis* prolyl bond, which allows turn stabilization by ring stacking effects [30,47].

cis/trans Isomers of proline-containing peptides also exist in gas phase ions produced by electrospray ionization mass spectrometry of tryptic digestion mixtures of common proteins. The peak splitting in drift time distributions indicate isomer-specific collision cross-sections and peptides that do not contain proline residues do not demonstrate this behavior. The ability to resolve different conformations for a given sequence appears to be related to the number of prolines present in the sequence and to the ion charge. Presently it is not clear whether the method can be used to monitor the conformational state of a prolyl bond of the respective protein in its native state [48,49].

9.1.4
Solvent and pH Effects

Unimolecular CTI proceeds via a single nonplanar transition state that can realize four different configurations [5]. Because the peptide bonds strongly interact with water molecules, solvation was thought to be an important factor determining CTI. However, using ^{17}O shielding constants, no isomer-specific differences in hydration was found in simple amides and proline-containing peptides selectively enriched in ^{17}O at the amide carbonyl group [50]. For acetylproline the hydration state of the acetyl carbon is essentially the same for both isomers. In the equilibrium the *trans* form is enthalpically favored ($\Delta H° = -5.14$ kJ mol^{-1}) but is disfavored in entropic term ($\Delta S° = -5.47$ J mol^{-1} K^{-1}) relative to the *cis* conformer. The enthalpy–entropy compensation was attributed to greater solvent immobilization in the *trans* isomer [51]. Thus, differential hydration of isomers does not contribute much to the factors determining ratio of isomers in water, but charged groups in close spatial proximity to the isomerizing bond might suspend the rule. Consequently, the *cis/trans* equilibrium of simple secondary amides was found to be almost completely insensitive to the solvent environment. This is expressed by $\Delta\Delta G°$ values of less than 0.5 kJ mol^{-1} when transferring N-methyl formamide from water to cyclohexane [52].

The results on the values of the *cis/trans* equilibrium of simple amides in different solvents confirm observations made with Gly-Pro and acetyl-Gly-Pro-OMe [53,54]. In all these cases the free energy difference for the isomers originates primarily from enthalpic differences. According to $\Delta H°$ values obtained in different solvents, it was concluded that the enthalpic component that differentiates the *cis* and *trans* isomer is similar in protic and aprotic solvents. Thus, the entropic term was found to favor the *cis* isomer in both aqueous buffers and aprotic organic solvents.

In fact, the fraction of *cis* isomer of proline-containing tetrapeptides was found to decrease only slightly on lowering of the temperature [34]. Solvent effects are much more pronounced in oligopeptides containing proline clusters and polyproline stretches. Polyproline stretches in segments of >four residues adopt the left-handed all-*trans* helix in aqueous solution termed polyproline II (PP II) helix. Intramolecular hydrogen bonds are missing in the polyproline II helix due to its constitution. However, in organic solvents, such as isopropanol, a dramatic conformational change occurs where the all-*trans* polyproline II helix completely converts to a secondary all-*cis* structure termed polyproline I helix (PP I). The conformational helical interconversion of polyproline has been suggested to occur by either a cooperative C- to N-terminal isomerization of the prolyl bonds or via a conformational intermediate composed of dispersed sequences of the different prolyl bond isomers [55].

Interaction of peptides with Li$^+$ ions in dry solvents, such as trifluorethanol (TFE) and tetrahydrofurane, can dramatically influence the free energy difference that discriminates the *cis* from the *trans* isomers for both prolyl bonds and secondary amide peptide bonds [56–59]. However, the Li$^+$-induced increase of the *cis* isomer population in linear oligopeptides depends on the nature of the amino acids preceding proline (G. Fischer, unpublished results).

The barriers to rotation of prolyl bonds are much more sensitive to solvents, and are generally increased on going to polar solvents. For acetyl-Gly-Pro-OMe the free energy of activation $\Delta G^{\ddagger}$ was found to be linearly correlated to the amide I vibrational mode of acetyl-Pro-OMe when both parameters were measured in different solvents [60]. For example, the rotational barrier in H$_2$O and TFE is increased about 50-fold when compared to toluene. Lipophilic solvents stabilize the nonpolar carbonylamine-like transition state of isomerization, whereas water stabilizes the C–N bond against twisting by preferentially solvating the ground-state carbonyl group [5]. Increased hydrogen bond donating capability of the solvent is another important factor that slows down isomerization rates.

Detailed analysis of *t*-butyloxycarbonyl (Boc)-protected peptides using infrared spectroscopy showed that intermolecular hydrogen bonds occurred within the *trans* conformation. Thus, the observed solvent effects on CTI rates originate from solvent interaction with the *trans* form [61]. Micelles of aqueous surfactants such as N-dodecyl-N,N-dimethyl-3-ammoniopropane-1-sulfonate (SB12) and phosphatidylcholine vesicles are described to increase the population of the *trans* isomer as well as the CTI rate for proline-containing peptides. For example, the *cis/trans* interconversion of succinyl-Ala-Ala-Pro-Phe-4-nitroanilide was 20-fold faster in micellar solutions compared to aqueous buffer [62]. This rate enhancement proved to be entirely enthalpy-driven, ruling out specific intramolecular effects. The finding that the isomerization rate of simple amides and uncharged proline-containing oligopeptides in buffered solutions is nearly pH independent between pH 3 and 10 indicates that general acid/base catalysis plays no role for the *cis/trans* transition under physiological conditions.

Specific acid catalysis is also restricted to biologically noncompatible conditions because amides are protonated in oxygen rather than nitrogen with pK_a values in the range of <0 [63,64]. The rarity of intermolecular acid/base effects on CTI was affirmed by the kinetic deuterium solvent isotope effects (KSIE) of about 1.0, indicating that the mechanism of spontaneous isomerization does not involve a proton-in-flight [29,42,65].

9.1.5
Sequence-Specific Effects

Intramolecular effects play an important role in the modulation of rates and equilibria of CTI in polypeptides. Both local and remote substituents, which are characterized by amino acid substitutions, charge differences, and posttranslational modifications, can be influential depending on the folding state of the polypeptide chain. The picture emerging from studies on dipeptides is that a charged backbone can considerably affect CTI. The highest *cis* fraction for the Ala-Tyr dipeptide (0.43%) occurs in the zwitterionic state. When altering the pH below the carboxyl pK_A (0.14% *cis*) or above the amino pK_A (0.26% *cis*), the *cis* population decreases. In contrast, anionic Xaa-Pro dipeptides usually show a high *cis/trans* ratio in the anionic state. A minimum value for the CTI rate constants exists for zwitterionic secondary amide peptide bonds in dipeptides but Xaa-proline CTI is especially slow in its anionic state [21,66,67].

Among a wealth of information on CTI from side-chain substitutions, those resulting from the minor hydrogen/deuterium isotopic replacement to the C_α atom of the amino acid preceding proline are especially clear, and indicate differences in the hyperconjugative stabilization of the prolyl isomers arising from a backbone substitution. An inverse equilibrium isotope effect of the *cis/trans* equilibrium of succinyl-Ala-Ala(C_α (H/D)-Pro-Phe-4-nitroanilide ($K_H/K_D = 0.989 \pm 0.009$) suggests that the deuterium is a *cis* isomer stabilizing substitution [68]. Furthermore, the ratio of rate constants of the CTI ($k_H/k_D = 1.05 \pm 0.03$) indicates that hyperconjugation is considerably enhanced in the transition state of isomerization. Thus, C-N bond rotation is far advanced in the transition state, and the state of the carbonyl group is ketone-like [68,69].

Changing the chirality at the C_α atom in proline-containing tetrapeptides showed position-dependent thermodynamic and kinetic effects on the CTI [70]. Positions adjacent to proline were found to be critical for spontaneous bond rotation. Compared with L-amino acids, D-amino acids in positions preceding and following proline, lead to an increased *cis* population. Considering the stereospecificity of rate constants C_α chirality affects CTI mainly due to destabilization of the planar *trans* state relative to the twisted transition state of rotation.

Early studies of model compounds, such as N-acetylproline-N'ethylamide (24% *cis* in D_2O) and N-acetyl-methylglycylproline-N'-methylamide (19% *cis*) [71] point to the remarkable effects of the residues preceding proline. This is also true for a single amino acid substitution in the unprotected Ser-Asn-Pro-Tyr-Asp-Val (7.3% *cis* isomer in neutral aqueous solution) resulted in 67.9% *cis* isomer for Ser-Trp-Pro-Tyr-Asp-Val [30]. In general, peptides of the constitution Yaa-Yaa-Xaa-Pro-Xaa-Yaa have the highest *cis* contents if Xaa is an aromatic residue, where Yaa represents any amino acid. The major stabilizing factor is the stacking interaction between the aromatic residues and the proline ring. It was noted too, that apart from the three aromatic residues, hydrophobic side-chains at position two are marginally better at promoting the formation of the *cis* peptide bond than hydrophilic ones [29,30].

The propensity of Xaa-Pro moieties to adopt the *cis* conformation in a pentapeptide (acetyl-Ala-Xaa-Pro-Ala-Lys-amide) was systematically determined along with CTI-relevant kinetic data [29]. Since there was no evidence for the existence of ordered structure, neither in *cis* nor in *trans* isomers of the pentapeptides, substitution-specific differences in CTI parameters can be exclusively attributed to local side-chain effects. With the sole exception of the side-chain protonated His, aromatic amino acids for Xaa slow down isomerization rates and enhance the population of *cis* isomers. Small aliphatic side-chains in Xaa represent the other extreme of low *cis* isomer probability and fast CTI rates. Proline clustering leads to a specific situation that can be characterized by very slow isomerization rates for certain isomers.

Oligopeptides where a histidine residue precedes the prolyl bond (-His-Pro-) show unusually high rates of CTI at physiological pH values [72]. The influence of pH on the prolyl isomerization rates indicated that the side-chain protonated molecules exhibited an up to 10-fold rate enhancement of CTI when compared with the same molecule in its unprotonated state. Arginine and lysine flanking proline do not show this effect, and for histidine the position preceding proline is essential. Intramolecular general acid catalysis by an interaction of the proton attached to an imidazole nitrogen of histidine and the imide nitrogen of the succeeding proline can be supposed. The kinetic solvent deuterium isotope effect (KSIE) of 2.0 ± 0.1 for CTI of angiotensin III (Arg-Val-Tyr-Ile-His-Pro-Phe) at low pH values, where the protonation state of the His side-chain approached 100%, fully agrees with intramolecular catalysis. For the unprotonated state of the side-chain no KSIE of CTI can be observed. Beside catalysis by enzymes this sequence-specific effect provides a convenient way to avoid folding phases in proteins limited in rate by prolyl isomerization [72]. Interestingly, side-chain protonation of histidine at this position is isomer-specific in that a series of oligopeptides exhibited pK_a (*cis*) values in the range of 6.3 ± 0.2 whereas pK_a (*trans*) values were found to be 6.6 ± 0.2. Intramolecular catalysis of prolyl bond isomerization was first observed during refolding of denatured dihydrofolate reductase. Proline 66 undergoes a *trans* to *cis* isomerization catalyzed by the guanidinium group of arginine 44, which interacts with the imide nitrogen group of the Gln65-Pro66 peptide bond [73]. The analysis of the RCSB database revealed that in almost 6% of all available structures at least one arginine guanidinium group is within 4 Å of the proline imide nitrogen (Wille, G. et al., in preparation). Under nonaqueous conditions intramolecular catalysis of CTI of peptide bonds has been frequently observed [74]. Disulfide-mediated cyclization of acetyl-Cys-Pro-Xaa-Cys-amide (Xaa = Phe, His, Tyr, Gly, and Thr) and acetyl-Cys-Gly-Pro-Cys-amide led to 2- to 13-fold enhanced rates of CTI in water that was thought to result from an intramolecular catalysis mechanism in which the NH proton of the Pro-Xaa peptide bond hydrogen binds to the proline nitrogen in the transition state.

Posttranslational modifications considerably extend the chemical diversity of the building blocks forming polypeptide chains. A good deal of effort has been directed toward understanding how the individual amino acid derivative affects the CTI of neighboring prolyl bonds in peptides and proteins.

The influence of phosphorylation at the residue preceding the critical proline on the *cis:trans* ratio is small in terms of the free energy differences $\Delta\Delta G°$ when compared with the unphosphorylated derivative [75]. Although, there is a significant increase of the *cis:trans* ratio for the pSer-Pro moiety in oligopeptides, which is reversed for pThr-Pro segments. More importantly, the rate constants for the *cis/trans* interconversion decrease 2- and 8-fold for the pSer-Pro and the pThr-Pro motif respectively [76]. This is in accordance with molecular dynamics simulations performed for the Thr-Ser-Pro-Ile/Thr-pSer-Pro-Ile segment where an higher free energy barrier to rotation upon phosphorylation along with a change in the distribution of the preferred backbone conformation has been observed [43]. The dianionic form of the side-chain phosphate residue represents the slow isomerizing species whereas the monoanion does not reveal electronic effects leading to deceleration of CTI [76]. Peptide bond CTI of Tyr-Pro segments are not affected upon side-chain phosphorylation [77]. Despite a long-range structure-destabilizing effect, threonine phosphorylation at the Thr44-Pro45 site of hirudin does not affect the isomeric state of the neighboring prolyl bond in solution [78].

Nature provided various patterns of modified prolines, including L-*trans*-4-hydroxyprolines (Hyp). The latter are often embedded in peptide and protein structures especially in collagens. The hydroxyl substituent of the pyrrolidine ring is able to tune bond angles and therefore it influences backbone orientation. Collagen, a structural protein of immense importance, consists of repeating Gly-Xaa-Yaa triads where Xaa is often proline and Yaa is often *trans*-4-hydroxyproline. Studies with *trans*-4-fluroproline (Flp) peptides showed that the triple helix formed by (Gly-Pro-Flp)$_{10}$ is more stable than (Gly-Pro-Hyp)$_{10}$, which is in turn more stable than (Gly-Pro-Pro)$_{10}$ [79]. This implied that the electron-withdrawing nature of fluoride is the driving force toward stability. Other detailed investigations confirm these findings and form a picture of electron-withdrawing 4R substitutions stabilizing the *trans* conformation of the preceding imide bond. Hydroxyl substitutions in the 4S orientation shift the equilibrium in the opposite direction [39,80]. The hydroxyl group in the 4-position influences the puckering of the pyrrolidine ring, which affects the conformational preference of the X-Pro peptide bond and thereby has an impact on the neighboring peptide backbone [81]. Introduction of a *trans*-4-hydroxy substituent leads to a stabilization of the *trans* peptide bond in acetyl-Phe-Hyp-NHMe (~83% *trans*, D$_2$O, 298 K) compared with acetyl-Phe-Pro-NHMe (~67% *trans*, D$_2$O, 298 K) [40]. Electron-withdrawing substituents in the 4th position inductively reduce electron density in the peptide bond, increasing N-pyramidalization, reducing the bond order of the C–N linkage and thereby facilitating the interconversion barrier to favor that which is lower in energy [82,83].

According to X-ray crystallographic data for *allo*-4-hydroxy-L-proline gives rise to a C$_\beta$-*exo* conformation of the Hyp pyrrolidine ring, but C$_\beta$-*exo* is also confirmed in other sequences [40,84]. An explanation could be that the C$_\beta$-*exo* and the C$_\gamma$-*exo* conformation of the pyrrolidine ring influence the φ and ψ dihedral angles in such way that it strengthens the backbone stereoelectronic interactions in the *trans* conformer. Recent theoretical studies suggest that hyperconjugative interactions are the driving force for adopting the C$_\beta$-*exo* and C$_\gamma$-*exo* conformations [85]. This ring pucker conformation would allow effective $n \rightarrow \pi^*$ electrostatic interactions between the carbonyl groups of the i–1 and i residues when the peptide bond adopts a *trans* conformation [80]. Therefore, hydroxylation of proline in the (Gly-Pro-Hyp)$_n$ sequence stabilizes the *trans* conformation of the Gly-Pro peptide bond by restricting the dihedral angles φ and ψ.

In an earlier publication, side-chain N-triglycosylation at the Asn-Pro moiety received credit as a substitution greatly decelerating CTI dynamics [86]. Interestingly, O-glycosylation is often found attached to Ser/Thr residues immediately preceding proline [87,88]. O-Glycosylations introduce bulky substituents to Ser/Thr side-chains analogous to those found in aromatic amino acids. However, it was found during studies with monoglycosylated peptides that O-glycosylation preceding proline does not stabilize the *cis* imide conformation [88]. O-Glycosylation of serine residues preceding l-*trans*-hydroxyproline was devoid of rate effects on prolyl isomerization [89].

9.1.6
Secondary Structure Formation and CTI

In general, the overall stability of secondary structures results from a number of rather weak noncovalent interactions. These interactions are important determinants of peptide folding motives. Secondary structure formation affects the *cis/trans* equilibrium by favoring either the *cis* or *trans* isomer [30,90]. If CTI occurs in secondary structure elements, it should cause peptide backbone reorientation, reconfiguration of hydrogen bonding patterns, hydrophobic interactions, and solvatization [91]. Additionally, the isomers have different hydration shells and interacting water molecules therefore play a crucial role in determining whether the *cis* or *trans* isomer is favored [92].

The strong influence of proline on the conformation of the preceding residue reflects steric clashes involving the pyrrolidine ring. Of proline peptide bonds in secondary structures, 38% are found in loops or random coils, 26% appear in helices, 23% in turns, and 13% in β-strands [93].

The right-handed a-helix (helical pitch of 5.4 Å/turn, 3.6 residues per turn) is characterized by a pattern of strong hydrogen bonds. These hydrogen bonds between the peptide C=O and the i+4 N–H group display nearly perfect N $\cdots$ O distances (2.8 Å) and are therefore quite stable. In addition, the dense packing within this helix brings the inner atoms into a van der Waals distance leading to maximized association energies [94]. The side-chains are extended towards the environment in a way that avoids sterical clashes. *cis*-Proline peptide bonds are prohibited in a-helical structures because sterical conflicts between the proline C_a and the nitrogen of the preceding residue occur [93]. If the amino acid preceding proline shows dihedral angles typical for an a-helix, then clashes between the C_d of the proline and both C_β and amide nitrogen of the preceding Xaa residue are also possible [93]. Hence, it is not surprising that ψ angles for residues preceding proline are displaced to a more negative value than usually occur in a-helical protein structures.

Theoretical calculations showed that in a-helices *trans* proline peptide bonds can best be accepted up to the fourth position within the helix because there is neither a disruption of the hydrogen bonding network in this region nor does the bulk of the pyrrolidine ring seriously interfere with the regular helix geometry. Analyses of the Brookhaven database in 1991 also showed that the highest frequency of *trans* Xaa-Pro bonds in α-helices can be observed at positions 3 and 4 from the N-terminus. Also, no proline residue could be found in position 5, where an amide hydrogen is necessary to stabilize the first turn by forming a hydrogen bond with the carbonyl at position 1 [93]. Proline at position 2 within an α-helix shows an overwhelming preference for Asp, Asn, Ser, Thr, and Gly as the preceding residue. There is a strong tendency for the side-chains of these residues (except Gly) to form an additional hydrogen bond with the exposed backbone NH of the residue that follows proline to compensate the untypical α-helical ψ angle. Glycine is the only natural amino acid that can occupy ψ values in the α-helical region of the Ramachandran plot when placed preceding proline.

Like α-helices, β-sheets are regular hydrogen-bonded structures. Amino acids in β-sheets require both the carbonyl oxygen and the amide nitrogen to take part in hydrogen bonds [95]. This eliminates proline from full participation due to the lack of the amide hydrogen. However, in β-sheets there is no conflict between the C_δ group of proline and the preceding residue. In both parallel and antiparallel sheets, proline is most frequently found between widely spaced hydrogen bonding patterns with the carbonyl oxygen not contributing to the hydrogen bonding network, or at the N-terminal end where the prolyl oxygen is involved in hydrogen bonding [93].

In contrast to α-helices, β-sheets do not involve interactions between amino acids close in sequence. Amino acids that interact within β-sheets are often found widely separated in the primary structure. Therefore, β-sheet formation needs structures that bring two polypeptide segments into close proximity. This is achieved via reverse turn structures [96]. Turns are aperiodic or nonrepetitive elements of secondary structure which mediate the folding of the polypeptide chain into a compact tertiary structure. Turns usually occur on the environment-exposed surface of proteins [97,98]. Reverse turns play an important role in polypeptide function, both as elements of structure as well as modulators of bioactivity [99]. Among the reverse turns found in proteins the β-turn is the most relevant [100]. β-Turns comprise four amino acid residues (*i* to *i*+3) forming an almost complete 180° turn in the direction of the peptide chain [101,102].

The type VI turn is a unique member of the β-turn family because it is the only turn that incorporates a *cis* peptide bond [103]. It follows that type VI β-turns always contain a proline residue at the *i*+2 position, since peptides incorporating this amino acid are the only ones that can exist in the *cis* conformation to any substantial extent [104]. Type VI turns are often found in polypeptides containing the sequence Xaa-Pro-Yaa, where Xaa and Yaa represent amino acids with an aromatic side-chain. The type VI turn has been estimated to be 8.4 kJ mol^{-1} more stable than the extended structure [30]. Richardson divided the type VI turns into subtypes VIa and VIb, with VIa structures exhibiting an intramolecular H-bond between the *i* carbonyl oxygen and the *i*+1 or the *i*+3 amide hydrogen [105]. In type VIb structures the orientation of the torsion angle ψ^{i+3} directs the C-terminus of the turn away from the N-terminal end so that hydrogen bonding groups are unable to interact. Studies with bicyclic lactam amino acid conjugates showed that the hydrogen bonded type VIa turn is more stable than the nonhydrogen bonded type VIb conformation in the absence of any other structural constraints [104]. Turn formation is a very rapid process. Interestingly, in a recent study only the proline-containing peptide (Xan-Ser-Pro-Ser-NAal-Ser-Gly, Xan = xanthone, NAla = naphthylalanine) showed double exponential kinetics with rate constants of 4.7×10^6 s^{-1} and 2.7×10^7 s^{-1} [46] where the slower rate constant was assigned to the loop formation of the *trans* prolyl conformer (*β*-II turn). Both rates are too fast to be ω and ψ angle rotation of the proline residue. Analyses of nonprolyl *cis* peptide bonds in protein structures showed that the conformational space of *cis* peptide bonds is restricted compared with *trans* peptide bonds [22]. The rapid loop formation around *cis* prolines seems to be largely caused by the restricted conformational space and the shorter end-to-end distances when compared with the *trans* isomer.

The polyproline II (PP II) conformation, as found in collagen and related proteins, has been accorded less attention than the classical protein secondary structures (*α*-helix, *β*-sheet, and *β*-turn). This is possibly due to the PP II conformation often assigned inappropriately to the "disordered" or "random" conformation classes [90,106]. The PP II structures have been implicated in amyloid formation [107] and nucleic acid binding [108], and have long been known as protein–protein interaction motifs [90,109,110]. PP II was originally observed in poly-L-proline peptides. Poly-L-proline basically adopts two different helical conformations dependent on the solvent environment. In aqueous solutions polyproline peptides form an extended "all-*trans*" left-handed helical structure with a helical pitch of 9.3 Å per turn, three residues per turn and ω, φ, and ψ dihedral angles around 180°, -80 ± 45°, and 142.5 ± 42.5° respectively [106]. In hydrophobic solvents polyprolines tend to form the right-handed polyproline I (PP I) helix with a helical pitch of 5.6 Å and 3.3 *cis* prolyl residues per turn and dihedral angles of ω, φ, and ψ of 0°, -75°, and $+160$° [111]. Usually, prolyl peptides are in the PP II conformation but even sequences without any proline can adopt this structure at least partially, e.g. polyglycine, polylysine [112–114], polyglutamate, polyaspartate [115], and polyalanine [116].

Theoretical and experimental studies on the PP II structure showed that in an aqueous environment, water molecules form carbonyl–water–carbonyl H-bonds within the chain [117–119], which seems to be the driving force for favoring the *trans* conformation. Interchain water bridges in PP II are not possible because carbonyls in the PP II helix are sterically quite crowded by the neighboring atoms [120]. The carbonyl–water–carbonyl clusters cannot be formed when the peptide adopts the PP I conformation. This explains why the PP I structure can be formed in hydrophobic solvents where this effect is negated. The PP I conformation has not yet been found in protein structures.

An analysis of the HOMSTRAD database showed that the PP II conformation represents 3% of the all peptide conformations and even occurs in 1.3% of peptides when only helices consisting of more than three residues are considered [106]. Proline is greatly favored in PP II helices, whereas Gly and aromatic amino acids have low propensity for the PP II structure. It is worth noting that although Gly generally disfavors the PP II helix, it is common in collagen. Collagen consists of three polypeptide chains, each in a PP II conformation. These three helices in turn are supercoiled around each other. This supercoil is right handed and there is a repeating pattern of H-bonds between them. Close packing of the chains near the central axis imposes the requirement that Gly should occupy every third position. Gly → Ala substitution results in subtle changes of the conformation, which leads to a local unwinding of the triple helix [121].

Additionally, several lines of evidence suggest that the "random coil" is a left-handed helix with a structure similar to the PP II helix, at least locally. For example, it was shown that for an a-helical polyalanine chain (21 residues), heating transforms this structure into a PP II conformation [122]. This is possibly due to the fact that PP II disrupts water organization less than β-sheet and a-helical conformation and is therefore favored entropically [123,124].

9.2
Amide Relevant Conformations in Proteins

Peptide bond CTI might be functionally relevant in the unfolded state, the folding intermediates, and the native state of a given protein. Despite the identity of the amino acid sequence distinct thermodynamic and kinetic parameters exist for the CTI of the different folding states. On this basis, the ratio of isomers and isomerization kinetics of unfolded proteins resembles those observed in N- and C-terminally blocked oligopeptides. In the native protein structural constraints may strongly favor (or disfavor) a certain isomeric state for a particular peptide bond. The peptide bond conformation in proteins is found to be *trans* in most cases [125]. In the last years, several analyses of protein structure databases showed that about ~99.7% of all peptide bonds were found to be in *trans* whereas ~0.3% assume the *cis* conformation, 0.27% are of the type Xaa-Pro, and 0.03% are secondary amide peptide bonds [22,126,127]. *Cis* secondary amide peptide bonds could simply have been overlooked in the course of structure determination. Unless specified explicitly, most of the refinement programs used today will refine any nonprolyl peptide bond in the *trans* conformation [126]. Neighboring residue effects appear to play an important role for the *cis* prolyl bond propensity in proteins, which approaches about 5% of all proteinaceous proline residues [93,128].

Recently, much effort has been put into predicting the isomeric state of prolyl bonds in proteins [129,130]. A program on the basis of a secondary structure information was developed that predicts for a given sequence whether a particular peptide bond is in either *cis* or *trans* conformation [131].

Since structural preconditions of CTI are rather different in the various folding states of a protein, the refolding and unfolding processes themselves were found to be limited in rate by slow peptide bond isomerizations. Obviously, multiple parallel pathways are realized for the refolding reaction since the starting population of unfolded chains is heterogeneous in terms of number and position of folding relevant CTI. Prolyl bond limited refolding/unfolding is frequently found in single domain and multidomain proteins where the refolding kinetics are limited by either *cis* to *trans* or *trans* to *cis* isomerization. However, in a proline-free variant of the *a*-amylase inhibitor tendamistat, for a small fraction of refolding molecules the relaxation time, which was 400 ms, could be assigned to CTIs of secondary amide peptide bonds [27]. Obviously, this type of isomerization during refolding must be universal in unfolded proteins.

By use of site-directed mutagenesis in positions covering *cis* prolyl bonds, the proline has been replaced by nonproline amino acids. It came as a surprise that the secondary amide peptide bond formed in the substitution still adopts the thermodynamically disfavored *cis* conformation in many cases [25,26,132–135]. Thus, to overcome the free energy costs of a *cis* secondary amide peptide bond of about 15 kJ mol^{-1} the structural consequences favoring the *trans* conformation must be absent in the folded protein variant. Consequently, the CTI is largely retained in these protein variants [133].

Folding reactions limited by prolyl isomerizations have the following CTI-specific characteristics [41]:

1. The activation energy for the refolding process is in the range of 80 kJ mol^{-1}.
2. In most cases, amplitudes of prolyl-limited folding phases do not much vary with initial folding conditions.
3. Double jump techniques reveal CTI by time-dependent amplitudes of folding phases.
4. Peptidyl prolyl *cis-trans* isomerases (PPIases) are able to catalyze the folding or unfolding reaction.
5. Under favorable conditions, changed folding parameters after replacement of the critical proline residue by site-directed mutagenesis are indicative of prolyl-limited processes.

Many folding intermediates typically exist under strongly native refolding conditions with the native state containing *cis* prolyl bonds. For example, the refolding reaction of GdmCl-denatured RNase T1, which is a single domain protein of 104 amino acid residues encompassing two *trans* and two *cis* prolyl bonds, is properly described by a kinetic model containing four unfolded species (2^n with *n* for the number of *cis* prolyl bonds) and three folding intermediates, all of which contain one non-native prolyl bond at least. For the three intermediates with non-native prolyl bonds relaxation rates of formation and decay ranging from 170 s to 6500 s have been calculated. From the chemical properties of the reaction, prolyl isomerizations kinetically uncoupled from chain rearrangements can be inferred. The relaxation time for the folding process involving molecules with an isomer composition already identical to the native state is 175 ms at pH 4.6 and 25 °C (Fig. 9.3) [136,137].

Interestingly, NMR investigations on the long-lived folding intermediate with a *trans* Tyr38-Pro39 bond revealed that the structural effect of the *trans* to *cis* isomerization, which forms the native protein, is not a local one but also involves other regions of the protein [140]. There is increasing evidence from refolding experi-

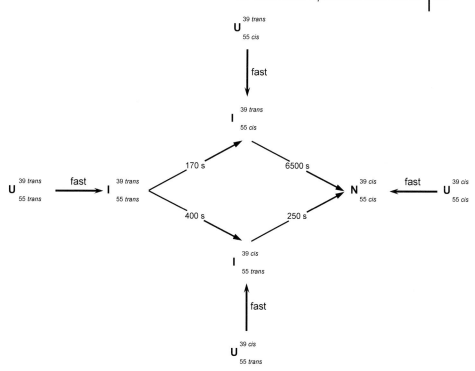

Fig. 9.3 The kinetic model for the slow prolyl isomerizations during refolding of RNase T1. U indicates the unfolded species, I the intermediates, and N is the native protein. Indices stand for the isomeric states of the prolines 39 and 55, respectively. Half-times given for the individual steps refer to folding conditions of 0.15 mol L^{-1} GdmCl, pH 5.0, 10 °C [138,139].

ments in vitro that the proper biological function of the refolding protein must await completion of the final prolyl isomerization [136,141,142].

9.3
Native State Peptide Bond Isomerization

The X-ray crystal structure database led us to believe that peptide bonds adopt either the *cis* or *trans* conformation in native proteins [22,128]. However, NMR spectroscopy [143], and in a few cases, crystal structure analysis [144], provide encouraging experimental evidence of conformational peptide bond polymorphism of folded proteins. Furthermore, conformational changes in response to ligand binding, crystallization conditions and point mutations at remote sites are frequent. Consequently, the three-dimensional protein structure database contains homologous proteins that have different native conformations for a critical prolyl bond [12].

Generally, multiple isomerization sites, low propensity of the thermodynamically disfavored isomer, the presence of the isomerization site in flexible protein segments, and poor dispersion of isomer-specific NMR chemical shifts illustrate the difficulties in the detection of native state peptide bond isomerization. However, methods have been developed to characterize native state prolyl isomerization in proteins that deviate much from equal partitioning of isomers [145].

In fact, there are a growing number of folded proteins that were shown to exhibit conformational heterogeneity about one or more peptide bonds and there is also growing evidence that PPIases modulate intracellular signaling events via native state isomerization [146]. Among the effects caused by prolyl isomerization the C_a atom displacement, which might propagate directed structural changes through the polypeptide backbone, may find widespread utility in the structural control of protein function [12].

Spontaneous peptide bond isomerization is relatively slow in respect to the NMR time-scale and, therefore, two resonance frequencies are observable. Nuclear Overhauser effect (NOE) patterns show the NMR couplings between neighboring atoms in space. That is why NOE spectra can be used to discriminate between the resonance patterns for the *cis* and *trans* conformers and to calculate the rate of exchange.

Regulation of biological functions can be achieved via catalytic and binding activities of cellular proteins. In recent years the potential of peptide bond CTI for a switch-like control of protein function beside amino acid side-chain modification or drastically reorientation of a whole polypeptide chains became apparent [147].

The best examined proteins that show conformationally heterogeneity in their native states are the sarc homology 2 domain of the interleukin-2 tyrosine kinase (ITK-SH2 domain), the HIV-1 coat protein Gag, the bacteriophage MS2 coat protein, and the transforming growth factor β-like domain (TBD) from human fibrillin-1 [143,148–160].

ITK is one of the native binding partners for the human PPIase Cyp18. Native state isomerization in the SH2 domain of ITK between the imide bond Asn286 and Pro287 has pronounced structural and functional consequences by causing long-range structural distortion. The population of the conformers are nearly equal (40% *cis*; 60% *trans*) [147]. In the *trans* form the loop (CD loop) preceding the N286-P287 peptide bond is extended away from the SH2 domain whereas in the *cis* conformer the CD loop bends down to the SH2 body (Fig. 9.4) [148].

Backbone dynamic experiments indicated a significant mobility for the *trans* peptide bond whereas the *cis* form is conformationally more restricted [148]. Each of the conformers binds ligands with different chemical properties. The *trans* conformer preferentially binds phosphotyrosine-containing peptides. When resuming the *cis* conformation the SH2 domain builds a specific intermolecular complex with the Src homology 3 (SH3) domain of ITK. This serves to restrict access to the catalytic domain. Both domains, the SH2 and the SH3, influence the conformational properties of the neighboring kinase domain and render the catalytic domain unable to carry out its physiological function [149]. This shows that pep-

Asn286-Pro287$_{trans}$ Asn286-Pro287$_{cis}$

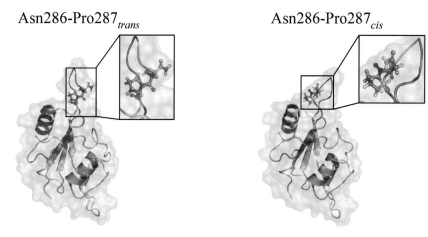

Fig. 9.4 Energy minimized structures of the *trans* (left, pdb: 1LUN) and *cis* (right, pdb: 1LUK) ITK-SH2 domain conformers. The Asn286-Pro287 segment in the CD-loop (red) is highlighted in green.

tide bond CTI is sufficient to alter ligand binding surfaces to an extent that can be distinguished between different ligands.

The bacteriophage MS2 coat protein lacks an terminal amino acid extension that in other virus coat proteins mediates assembly of the viral coat by switching between ordered and disordered forms [156]. Interestingly CTI around the imide bond preceding Pro78 is responsible for switching between two distinct structures in this region of the protein. As observed for ITK the *cis* imide bond-containing species is less flexible than the corresponding *trans* form [157]. A loop in the "open" *trans*-Pro-containing species protrudes away from the surface of the protein, allowing access to the active site [158]. In the *cis* ("closed") conformation, this loop makes hydrophobic contacts to the ITK body and thereby shields the active site from solvent [159].

In the given examples of native state isomerization for ITK and MS2 the *trans* form corresponds to an extended, solvent-exposed loop conformation. *Cis* forms are characterized by extensive contacts between the loops and the surface of the proteins compared to the *trans* conformers, where those contacts are weak or not present. It is speculated that such contacts provide the necessary energy needed to stabilize the inherently less stable *cis* conformation [147].

Human cyclophilin 18 (Cyp18) was found to be incorporated into the human immunodeficiency type 1 virus (HIV) by interacting with the HIV-1 coat protein Gag [150]. The Cyp18/Gag complex is essential for subsequent packing of multiple copies of Cyp18 into each HIV-1 viron [151]. Virons lacking Cyp18 are less infectious [152]. Available NMR [143] and crystal [153] structures of this complex provide valuable insights into the nature of the formed complex. The NMR structure of the capsid protein shows conformational heterogeneity in one location, the Gly89-Pro90 peptide bond. The minor 14% of the Gag population correspond to a *cis* conformation whereas 86% are in *trans*. Several X-ray structures showed that the human Cyp18 specifically binds to the G89-P90 motif of Gag, which is situated within a solvent-exposed loop of the capsid protein [154]. These structures reveal that the G89-P90 peptide bond adopts exclusively the *trans* conformation upon binding to Cyp18. This finding is in contrast to previously solved structures of Cyp18 in complex with short peptides, where the *cis* conformer was adopted [161,162]. Additionally it was shown that Cyp18 is able to accelerate the isomerization of the G89-P90 peptide bond in vitro but structural perturbations in the overall structure of HIV-1 capsid are very limited [155]. However, it is unclear whether the primary role of Cyp18 is of enzymatic nature or if it is a protein binding module that mediates specific protein–protein interactions [163].

In the case of TBD (transforming growth factor β-like domain of human fibrilin-1) two stable conformers are observable that differ with respect to the isomerization state of Pro22. The conformational heterogeneity is embedded in a region of stable β-sheet structure and in both conformers the region surrounding Pro22 is well ordered. In this case CTI around Pro22 causes measurable changes in the secondary structure topology of the concerned domain [160]. Molecular dynamic calculations suggest that the backbone ψ angle of the proline residue is predominantly around $-30°$ (α-helical region) for the *cis* isomer whereas the *trans* form can accept both $-30°$ and $150°$ (β-sheet region) for ψ [43]. This example shows that native state isomerization not only occurs in flexible loop regions but can also appear in more stable secondary structures.

The Lqh-8/6 (*Leiurus quinquestriatus hebracus* toxin-like peptide) can be isolated from scorpion venom. This 38-residue oligopeptide exhibits four disulfide bonds and shows nearly equal amounts of *cis* and *trans* conformers around the Ala36-Pro37 imide bond. The structural differences between these conformers are locally limited. In the *cis* form the tyrosine 38 is clearly orientated towards the protein core and is additional stabilized by two hydrogen bonds between the tyrosine side-chain OH and the protein body. In the *trans* conformer tyrosine 38 is completely solvent exposed. Interestingly, no interconversion between these conformers was detectable [164]. The question of whether these two conformationally fixed forms can direct different toxic actions has not yet been answered.

In conclusion, native state isomerizations around imidic peptide bonds point to a general role for proline as a molecular switch that can control protein–protein interaction.

A well-studied case of native state isomerization involving a secondary amide peptide bond is the saccharide binding protein concanavalin A (ConA). Concanavalin A is a metal ion-dependent lectin isolated from *Canavalia ensiformis* (jack bean) which can interact with cell surfaces by binding specific carbohydrates. The activity of the protein can be switched on and off by calcium-induced CTI of the Ala207-Asp208 bond [165]. In solution 13% of metal-free ConA exhibit the Ala207-Asp208 bond in the *cis* (locked) conformation. Previously it was known that the two ConA conformers are separated by an activation barrier of $\sim$92 kJ mol^{-1} [166]. The equilibrium of ConA is dependent on the calcium ion concentration [167]. Binding of Ca^{2+} ions to the S2 metal binding site of ConA forces several amino acids to reorientate [168]. At the end of this multistep process the side-chain of Thr11 is pushed against the Ala207-Asp208 peptide bond where it makes a clash with Asp208. A consequence of this movement is that the population in the locked conformation (*cis* Ala207-Asp208) rises to nearly 100%. It seems that the release of steric stress between Thr11 and the Ala207-Asp208 peptide backbone drives the *trans* to *cis* isomerization, resulting in an active ConA protein [169]. Going from the locked to the unlocked state, metal ions must be released in a pH-dependent reaction. This opens the conformational space around the Ala207-Asp208 peptide bond and the energetically demanding *cis* bond can refold to the usual *trans* conformer. Such an activation/inactivation mechanism can possibly control the gain and loss of ligand binding activity. The slow kinetics of the nonproline CTI could withhold the lectins from rebinding the storage glycoprotein's before diffusing out of the seedlings. In the outer environment this may contribute to protection against pathogens [170] after regaining activity in the less acidic and calcium-rich soil.

9.4
Biological Consequences

The relevance of catalyzed and spontaneous peptide bond CTI to the biological function of peptides and proteins has inspired considerable effort in research. Thus, distinct pathways have been identified that allow peptide bond isomers to affect physiological signaling differently. To fully understand isomer specificity of bioreactions at the molecular level, it is essential to characterize the structural and electronic differences between *cis* and *trans* peptide bond isomers. Most importantly, both isomers cannot sample the same conformational space around proline [30,171], thus presenting an isomer-specific topography to interacting molecules.

It is just three decades since evidence of the essential role of prolyl bond isomerization in protein chemistry was first obtained. It appeared that the refolding kinetics of denatured proteins could be described in terms of conformational polymorphism of the unfolded polypeptide chain [33]. Briefly, the observations indicated that *cis* prolyl bonds in a folding chain might be able to block the final stage of the folding reaction. A similar conclusion for *cis* secondary amide peptide bonds had to await first direct experimental demonstration in 2001 [59]. Presently, physiologically relevant folding effects of peptide bond CTI have been detected for both intramolecular single chain recognition and intermolecular subunit assembly. Folding blockade is shown to occur on the level of both *cis* and *trans* conformations depending on which of the two isomers is present in the native state. Next, the conformational specificity of enzyme catalysis became apparent in that nonreactive prolyl bond isomers of substrates have been detected in the reaction for many different proteases [172]. Interestingly, conformational specificity of enzymes is not limited to reactions of the isomerizing peptide bond itself but might play a crucial role in modulating reactivity of bonds remote to the isomerizing peptide bond [173].

Currently isomer specificity is known as an important property in the action of proteases, protein phosphatases, and proline-specific protein kinases on their proteinaceous substrates in vitro and in vivo [174–176].

Principally, both interacting protein and interacting ligand/substrate can undergo biologically relevant peptide bond CTI. Among the proteins prone to prolyl isomerization, enzymes and receptors as well as cytosolic binding proteins have been identified [177–183].

Even in complex biological processes such as phage infectivity CTI of a single prolyl bond has been found to be an essential reaction step. In this case the *trans* Gln212-Pro213 moiety of the gene-3-protein in the coat of phage fd, which is located in the hinge region between both domains of this protein, forms a kinetic block of domain reassembly ensuring interaction of the gene-3-protein N1 domain with TolA [41].

Furthermore, long-lived meta stable energy minima might exist in relation to the isomeric state of a critical peptide bond [184]. In the metastable state in which the folded forms of a polypeptide chain have similar structural characteristics but differ in their free energy level, the kinetically trapped species could demonstrate properties of a high energy peptide bond isomer [185].

Furthermore, the recognition of the peptide hormone angiotensin II by the respective receptor is only possible if the Xaa-Pro bond within the angiotensin II sequence assumes the *trans* conformation [186]. In contrast, the muscle selective μ-conotoxin GIIIB requires the *cis* conformation between two hydroxyprolines to become biologically active [187]. Direct evidence for an isomer-specific effect in vivo has been found in lung perfusion studies demonstrating isomer-specificity of angiotensin-converting enzyme [188]. In the case of oxytocin the isomers seem to have an antagonistic function [189]. Isomer-specific recognition power of the μ-receptor has been found for morphiceptin and endomorphin-2 mediated signaling because both ligand peptides are active only as the *cis* isomer, involving isomerization at the Tyr-Pro bond [190].

References

1 M. Ramek, C.-H. Yu, J. Sakon, L. Schafer, *J. Phys. Chem.* **2000**, *A104*, 9636–9645.

2 A. K. Dasgupta, R. Majumdar, D. Bhattacharyya, *Ind. J. Biochem. Biophys.* **2004**, *41*, 233–240.

3 L. Esposito, A. De Simone, A. Zagari, L. Vitagliano, *J. Mol. Biol.* **2005**, *347*, 483–487.

4 C. Dugave, L. Demange, *Chem. Rev.* **2003**, *103*, 2475–2532.

5 G. Fischer, *Chem. Soc. Rev.* **2000**, *29*, 119–127.

6 M. Avalos, R. Babiano, J. L. Barneto, J. L. Bravo, P. Cintas, J. L. Jimenez, J. C. Palacios, *J. Org. Chem.* **2001**, *66*, 7275–7282.

7 A. S. Edison, *Nat. Struct. Biol.* **2001**, *8*, 201–202.

8 K. B. Wiberg, *Acc. Chem. Res.* **1999**, *32*, 922–929.

9 L. Pauling, *The Nature of the Chemical Bond.* Cornell University Press, New York, **1939**.

10 T. Drakenberg, K. I. Dahlqvist, S. Forsen, *J. Phys. Chem.* **1972**, *76*, 2178–2183.

11 A. Avenoza, J. H. Busto, J. M. Peregrina, F. Rodriguez, *J. Org. Chem.* **2002**, *67*, 4241–4249.

12 U. Reimer, G. Fischer, *Biophys. Chem.* **2002**, *96*, 203–212.

13 O. Tchaicheeyan, *FASEB J.* **2004**, *18*, 783–789.

14 G. Fischer, T. Aumuller, *Rev. Physiol. Biochem. Pharmacol.* **2003**, *148*, 105–150.

15 K. B. Wiberg, M. W. Wong, *J. Am. Chem. Soc.* **1993**, *115*, 1078–1084.

16 D. Lauvergnat, P. C. Hiberty, *J. Am. Chem. Soc.* **1997**, *119*, 9478–9482.

17 M. P. Hinderaker, R. T. Raines, *Protein Sci.* **2003**, *12*, 1188–1194.

18 S. Ilieva, B. Hadjieva, B. Galabov, *J. Org. Chem.* **2002**, *67*, 6210–6215.

19 G. N. Ramachandran, A. K. Mitra, *J. Mol. Biol.* **1976**, *107*, 85–92.

20 P. Li, P. Li, X. G. Chen, E. Shulin, S. A. Asher, *J. Am. Chem. Soc.* **1997**, *119*, 1116–1120.

21 G. Scherer, M. L. Kramer, M. Schutkowski, U. Reimer, G. Fischer, *J. Am. Chem. Soc.* **1998**, *120*, 5568–5574.

22 A. Jabs, M.S. Weiss, R. Hilgenfeld, *J. Mol. Biol.* **1999**, *286*, 291–304.

23 B. L. Stoddard, A. S. Pietrokovski, *Nat. Struct. Biol.* **1998**, *5*, 3–5.

24 C. Odefey, L. M. Mayr, F. X. Schmid, *J. Mol. Biol.* **1995**, *245*, 69–78.

25 R. W. Dodge, H. A. Scheraga, *Biochemistry* **1996**, *35*, 1548–1559.

26 M. Vanhove, X. Raquet, T. Palzkill, R. H. Pain, J-M. Frere, *Struct. Funct. Genetics* **1996**, *25*, 104–111.

27 G. Pappenberger, J. Aygun, J. W. Engels, U. Reimer, G. Fischer, T. Kiefhaber, *Nat. Struct. Biol.* **2001**, *8*, 452–458.

28 C. Schiene-Fischer, J. Habazettl, F. X. Schmid, G. Fischer, *Nat. Struct. Mol. Biol.* **2002**, *9*, 419–424.

29 U. Reimer, G. Scherer, M. Drewello, S. Kruber, M. Schutkowski, G. Fischer, *J. Mol. Biol.* **1998**, *279*, 449–460.

30 J. Yao, H. J. Dyson, P. E. Wright, *J. Mol. Biol.* **1994**, *243*, 736–753.

31 J. Zhao, D. Wildemann, M. Jakob, C. Vargas, C. Schiene-Fischer, *Chem. Commun.* **2003**, 2810–2811.

32 R. Frank, M. Jakob, F. Thunecke, G. Fischer, M. Schutkowski, *Angew. Chem. Int. Ed. Engl.* **2000**, *39*, 1120–1122.

33 J. F. Brandts, H. R. Halvorson, M. Brennan, *Biochemistry* **1975**, *14*, 4953–4963.

34 G. Fischer, H. Bang, E. Berger, A. Schellenberger, *Biochim. Biophys. Acta.* **1984**, *791*, 87–97.

35 D. Kern, M. Schutkowski, T. Drakenberg, *J. Am. Chem. Soc.* **1997**, *119*, 8403–8408.

36 M. L. DeRider, S. J. Wilkens, M. J. Waddell, L. E. Bretscher, F. Weinhold, R. T. Raines, J. L. Markley, *J. Am. Chem. Soc.* **2002**, *124*, 2497–2505.

37 Y. Che, G. R. Marshall, *J. Org. Chem.* **2004**, *69*, 9030–9042.

38 M. Keller, C. Sager, P. Dumy, M. Schutkowski, G. S. Fischer, M. Mutter, *J. Am. Chem. Soc.* **1998**, *120*, 2714–2720.

39 C. Renner, S. Alefelder, J. A. Bae, N. Budisa, R. Huber, L. Moroder, *Angew. Chem. Int. Ed. Engl.* **2001**, *40*, 923–925.

40 C. M. Taylor, R. Hardre, P. J. Edwards, *J. Org. Chem.* **2005**, *70*, 1306–1315.

41 F. X. Schmid, J. Buchner, T. Kiefhaber, *Protein Folding Handbook*. Wiley-VCH, Weinheim, **2005**, 916–945.

42 R. L. Stein, *Adv. Protein Chem.* **1993**, *44*, 1–24.

43 D. Hamelberg, T. Shen, J. A. McCammon, *J. Am. Chem. Soc.* **2005**, *127*, 1969–1974.

44 S. Fischer, R. L. Dunbrack, J. M. Karplus, *J. Am. Chem. Soc.* **1984**, *116*, 11931–11937.

45 D. Q. McDonald, W. C. Still, *J. Org. Chem.* **1996**, *61*, 1385–1391.

46 F. Krieger, A. Moglich, T. Kiefhaber, *J. Am. Chem. Soc.* **2005**, *127*, 3346–3352.

47 S. S. Zimmerman, H. A. Scheraga, *Macromolecules* **1976**, *9*, 408–416.

48 A. E. Counterman, D. E. Clemmer, *Anal. Chem.* **2002**, *74*, 1946–1951.

49 N. P. Ewing, X. Zhang, C. J. Cassady, *J. Mass Spectrom.* **1996**, *31*, 1345–1350.

50 I. P. Gerothanassis, C. Vakka, A. Troganis, *J. Magn. Reson. B* **1996**, *111*, 220–229.

51 A. Troganis, I. P. Gerothanassis, Z. Athanassiou, T. Mavromoustakos, G. E. Hawkes, C. Sakarellos, *Biopolymers* **2000**, *53*, 72–83.

52 A. Radzicka, L. Pedersen, R. Wolfenden, *Biochemistry* **1988**, *27*, 4538–4541.

53 E. S. Eberhardt, S. N. Loh, R. T. Raines, *Tetrahedron Lett.* **1993**, *34*, 3055–3056.

54 H. N. Cheng, F.A. Bovey, *Biopolymers* **1977**, *16*, 1465–1472.

55 E. Beausoleil, W. D. Lubell, *Biopolymers* **2000**, *53*, 249–256.

56 J. L. Kofron, P. Kuzmic, V. Kishore, E. Colon-Bonilla, D.H. Rich, *Biochemistry* **1991**, *30*, 6127–6134.

57 J. L. Kofron, P. Kuzmic, V. Kishore, G. Gemmecker, S. W. Fesik, D. H. Rich, *J. Am. Chem. Soc.* **1992**, *114*, 2670–2675.

58 M. Köck, H. Kessler, D. Seebach, A. Thaler, *J. Am. Chem. Soc.* **1992**, *114*, 2676–2686.

59 G. Pappenberger, J. Aygun, J. W. Engels, U. Reimer, G. Fischer, T. Kiefhaber, *Nat. Struct. Biol.* **2001**, *8*, 452–458.

60 E. S. Eberhardt, et al., *J. Am. Chem. Soc.* **1992**, *114*, 5437–5439.

61 B. Ishimoto, K. Tonan, S. Ikawa, *Spectrochim. Acta. A Mol. Biomol. Spectrosc.* **2000**, *56A*, 201–209.

62 M. L. Kramer, G. Fischer, *Biopolymers* **1997**, *42*, 49–60.

63 R. B. Martin, W. C. Hutton, *J. Am. Chem. Soc.* **1973**, *95*, 4752–4754.

64 F. X. Schmid, R. L. Baldwin, *Proc. Natl Acad. Sci. USA* **1978**, *75*, 4764–4768.

65 C. S. G. Scholz, L. M. Mayr, T. Schindler, G. Fischer, FX. Schmid, *Biol. Chem.* **1998**, *379*, 361–365.

66 C. Grathwohl, K. Wuthrich, *Biopolymers* **1976**, *15*, 2043–2057.

67 C. Schiene-Fischer, G. Fischer, *J. Am. Chem. Soc.* **2001**, *123*, 6227–6231.

68 G. Fischer, E. Berger, H. Bang, *FEBS Lett.* **1989**, *250*, 267–270.

69 R. K. Harrison, R. L. Stein, *J. Am. Chem. Soc.* **1992**, *114*, 3464–3471.

70 C. Schiene, U. Reimer, M. Schutkowski, G. Fischer, *FEBS Lett.* **1998**, *432*, 202–206.

71 E. R. Stimson, S. S. Zimmerman, H. A. Scheraga, *Macromolecules* **1977**, *10*, 1049–1060.

72 U. Reimer, N. El Mokdad, M. Schutkowski, G. Fischer, *Biochemistry* **1997**, *36*, 13802–13808.

73 F. L. Texter, B. D. Spencer, R. Rosenstein, C. R. Matthews, *Biochemistry* **1992**, *31*, 5687–5691.

74 C. Cox, T. Lectka, *Org. Lett.* **1999**, *1*, 749–752.

75 R. Hoffmann, N. F. Dawson, J. D. Wade, L. Otvos, Jr., *J. Pept. Res.* **1997**, *49*, 163–173.

76 M. Schutkowski, A. Bernhardt, X. Z. Zhou, M. Shen, U. Reimer, J-U. Rahfeld, K. P. Lu, G. Fischer, *Biochemistry* **1998**, *37*, 5566–5575.

77 M. B. Yaffe, M. Schutkowski, M. Shen, X. Z. Zhou, P. T. Stukenberg, J. U. Rahfeld, J. Xu, J. Kuang, M. W. Kirschner, G. Fischer, L. C. Cantley, K. P. Lu, Science **1997**, *278*, 1957–1960.

78 M. Kipping, T. Zarnt, S. Kiessig, U. Reimer, G. Fischer, P. Bayer, *Biochemistry* **2001**, *40*, 7957–7963.

79 S. K. Holmgren, K. M. Taylor, L. E. Bretscher, T. R. Raines, *Nature* **1998**, *392*, 666–667.

80 L. E. Bretscher, C. L. Jenkins, K. M. Taylor, M. L. DeRider, R. T. Raines, *J. Am. Chem. Soc.* **2001**, *123*, 777–778.

81 P. Chakrabarti, D. Pal, *Prog. Biophys. Mol. Biol.* **2001**, *76*, 1–102.

82 N. Panasik, Jr., E. S. Eberhardt, A. S. Edison, D. R. Powell, R. T. Raines, *Int. J. Pept. Protein Res.* **1994**, *44*, 262–269.

83 E. S. Eberhardt, N. J. Panasik, R. T. Raines, *J. Am. Chem. Soc.* **1996**, *118*, 12261–12266.

84 N. Shamala, R. Nagaraj, P. Balaram, *Biochem. Biophys. Res. Commun.* **1977**, *79*, 292–298.

85 F. Weinhold, *Nature* **2001**, *411*, 539–541.

86 T. Takeda, K. Kojima, Y. Ogihara, *Chem. Pharm. Bull* **1991**, *39*, 2699–2701.

87 G. D. Holt, G. W. Hart, *J. Biol. Chem.* **1986**, *261*, 8049–8057.

88 Y. L. Pao, M. R. Wormarld, R. A. Dwek, A. C. Lellouch, *Biochem. Biophys. Res. Commun.* **1996**, *219*, 157–162.

89 E. Beausoleil, R. Sharma, S. W. Michnick, W. D. Lubell, *J. Org. Chem.* **1998**, *63*, 6572–6578.

90 G. Siligardi, A.F. Drake, *Biopolymers* **1995**, *37*, 281–292.

91 A. Troganis, I. P. Gerothanassis, Z. Athanassiou, T. Mavromoustakos, G. E. Hawkes, C. Sakarellos, *Biopolymers* **2000**, *53*, 72–83.

92 E. S. Eberhardt, R. T. Raines, *J. Am. Chem. Soc.* **1994**, *116*, 2149–2150.

93 M. W. MacArthur, J. M. Thornton, *J. Mol. Biol.* **1991**, *218*, 397–412.

94 J. Voet, D. Voet, *Biochemistry.* Wiley-VCH, Weinheim, **1994**.

95 S. Deechongkit, H. Nguyen, E. T. Powers, P. E. Dawson, M. Gruebele, J. W. Kelly, *Nature* **2004**, *430*, 101–105.

96 B. L. Sibanda, J. M. Thornton, *J. Mol. Biol.* **1993**, *229*, 428–447.

97 I. D. Kuntz, *Protein folding. J. Am. Chem. Soc.* **1972**, *94*, 4009–4012.

98 J. A. Smith, L. G. Pease, *CRC Crit. Rev. Biochem.* **1980**, *8*, 315–399.

99 T. Creighton, *Proteins: Structure and Molecular Properties.* W. H. Freeman, New York, **1993**.

100 G. D. Rose, L. M. Gierasch, J. A. Smith, *Adv. Protein. Chem.* **1985**, *37*, 1–109.

101 E. Vass, M. Hollosi, F. Besson, R. Buchet, *Chem. Rev.* **2003**, *103*, 1917–1954.

102 W. J. Zhang, et al., *J. Am. Chem. Soc.* **2003**, *125*, 1221–1235.

103 P. N. Lewis, F. A. Momany, H. A. Scheraga, *Biochim. Biophys. Acta* **1973**, *303*, 211–229.

104 K. Kim, J. P. Germanas, *J. Org. Chem.* **1997**, *62*, 2847–2852.

105 J. S. Richardson, *Adv. Protein. Chem.* **1981**, *34*, 167–339.

106 M. V. Cubellis, F. Caillez, T. L. Blindell, S. C. Lovell, *Proteins Struct. Funct. Bioinfo.* **2005**, *58*, 880–892.

107 E. W. Blanch, L. A. Morozova-Roche, D. A. E. Cochran, A. J. Doug, L. Hecht, L. D. Barron, *J. Mol. Biol.* **2000**, *301*, 553–563.

108 J. M. Hicks, V. L. Hsu, *Proteins* **2004**, *55*, 316–329.

109 M. A. Kelly, B. W. Chellgren, A. L. Rucker, J. M. Troutman, M. G. Fried, A.-F. Miller, T. P. Creamer, *Biochemistry* **2001**, *40*, 14376–14383.

110 J. T. Nguyen, C. W. Turck, F. E. Cohen, R. N. Zuckermann, W. A. Linn *Science* **1998**, *282*, 2088–2092.

111 U. Shmueli, W. Traub, *J. Mol. Biol.* **1965**, *12*, 205–214.

112 A. F. Drake, G. Siligardi, W. A. Gibbons, *Biophys. Chem.* **1988**, *31*, 143–146.

113 S. C. Yasui, T. A. Keiderling, *J. Am. Chem. Soc.* **1986**, *108*, 5576–5581.

114 M. G. Paterlini, T. B. Freedman, L. A. Nafie, *Biopolymers* **1986**, *25*, 1751–1765.

115 P. A. Osumi-Davis, M. C. de Aguilera, R. W. Woody, A. Y. Woody, *J. Mol. Biol.* **1992**, *226*, 37–45.

116 Z. Shi, C. A. Olson, G. D. Rose, R. L. Baldwin, N. R. Kallenbach, *Proc. Natl Acad. Sci. USA* **2002**, *99*, 9190–9195.

117 N. Sreerama, R. W. Woody, *Proteins* **1999**, *36*, 400–406.

118 B. J. Stapley, T. P. Creamer, *Protein Sci.* **1999**, *8*, 587–595.

119 A. A. Makarov, V. M. Lobachov, I. A. Adzubei, N. G. Esipova, *FEBS Lett.* **1992**, *306*, 63–65.

120 C. A. Gough, R. W. Anderson, R. S. Bhatnagar, *J. Biomol. Struct. Dyn.* **1998**, *15*, 1029–1037.

121 J. Bella, M. Eaton, B. Brodsky, H. M. Berman, *Science* **1994**, *266*, 75–81.

122 S. A. Asher, A. V. Mikhonin, S. Bykov, *J. Am. Chem. Soc.* **2004**, *126*, 8433–8440.

123 A. Kentsis, M. Mezei, T. Gindin, R. Osman, *Proteins* **2004**, *55*, 493–501.

124 M. Mezei, P. J. Fleming, R. Srinivasan, G. D. Rose, *Proteins* **2004**, *55*, 502–507.

125 G. N. Ramachandran, and V. Sasisekharan, *Adv. Protein Chem.* **1968**, *23*, 283–438.

126 A. Weiss, A. Jabs, R. Hilgenfeld, *Nat. Struct. Mol. Biol.* **1998**, *5*, 676.

127 R. E. London, D. G. Davis, R. J. Vavrek, J. M. Stewart, R. E. Handschumacher, *Biochemistry* **1990**, *29*, 10298–10302.

128 D. E. Stewart, A. Sarkar, J. E. Wampler, *J. Mol. Biol.* **1990**, *214*, 253–260.

129 M. L. Wang, W. J. Li, W. B. Xu, *J. Pept. Res.* **2004**, *63*, 23–28.

130 S. Lorenzen, B. Peters, A. Goede, R. Preissner, C. Froemmel, *Proteins* **2005**, *58*, 589–595.

131 D. Pahlke, D. Leitner, U. Wiedeman, D. Labudde, *Bioinformatics* **2005**, *21*, 685–686.

132 N. B. Tweedy, A. K. Nair, S. A. Paterno, C. A. Fierke, D. W. Christianson, *Biochemistry* **1993**, *32*, 10944–10949.

133 L. M. Mayr, O. Landt, U. Hahn, F. X. Schmid, *J. Mol. Biol.* **1993**, *231*, 897–912.

134 W. F. Walkenhorst, S. M. Green, H. Roder, *Biochemistry* **1997**, *36*, 5795–5805.

135 C. Cheesman, R.B. Freedman, L.W. Ruddock, *Biochemistry* **2004**, *43*, 1618–1625.

136 T. Kiefhaber, F. X. Schmid, K. Villaert, Y. Engelsborghs, A. Chaffote, *Protein Sci.* **1992**, *1*, 1162–117.

137 L. M. Mayr, C. Odefey, M. Schutkowski, F. X. Schmid, *Biochemistry* **1996**, *35*, 5550–5561.

138 T. Kiefhaber, R. Quass, U. Hahn, F. X. Schmid, *Biochemistry* **1990**, *29*, 3053–3061.

139 T. Kiefhaber, R. Quass, U. Hahn, F. X. Schmid, *Biochemistry* **1990**, *29*, 3061–3070.

140 J. Balbach, C. Steegborn, T. Schindler, F. X. Schmid, *J. Mol. Biol.* **1999**, *285*, 829–842.

141 F. Chiti, N. Taddei, E. Giovannoni, N. A. J. Van Nuland, G. Ramponi, C. M. Dobsson, *J. Biol. Chem.* **1999**, *274*, 20151–20158.

142 M. J. Thies, J. Mayer, J. G. Augustine, C. A. Frederick, L. Hauke, J. Buchner, *J. Mol. Biol.* **1999**, *293*, 67–79.

143 R. K. Gitti, M. B. Lee, J. Walker, M. F. Summers, S. Yoo, W. I. Sundquist, *Science* **1996**, *273*, 231–235.

144 J. L. Smith, W. A. Hendrickson, R. B. Honzatko, S. Sheriff, *Biochemistry* **1986**, *25*, 5018–5027.

145 M. Weiwad, A. Werner, P. Rucknagel, A. Schierhorn, G. Kullertz, G. Fischer, *J. Mol. Biol.* **2004**, *339*, 635–646.

146 M. Arevalo-Rodriguez, M. E. Cardenas, X. Wu, S. D. Hanes, J. Heitman, *EMBO J.* **2000**, *19*, 3739–3749.

147 A. H. Andreotti, *Biochemistry* **2003**, *42*, 9515–9524.

148 R. J. Mallis, K. N. Brazin, D. B. Fulton, A. H. Andreotti, *Nat. Struct. Biol.* **2003**, *9*, 900–905.

149 K. N. Brazin, D. B. Fulton A. H. Andreotti, *J. Mol. Biol.* **2000**, *302*, 607–623.

150 J. Luban, K. L. Bossolt, E. K. Franke, G. V. Kalpana, S. P. Goff, *Cell* **1993**, *73*, 1067–1078.

151 D. Braaten, E. K. Franke, J. Luban, *J. Virol.* **1996**, *70*, 3551–3560.

152 D. Braaten, E. K. Franke, J. Luban, *J. Virol.* **1996**, *70*, 4220–4227.

153 T. R. Gamble, F. F. Vajdos, S. Yoo, D. K. Worthylake, M. Houseweart, W. I. Sundquist, C. P. Hill, *Cell* **1996**, *87*, 1285–1294.

154 Y. Zhao, Y. Chen, M. Schutkowski, G. Fischer, H. Ke, *Structure* **1997**, *5*, 139–146.

155 D. A. Bosco, E. Z. Eisenmasser, S. Pochapsky, W. I. Sundquist, D. Kern, *Proc. Natl Acad. Sci. USA* **2002**, *99*, 5247–5252.

156 T. Stehle, S. C. Harrison, *EMBO J.* **1997**, *16*, 5139–5148.

157 R. Golmohammadi, K. Valegaard, K. Fridborg, L. Lilijas, *J. Mol. Biol.* **1993**, *234*, 620–639.

158 P. Grochulski, Y. Li, J. D. Schrag, F. Bouthillier, P. Smith, D. Harrison, B. Kubin, M. Cygler, *J. Biol. Chem.* **1993**, *268*, 12843–12847.

159 P. Grochulski, Y. Li, J. D. Schrag, M. Cygler, *Protein Sci.* **1994**, *3*, 82–91.

160 X. Yuan, J. M. Werner, V. Knott, P. A. Handford, I. D. Campbell, K. Douning, *Protein Sci.* **1998**, *7*, 2127–2135.

161 Y. Zhao, H. Ke, *Biochemistry* **1996**, *35*, 7356–7361.

162 J. Kallen, C. Spitzfaden, M. G. Zurini, G. Wider, H. Widmer, K. Wüthrich, M. D. Walkinshaw, *Nature* **1991**, *353*, 276–279.

163 S. L. Schreiber, G. R. Crabtree, *Immunol. Today* **1992**, *13*, 136–142.

164 E. Adjadj, V. Naudat, E. Quiniou, D. Wouters, P. Santiere, C. T. Craescu, *Eur. J. Biochem.* **1997**, *246*, 218–227.

165 J. Bouckaert, F. Poortmans, L. Wyns, R. Loris, *J. Biol. Chem.* **1996**, *271*, 16144–16150.

166 R. D. Brown, S. H. Koenig, C. F. Brewer, *Biochemistry* **1982**, *21*, 465–469.

167 C. F. Brewer, R. D. Brown, S. H. Koenig, *Biochemistry* **1983**, *22*, 3691–3702.

168 J. Bouckaert, R. Loris, F. Poortmans, L. Wyns, *Proteins* **1995**, *23*, 510–524.

169 J. Bouckaert, Y. Dowalelef, F. Poortmans, L. Wyns, R. Loris *J. Biol. Chem.* **2000**, *275*, 19778–19787.

170 W. J. Peumans, E. J. Van Damme, *Plant. Physiol.* **1995**, *109*, 347–352.

171 J. B. Ball, R. A. Hughes, P. F. Alewood, P. R. Andrews, *Tetrahedron* **1993**, *49*, 3467–3478.

172 L. N. Lin, J. F. Brandts, *Biochemistry* **1979**, *18*, 43–47.

173 G. Fischer, J. Heins, A. Barth, *Biochim. Biophys. Acta* **1983**, *742*, 452–462.

174 J. F. Brandts, L. N. Lin, *Methods Enzymol.* **1986**, *131*, 107–126.

175 M. Weiwad, G. Küllertz, M. Schutkowski, G. Fischer, *FEBS Lett.* **2000**, *478*, 39–42.

176 X. Z. Zhou, O. Kops, A. Werner, P. J. Lu, M. H. Shen, G. Stoller, G. Küllertz, M. Stark, G. Fischer, K. P. Lu, *Mol. Cell.* **2000**, *6*, 873–883.

177 S. D. Betts, J. R. Ross, E. Pichersky, C. F. Yocum, *Biochemistry* **1996**, *35*, 6302–6307.

178 P. J. Breheny, A. Laederach, D. B. Fulton, A. H. Andreotti, *J. Am. Chem. Soc.* **2003**, *125*, 15706–15707.

179 A. B. Brauer, G. J. Domingo, R. M. Cooke, S. J. Matthews, R. J. Leatherbarrow, *Biochemistry* **2002**, *41*, 10608–10615.

180 K. K. Ng, W. I. Weis, *Biochemistry* **1998**, *37*, 17977–17989.

181 M. Lopez-Ilasaca, C. Schiene, G. Küllertz, T. Tradler, G. Fischer, R. Wetzker, *J. Biol. Chem.* **1998**, *273*, 9430–9434.

182 G. Coaker, A. Falick, B. Staskawicz, *Science* **2005**, *308*, 548–550.

183 L. Perera, T. A. Darden, L.G. Pedersen, *Biochemistry* **1998**, *37*, 10920–10927.

184 L. D. Cabrita, S. P. Bottomley, *Eur. Biophys. J.* **2004**, *33*, 83–88.

185 C. Schiene-Fischer, C. Yu, *FEBS Lett* **2001**, *495*, 1–6.

186 P. Juvvadi, D. J. Dooley, C. C. Humblet, G. H. Lu, E. A. Lunney, R. L. Panek, R. Skeean, G. R. Marshall, *J. Pept. Res.* **1992**, *40*, 163–170.

187 K. J. Nielsen, M. Watson, D. J. Adams, A. K. Hammarstrom, P. W. Gage, J. M. Hill, D. J. Craik, L. Thomas, D. Adams, P. F. Alawood, R. J. Lewis, *J. Biol. Chem.* **2002**, *277*, 27247–27255.

188 M. P. Merker, I. M. Armitage, S. H. Audi, L. T. Kakalis, J. H. Linehau, J. R. Maehl, D. L. Roehrig, *Am. J. Physiol.* **1996**, *270*, L251–L259.

189 L. Halab, W.D. Lubell, *J. Org. Chem.* **1999**, *64*, 3312–3321.

190 M. Keller, C. Boissard, L. Patiny, N. N. Chung, C. Lemieux, M. Mutter, P. W. Schiller, *J. Med. Chem.* **2001**, *44*, 3896–3903.

10
Enzymes Catalyzing Peptide Bond *Cis-Trans* Isomerizations

Gunter Fischer

10.1
Introduction

The arsenal of enzymes that control posttranslational modifications of proteins includes peptide bond *cis-trans* isomerases whose corresponding modification reaction is unique among posttranslational events. Unlike other modifying enzymes, peptide bond *cis-trans* isomerases catalyze conformational interconversions (folding) of the polypeptide chain of a substrate while leaving unaltered its chemical constitution. In the spontaneous process a substrate polypeptide chain constantly oscillates between the two alternative conformational states of a particular peptide bond. According to the theory of enzyme catalysis of reversible reactions this oscillatory movement is faster in the presence of a peptide bond *cis-trans* isomerase. But a gross chemical change in the reactants is difficult to detect under these conditions. However, these enzymes were also shown to facilitate the catalyzed re-equilibration of a disturbed equilibrium between peptide bond *cis-trans* isomers, leaving the equilibrium constant unchanged. This property served as a probe to discover the first peptide bond *cis-trans* isomerase in pig kidney in 1984 [1]. These enzymes are thus far the only cellular catalysts known to be directly targeted for a conformational interconversion in a protein.

A general effect of catalysis by peptide bond *cis-trans* isomerases is to allow folding reactions to proceed under thermodynamic control escaping from kinetic control. Thus, the advantages of thermodynamic control of global folding and local rearrangements of proteins may exert evolutionary pressure from which folding helper enzymes could have evolved.

Thus far no enzymes have been reported that actively produce a particular peptide bond isomer by means of an energy-consuming reaction, thereby allowing the *cis/trans* equilibrium to be shifted relative to that of the original state of the substrate.

The functional dynamics of proteins in cells depends on how fast those macromolecules adopt a unique three-dimensional structure and, in addition, how spatial–temporal control of the backbone conformation by external molecules underlies the mechanism of folding and chain recognition in vivo. The numerous con-

cis-trans Isomerization in Biochemistry. Edited by Christophe Dugave
Copyright © 2006 WILEY-VCH Verlag GmbH & Co. KGaA, Weinheim
ISBN: 3-527-31304-4

tact points formed and released in the course of intra- and intermolecular bio-recognition are based on countless rotational movements. Potentially, rotational rates could be externally influenced by a wide variety of physical and chemical means such as enzyme catalysis, catalysis by low molecular mass compounds, heat, mechanical forces, and supportive microenvironments. In polypeptides, most rotations of covalent bonds proved intrinsically to be very fast and do not need further rate acceleration by either external factors or intramolecular assistance. Thus, the continuous stretch of single bonds in the graphical representation of the protein backbone conveys a feeling of mobility and flexibility. However, a considerable degree of rigidity exists in the backbone that is nearly independent of its three-dimensional fold and resides in the electron distribution within the C(=O)–N linkage. In fact, very slow conformational interconversions can be detected for backbone rearrangements in native and unfolded proteins in the course of chain folding and assembly, protein–ligand interactions, enzyme catalysis, and transport processes. These slow processes have been characterized molecularly and have frequently been found to be *cis-trans* isomerizations (CTIs) of specific peptide bonds of a protein [2].

By virtue of having two thermodynamically stable chain arrangements of different biological reactivity separated by a rather high energetic barrier, peptide bond CTI resembles a molecular switch that can control cellular signals [3]. Site-specific catalysts of peptide bond CTI would then serve as "liquifying" agents for narrow regions of the backbone that, by reducing the free energy barrier to C–N bond rotation, act as a means of coupling the frequency of mechanical bond movement to the rate of an enzyme-catalyzed biochemical reaction. Provided that the rate-limiting step does not change, intermolecular catalysts concentration-dependently influence resistance to switching. Clearly, enzymes combine favorably two important aspects of catalysis: specificity and rate enhancement. As a consequence, organisms have evolved an array of enzymes that simultaneously accelerate CTI of specific peptide bonds in both directions, *cis* to *trans* and *trans* to *cis*.

Originally, peptide bond *cis-trans* isomerase activity was discovered using a protease-coupled assay for CTI of the imidic peptide bond of proline, commonly termed the prolyl bond [1]. Magnetization transfer in ^{1}H NMR-based assays permitted kinetic characterization of enzymes catalyzing CTI of secondary amide peptide bonds in *Escherichia coli* about 20 years later [4]. The current state of knowledge is that peptide bond *cis-trans* isomerases are present in all organisms and function both inside and outside the cell. To date, two enzyme classes of different primary specificity profile, the peptidyl prolyl *cis-trans* isomerases (PPIases, E.C.5.2.1.8) and the secondary amide peptide bond *cis-trans* isomerases (APIases) have been identified and characterized. They are targeted to imidic (-Xaa-Pro-) and the secondary amide peptide bonds (Xaa-Yaa), respectively (where Yaa stands for all gene-encoded amino acids except Pro, and Xaa is for all gene-encoded amino acids), some of which might be important for the folding of newly synthesized proteins and others, in particular, for the timing of conformational changes in segments of native proteins. In both cases, the timely appearance/disappearance of biological activity of substrate proteins is thought to be affected enzymatically.

In most organisms, including humans, multiple members of each subtype of peptide bond *cis-trans* isomerases are expressed that differ in their molecular masses, domain composition, intracellular localization, and substrate specificity. Only a few investigations exist about the cooperation of the extra modules with the catalytic core.

The PPIases can be classified into three distinct families according to amino acid sequence homology. The high degree of structural relatedness and amino acid sequence similarity define membership in each individual family. Two families, the cyclophilins (Cyp) [5,6] and the FK506 binding proteins (FKBP) [7,8], implicated in the cellular action of the immunosuppressive drugs cyclosporin A (CsA) and FK506 (tacrolimus) respectively are sometimes termed immunophilins. Chimeric proteins containing a N-terminal FKBP domain and a C-terminal cyclophilin domain, or these domains in a reversed orientation, have also been observed in microorganisms [9]. One of the most fascinating aspects of immunophilins is that the immunophilin/drug complexes of 1:1 stoichiometry display discrete reactivity changes when compared with the complex constituent in isolation. Only in its Cyp18 complex does CsA exhibit high-affinity, saturable, specific binding to another cytosolic enzyme, the pSer(pThr)-specific protein phosphatase 2B calcineurin [10,11]. In essence, immunophilin/drug complexes are able to recruit a set of proteins unable to exhibit affinity for the complex constituents when offered individually. This gain of function, which also leads to the recruitment of calcineurin by the FKBP12/FK506 complex, is presumably independent of PPIase-mediated CTI catalysis. However, it should be noted that partitioning of these drugs into peptide bond *cis-trans* isomers plays a role for the drug activities in aqueous solutions. Considering PPIase inhibition, a single imidic peptide bond of the drugs, the N-MeLeu^9-N-MeLeu^{10} moiety and the pipecolinyl amide bond of CsA and FK506, respectively, separates into an inhibitory and a noninhibitory isomer [12–14].

The third family of PPIases, the parvulins, does not belong to immunophilins because family members do not show affinity to immunosuppressive drugs. In all cases the prototypic family members are small, single-domain proteins expressed in high abundance predominantly in the cytosol of mammalian and bacterial cells.

Among potential substrates of PPIases oligopeptides, unfolded polypeptide chains, folding intermediates as well as native proteins have been identified. In keeping with the high cellular PPIase concentrations, these enzymes can be present in stoichiometric amounts with potential substrates in vivo. A simple four-species reaction model accounting for PPIase catalysis involving both isomers in the unbound and the enzyme-bound state matches well the observed kinetic parameter of oligopeptide substrates. It is evident from this model that an internal equilibrium constant K_{int} = [E*trans]/[E*cis] for the Michaelis complexes exists and can deviate strongly from the reaction equilibrium constant in the absence of the enzyme. Under these conditions, of the *cis-trans* isomers of a substrate in solution, a PPIase was found to be able to sequester more *cis* leading to an increased *cis:trans* ratio of the Michaelis complexes near unity [15]. It has been suggested that for an evolutionarily optimized enzyme catalysis K_{int} depends on how close to reaction equilibrium the enzyme can maintain its catalysis under physiological conditions [16]. According to this model, the magnitude of the internal equilibrium constant $K_{int,}$ which approaches unity for Cyp18, thus points to a physiological function of this PPIase for catalytic interconversions involving native state CTI of proteins. However, catalytic amounts of peptide bond *cis-trans* isomerases ($[E]_0 \ll [S]_0$) cannot markedly shift the *cis:trans* ratio of a substrate. It follows that both directions of the conformational interconversion are catalyzed in a reversible manner unless CTI is coupled to an irreversible reaction, such as protein folding under strongly native conditions and proteolytic cleavage [17,18].

To form catalytically productive enzyme/substrate complexes, many peptide bond *cis-trans* isomerases essentially require the location of the reactive bond of the substrate in the context of secondary binding sites or a specific spatial organization of the polypeptide chain thus creating features of stereo- and regiospecificity [19,20]. As in the case of many endoproteases, PPIases can utilize an extended array of catalytic subsites to enhance catalytic efficiency and substrate specificity. These properties precondition peptide bond *cis-trans* isomerases toward a complex reaction pattern. Consequently, biochemical investigations have led to the elucidation of three distinct molecular mechanisms that might be operative either in isolation or collectively in the cellular action of both prototypical and multidomain peptide bond *cis-trans* isomerases:

1. catalysis of peptide bond CTI,
2. an isomer-specific, site-directed holding function for unfolded polypeptide chains in the Michaelis complex, and
3. a presenter protein function for an unknown number of physiological ligands (gain of function).

Mechanistic pathways 1 and 2 involving isomerization catalysis and holding of unfolded polypeptide chains can be discussed in relation to all subfamilies of peptide bond *cis-trans* isomerases. In contrast, only members of the two families of cyclophilins and FKBP, were found to play a role in presenting physiological ligands to further cellular constituents.

Based on current knowledge, catalysis of peptide bond CTI (mechanism 1) plays a major role in the physiological function of peptide bond *cis-trans* isomerases. Mechanisms 2 and 3 might be utilized as functional auxiliaries. Currently, natural molecules using mammalian PPIases as presenter proteins (mechanism 3) are all of microbial origin, and are only present in mammals under artificial conditions. It follows that peptide bond *cis-trans* isomerases have evolved to bind the transition state of CTI tightly and thus catalyze CTI very efficiently. For instance, a level of fewer than 400 molecules of catalytically active ESS1 (a parvulin-like PPIase) per yeast cell is sufficient for growth [21]. The complete loss of ESS1 in a gene knockout mutant resulted in a lethal phenotype and showed the essential character of ESS1 for survival of *Saccharomyces cerevisiae*.

In conclusion, among all known enzymes, peptide bond *cis-trans* isomerases exhibit a unique reaction profile with the potential of targeting a large number of rather similar reactive sites in a polypeptide chain, and a low degree of chemical differences that separates the reactant and product state of the enzyme reaction.

For PPIases, oligopeptides containing -Xaa-Pro- moieties mostly serve the role of assay substrates in vitro but native proteins presumably represent the substrates targeted in the cell. In order to determine catalytic efficacy, re-equilibration kinetics following rapid equilibrium perturbation by physical and chemical means have to be measured for both the spontaneous and enzyme-catalyzed CTI. Using dynamic ^{1}H NMR spectroscopy isomer-specific chemical shift dispersion is the basis of a sensitive PPIases assay using 2D-NOESY experiments [22,23]. Transferred magnetization to *cis-trans* isomers substitutes for chemically induced equilibrium perturbation. Visualization and direct monitoring of PPIase catalysis in living cells is not possible as of yet.

10.2
Cyclophilins

The prototypical member of the cyclophilins family, cytosolic cyclophilin 18 (Cyp18, CypA) of mammalian cells specifically catalyzes prolyl CTIs but remains inert toward secondary amide peptide bonds [1]. Despite the existence of members of the PPIase class of enzymes in virtually all organisms, some prokaryotic and archaea genomes do not encode enzymes of the cyclophilin family [24]. In larger cyclophilins, additional polypeptide segments complete the full-length proteins. They were found to be located N-terminally and C-terminally to the catalytic core domain. Functionally, the extra domains and segments are coupled with intracellular targeting, RNA recognition, RING finger motif mediated metal ion binding, prenylation motifs, tetratricopeptide repeat (TPR)-mediated protein–protein interactions, RAN GTPase binding and calmodulin binding. Furthermore, a Cyp18 fusion represents the first vertebrate example of a chimeric protein, TRIM5-Cyp18, the expression of which is generated by insertion of Cyp18 complementary DNA into the locus of the TRIM5-*a* restriction factor of Old World primates. The presence of either TRIM5-*a* or TRIM5-Cyp18 in these animals is a potent impediment to HIV-1 infection [25].

A schematic representation of the diverse group of human cyclophilins is given in Fig. 10.1.

In their catalytic domain, larger human cyclophilins share high sequence similarity to prototypical human Cyp18 (hCyp18), which has a molecular mass of about 18 kDa. Furthermore, amino acid sequences of cyclophilin-like domains are highly conserved from yeast to humans with a smaller degree of conservation across prokaryotic taxa [26]. The three-dimensional fold of hCyp18 consists of eight antiparallel β-strands supplemented by three a-helices that make up an antiparallel β-barrel structure. Cellular concentrations are variable among the cyclophilins. The Cyp18 concentration approaches high levels in the brain but is also

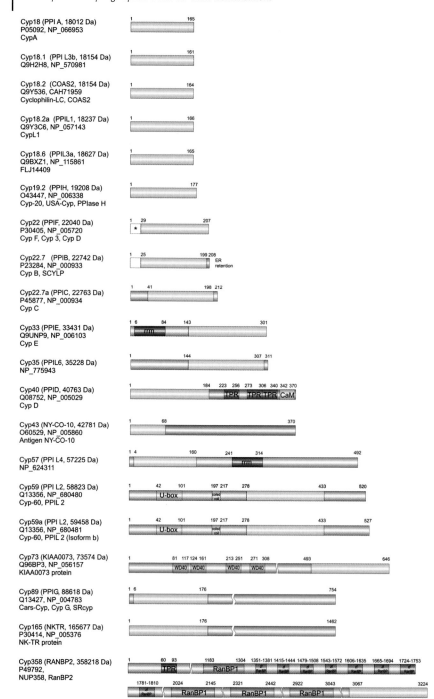

◄ **Fig. 10.1** Human cyclophilins. Protein nomenclature is according to Fischer [3]. The gene name and the molecular mass of the unprocessed proteins are shown in the brackets. In the second row, the accession number of the SWISS-PROT/TREMBL database and an example of an alternative name of the protein is given. The amino acid residues that border the protein domains or functional modules are designated according to SWISS-PROT or Pfam databases. The cyclophilin domain is depicted in yellow. Signal sequence regions are shown as colorless boxes. CaM, Ca^{2+}/calmodulin binding motif; rrm, RNA recognition motif; TPR, tetratricopeptide repeat; RanBP1, Ran binding protein 1 domain; U-box, U-box domain; WD40, WD40 repeats; zf RnaBP, Zn-finger, Ran binding. * indicates "potential."

remarkable in other tissues and cells, as could be exemplified by measuring 5–10 µg mg^{-1} total protein in kidney tubules and endothelial cells [27–29]. These high cytosolic concentrations enable Cyp18-binding drugs, notably high-affinity cyclosporins, to accumulate in the cell at high intracellular levels where they could translate into potent physiological signals. This is an important determinant of cyclosporin efficacy because cells lacking high levels of cyclophilins, as could be achieved in gene deletion experiments, might become resistant to the drug [30]. Despite the fact that the Cyp18 gene is regarded as a housekeeping gene, the regulated expression of Cyp18 mRNA in rat brain points to stress control [31]. Many larger cyclophilins are subject to stress regulation as well [32–34]. Similarly, proteome analyses detected upregulation of Cyp18 in the higher passages of fetal skin cells and downregulation in the fibroblasts of higher aged adults [35,36]. Autoantibodies against cyclophilins are commonly found in autoimmune diseases, and prototypic cyclophilins of various organisms known to cause respiratory allergies, are major allergens [37]. Most strikingly, all genes encoding cyclophilins could be deleted without seriously affecting the viability of *S. cerevisiae* [38]. This is a clear indication for the fact that optimal growth conditions inherently test only a small portion of essential gene function. Consequently, dramatic phenotypes can be observed in deletion mutants of the Cpa1 and Cpa2 proteins, the Cyp18 homologs in *Cryptococcus neoformans*. A Cpa1/cpa2 double mutant exhibited severe synthetic defects in growth and virulence, indicating strong species-specific differences in cyclophilin-mediated folding control [39].

hCyp18 represents a thermally and chemically rather stable protein containing a well-defined active site comprising side-chains of residues Arg55, His126, Phe60, Gln63, and Phe113. The cyclic undecapeptide CsA represents a tight-binding inhibitor for the PPIase activity of most human cyclophilins. The catalytic site is coincident with the CsA binding site. Typically the mode of inhibition of human Cyp18 is competitive with a K_i-value in the range of 2 nmol L^{-1} [5]. A Trp residue at position 121 is a key element of cyclophilins for high-affinity binding of CsA [40]. Cyp18 is a rather promiscuous catalyst as indicated by comparable k_{cat}/K_m values for a series of Xaa-Pro-containing tetrapeptides while the Xaa position varied within the natural amino acids [1,20,41]. Specificity constants approach high values in the range $>10^7$ M^{-1} s^{-1}, implying a nearly diffusion controlled bimolecular reaction; turnover numbers are also impressive (in the range of 10^2–10^4 s^{-1}) for oligopeptide substrates. From these data, Cyp18 appears to belong to the perfectly evolved enzymes [16]. It follows that prediction of physiological substrates of PPIases that is solely based on libraries selecting oligopeptides for preferred ground state binding affinity is misleading.

The energy barrier of a Cyp18-catalyzed isomerization is determined almost entirely by the activation entropy. This contrasts with the spontaneous reaction which is enthalpy driven [20]. PPIase-mediated catalysis of prolyl isomerization was observed for oligopeptides, unfolded polypeptide chains and native proteins. As with the RNase T1 substrate, rate-limiting prolyl isomerization in protein folding can be assessed by monitoring rate acceleration in the refolding kinetics of denatured proteins in the presence of various PPIases [42]. Kinetic control of slow steps of protein folding by cyclophilins may avoid accumulation of reaction intermediates prone to formation of non-native protein conformations in vitro and in vivo. In addition, intermolecular association reactions, exemplified by homo- and hetero-oligomer formation, were found to be sensitive to Cyp18 catalysis in vitro. For example, Cyp18 is able to facilitate the CTI during refolding and assembly of the antibody domain C(H)3. Thus, the rate-limiting step of protein dimerization changed from slow prolyl isomerization to rapid chain collapse in the presence of the PPIase [43].

The calcium/magnesium-dependent nuclease activity reported for cyclophilins [44,45] might originate in an impurity present in cyclophilin preparations [46,47].

Interestingly, clinicians have long utilized low-molecular-mass inhibitors of hCyp18 such as CsA and its derivatives in the prevention of allograft rejection, for treatment of autoimmune diseases and for prophylaxis of graft-versus-host diseases [48,49]. The results of the structure–activity relationship studies for cyclosporin derivatives suggest that a causal relationship between inhibition of the PPIase activity of Cyp18 and T cell replication does not exist [50,51]. Functional analyses revealed that the binary Cyp18/CsA complex might form a composite surface that interacts with calcineurin, whose protein phosphatase activity on transcription factors such as nuclear factor of activated T cell (NFAT) is inhibited in the ternary complex [52,53]. It became evident that several hundred polypeptides were found to be up- or downregulated when comparing CsA-treated and untreated ConA-activated T lymphocytes [54]. Obviously, differentially expressed proteins include those affected by inhibition of the PPIase activity of many cellular cyclophilins as well as inhibition of protein dephosphorylation by calcineurin.

Strikingly, cyclosporin derivatives exist that inhibit calcineurin, circumventing prior binding to hCyp18. For instance, the complex between [dimethylaminoethylthiosarcosine]^3-CsA and Cyp18 does not lead to calcineurin inhibition, but this derivative potently inhibits the protein phosphatase activity on its own [55]. In the Cyp18/CsA complex the bioactive surface is mainly determined by the portion of the cyclosporin A macrocycle protruding out of the hCyp18 binding cleft, and [dimethylaminoethylthiosarcosine]^3-CsA does not require Cyp18 to adopt the bioactive conformation of the protruding segment [56]. Generally, modification of CsA residues at positions 9 to 2 (collectively termed the Cyp18 binding domain) reduces the affinity to Cyp18, while calcineurin inhibition by the Cyp18/CsA derivative complex could be impaired by changing the residues at positions 4 to 7 (collectively termed the calcineurin binding domain). On the other hand, strong Cyp18 inhibitors of the sanglifehrin type, which are structurally distinct from cyclosporins, also offer the capability for attenuation of clonal T cell expansion. Sanglifehrin A blocks T cell proliferation in response to interleukin-2 (IL-2) by inhibiting the appearance of activity of the cell cycle kinase cyclin E-Cdk2; calcineurin activity is nevertheless left unaltered [57,58]. There is no indication that the Cyp18/sanglifehrin complex is able to recruit other cell constituents using Cyp18 as presenter protein for this drug.

To add to the puzzle, the estimation of calcineurin activity in CsA-treated renal transplant patients raised doubts about the direct relationship between calcineurin inhibition and immunosuppression. In circulating lymphocytes of the immunosuppressed patients, calcineurin activity is only partially reduced (50% to 85% of the control) [59]. It was hypothesized that partial calcineurin inhibition might account for both the immunosuppression and the immunocompetence of CsA-treated patients [60]. Altogether, the data demonstrate that cyclophilins other than Cyp18 might contribute directly to immunosuppression through the PPIase-inhibitory effect of CsA. Unfortunately, few data are available quantifying the magnitude of inhibition of CsA derivatives for the different cyclophilins. CsA derivatives that do not show a gain of function in their Cyp18 complexes and thus avoid calcineurin inhibition but still inhibit the PPIase activity of Cyp18 could facilitate the studies of the distinct physiological functions of cyclophilins [61–63]. For example, side-chain substitution in position 6 of CsA ([N-methylalanine]^6-CsA) leads to a nearly unchanged Cyp18 inhibition, but a 25-fold decrease in calcineurin inhibition and about 250-fold decrease in immunosuppressive activity as compared with CsA. Similarly, [N-methylisoleucine]^4-CsA (SDZ NIM811) and [N-methylvaline]^4-CsA have proved to be potent Cyp18 inhibitors with only minor effects on calcineurin activity and clonal T cell expansion. In addition, non-cyclosporin-like inhibitors of moderate potency have been published but biological data are still lacking [64,65].

The methods used to identify physiological functions of an individual cyclophilin or a group of related cyclophilins include combined depletion and complementation in cells and organisms, the application of monofunctional cyclophilin inhibitors that lack inhibition of calcineurin and the effects of wild-type cyclophilins and enzymatically inactive cyclophilin variants overexpressed in certain cell types. These studies suggest the major involvement of cyclophilins in viral and parasitic infections, glucocorticoid response, malignancies, ischemia/reperfusion injury via activation of the mitochondrial permeability transition pore, inflammation, stress response, and allergies. The pharmacologically relevant profile of cyclophilins has been extensively reviewed [66].

For instance, the putative involvement of Cyp18 in the elimination of damaged and misfolded proteins produced by oxidative stress from cells has been found when exploring the cause of familial amyotrophic lateral sclerosis [67]. Cyp18 acts catalytically on still unknown prolyl bonds either by direct participation in protein degradation or indirectly by supporting the functional structure of a Cu/Zn superoxide dismutase-1. A mutant protein of the latter enzyme is responsible for the fatal apoptosis of neurons in this disease. Overexpression of wild-type Cyp18 in cells containing the mutant dismutase rescued the cell response from apoptotic death. In another example, native state isomerization of the T cell-specific protein tyrosine kinase (Itk) SH2 domain causes a proline-dependent conformational switch that regulates substrate recognition and mediates regulatory interactions with the active site of Cyp18. Both proteins, Cyp18 and Itk form a stable complex in Jurkat T cells that is disrupted by treatment with CsA [68,69]. It adds to the story that mutation of a conformationally heterogeneous proline residue in the SH2 domain specifically increased Th2 cytokine production from wild-type CD4+ T cells. Consequently, the enhanced CD4+ T cell response in the absence of Cyp18 leads to allergic disease, with elevated IgE and tissue infiltration by mast cells and eosinophils [70].

Cyp18 itself has pronounced biochemical and physiological effects on living cells. When Cyp18 is secreted by vascular smooth muscle cells in response to oxidative stress, it mediates typical effects of reactive oxygen species the activation of extracellular signal-regulated kinase (ERK1/2) and stimulation of cell growth [71]. The host cell response against parasites also involves secreted and host cell cyclophilins [72,73]. Presumably, cyclophilin-catalyzed prolyl isomerizations do not play a role in the de novo synthesis of proteins.

10.3
FK506 Binding Proteins (FKBPs)

Among the PPIases, FKBPs constitute a large family discovered in 1989 [7,8]. At this time, the prototypic human enzyme (FKBP12) was identified as a FK506 binding molecule; the term FKBP includes all proteins that share amino acid sequence homology with FKBP12, although high-affinity FK506 binding is not a common property of FKBP. Our current thinking is that cellular conditions play a decisive role in the affinity of FKBP for FK506, and FK506-mediated enzyme inhibition may occur in a large part of human FKBP. Using in vitro assays, K_i values range from high picomolar concentrations to no inhibition at all. FKBPs are widely distributed throughout all living organisms, being present even in the very small genome of *Mycoplasma genitalum* as sole PPIase. At least 17 FKBP have been identified, mostly at the protein level in human cells and cover a wide range of molecular masses from 12 kDa to about 135 kDa (Fig. 10.2). The PPIase properties of the large FKBPs are rooted in one or in multiple FKBP domains. As already

Fig. 10.2 Human FK506-binding proteins. Protein nomenclature is according to Fischer [3]. The gene name and the molecular mass of the unprocessed proteins are shown in the brackets. In the second row, the accession number of the SWISS-PROT/TREMBL database and an example of an alternative name of the protein is given. The amino acid residues that border the protein domains or functional modules are designated according to SWISS-PROT or Pfam databases. The FKBP domain is depicted in yellow. Signal sequence regions are shown as colorless boxes. CaM, Ca^{2+}/calmodulin binding motif; EF, EF hand; TPR, tetratricopeptide repeat; RanBP1, Ran binding. * indicates "potential."

FKBP12 (FKBP1A, 11950 Da)
P62942, NP_463460
FK506-binding prot.1A

FKBP12.6 (FKBP1B, 11782 Da)
P68106, NP_004107
FK506-binding prot.1B

FKBP15.9 (FKBP1C, 15903Da)
NP_001011510
FK506-binding prot.1C

FKBP15.6 (FKBP2, 15649 Da)
P26885, NP_476433
FK506-binding prot.2, FKBP13

FKBP22 (FKBP11, 22180 Da)
Q9NYL4, NP_057678
FK506-binding prot.11, FKBP19

FKBP24 (FKBP14, 24172 Da)
Q9NWM8, NP_060416
FK506-binding prot.14, FKBP22

FKBP25 (FKBP3, 25177 Da)
Q00688, NP_002004
FK506-binding prot. 3

FKBP30 (FKBP7, 30009 Da)
Q9Y680, NP_057189
FK506-binding prot. 7, FKBP23

FKBP36 (FKBP6, 37214 Da)
O75344, NP_003593
FK506-binding prot. 6

FKBP37 (AIP, 37664 Da)
O00170, NP_003968
AH receptor-interacting protein

FKBP38 (FKBP8, 38408 Da)
Q14318, NP_036313
FK506-binding prot. 8, FKBPR8

FKBP44 (AIPL1, 43903 Da)
Q9NZN9, NP_055151
Aryl-hydrocarbon
interacting protein-like 1

FKBP51 (FKBP5, 51212 Da)
Q13451, NP_004108
FK506-binding prot. 5, FKBP54

FKBP52 (FKBP4, 51804 Da)
Q02790, NP_002005
FK506-binding prot. 4, FKBP59

FKBP63 (FKBP9, 63084 Da)
O95302, NP_009201
FK506-binding prot. 9

FKBP65 (FKBP10, 64245 Da)
Q96AY3, NP_068758
FK506-binding prot. 10

FKBP135 (KIAA0674, 135178 Da)
Q9Y4D0
Hypothetical protein KIAA0674
Fragment

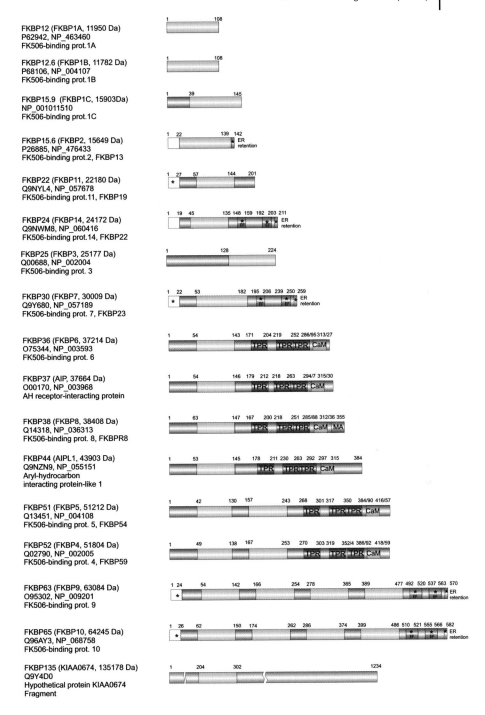

determined for cyclophilins, the catalytic domains are supplemented by C-terminal and N-terminal extensions known for other biochemical functions. The concentration of FKBPs can approach high levels (about 20 μmol L^{-1}), and human FKBP locates to nuclei, the endoplasmic reticulum, the cytosol, and can be secreted [3].

The major structural elements surrounding the hydrophobic catalytic cleft of FKBP12 are a concave five-stranded antiparallel β-sheet wrapping around a short α-helix and a large flap region. In preferred substrates, hydrophobic side-chains are positioned toward the N-terminus of the reactive prolyl bond [20]. Unlike hCyp18 and *E. coli* parvulin 10 (Par10), the prototypic members of other PPIase families, k_{cat}/K_m values of FKBP12 toward oligopeptide substrates do not achieve the diffusion controlled limit. Similarly, turnover numbers of FKBP are also about 10-fold lower than those observed with cyclophilins. Comparison of the effects of point mutations on k_{cat}/K_m values for tetrapeptide substrates revealed catalytically important side-chains of FKBP12 [74,75]. The point mutations of FKBP12 variants associated with a single amino acid substitution (Tyr82Phe, Phe99Tyr, Trp59Ala, Asp37Val, Asp37Leu) experienced marked reduction (about 10-fold) in the specificity constant k_{cat}/K_m, with the k_{cat} value being the most severely affected parameter. Provided that the native tertiary structure remained unchanged, none of the FKBP variants resulting from point mutations completely blocked the catalytic machinery.

Knowledge of FKBP12 and FKBP12.6 structures either alone or complexed with ligands exists but a three-dimensional model of a FKBP12/oligopeptide substrate complex on the basis of a resolved X-ray structure is still missing. Thus, the FK506 binding site has been assumed to be identical with the catalytic site.

Native FKBP12 contains seven *trans* prolyl peptide bonds, and the CTIs of some or all of them constitute a slow, rate-limiting event in folding. Its refolding process from a chemically denatured state constituted the first example of an autocatalytic formation of a native protein from kinetically trapped intermediates with non-native prolyl isomers [76,77].

The genomes of lower eukaryotes, eubacteria, and archaea encode homologs that have, in some cases, a bulge insertion in the flap region of FKBP12, conferring holding chaperone properties to the prototypic enzymes [78,79]. Chaperoning unfolded chains can be explained with a largely extended catalytic subsite required for optimal recognition of the cellular substrates in these organisms and is a biochemical feature of many polypeptide chain interconverting enzymes including proteases [80,81]. Similarly, FKBP are likely to have evolved to help establish accelerated prolyl isomerization by utilizing ground state polypeptide binding as a necessary auxiliary.

Several studies have highlighted the importance of multidomain FKBPs in the control of signal transduction pathways. For example, hFKBP51 and hFKBP52 have been found in steroid hormone receptor complexes and are thought to play an important role in complex formation and translocation of receptor–ligand complexes from the cytosol into the nucleus [82,83]. Notably, these multidomain FKBPs share structural characteristics as demonstrated by an N-terminal PPIase domain followed by three tetratricopeptide repeats (TPR) and a putative calmodulin (CaM) binding motif (Fig. 10.2). In most FKBPs the active site adopts a substrate-exposed, functional state that does not require prior activation by cofactors for substrate and inhibitor binding. However, several larger FKBP, such as FKBP38, are suspected of not conforming to this rule.

Extensive studies have revealed that hFKBP38 (FKBP8), a multidomain protein, is characterized by a pronounced expression of its corresponding mRNA in human brain tissue and its specific antitumor effects in mice, caused by upregulation of the anti-invasive syndecan 1 gene expression and suppression of the proinvasive MMP9 gene [84,85]. FKBP38 belongs to the FKBP family in terms of sequence similarity, but neither PPIase activity nor FK506 binding could be detected for the isolated protein. Surprisingly, the PPIase activity of FKBP38 was found to be controlled by its association with Ca^{2+}/calmodulin. Under conditions of saturating calmodulin concentrations, enzymatic activity was observed at calcium concentrations below 1 µmol L^{-1}. Simultanously, appearance of a high-affinity FK506 binding site in the heterodimeric complex was observed [86]. This identification of the first FKBP functionally controlled by recruitment of a second messenger also raises the question of whether the physiological processes involving similarly structured multidomain FKBPs, such as the steroid receptor constituents FKBP51, FKBP52, FKBP37, and FKBP44, might also respond to calcium signals.

The formation of tight complexes of the peptidomacrolides FK506 (also known as tacrolimus) and rapamycin (also known as sirolimus) with endogenous FKBP12 inhibits T cell proliferation via different pathways. Similar to CsA in the Cyp18/CsA complex, FK506 when bound to FKBP12 experiences gain of function that leads to calcineurin inhibition. The FKBP12/FK506 complex inhibits calcineurin more powerful when compared to Cyp18/CsA. Dissociation constants for ternary calcineurin/FKBP/FK506 complexes containing different FKBP range from 88 nmol L^{-1} to 27 µmol L^{-1} [87]. Recently it was found that FKBP12 is the only FKBP family member to play a key role in FK506-mediated immunosuppression [88]. A topically active FK506 derivative (pimecrolimus) with reduced system exposure and thus increased immunological safety has been launched for therapeutic application in atopic dermatitis, psoriasis, and allergic contact dermatitis [89].

By contrast, the FKBP12/rapamycin complex targets mTOR kinase, a downstream effector of the phosphatidylinositol 3-kinase (PI3K)/Akt (protein kinase B) signaling pathway. mTOR kinase plays an important role in RNA stability and transcription, and controls the translation machinery in response to amino acids and growth factors via activation of p70 ribosomal S6 kinase and inhibition of the eIF-4E binding protein [90–92]. For rapamycin, the effect of gain of function achieved by omitting FKBP12 from mTOR binding assay was approximately 2000-fold in favor of higher affinity for the FKBP12/rapamycin complex [93].

Identification of signaling pathways solely affected by inhibition of the PPIase activity is critical for obtaining a complete picture of cellular responses mediated by FKBP-sensitive prolyl isomerizations.

Knowledge of these processes has been facilitated by studying in vivo effects of nonimmunosuppressive FKBP inhibitors such as GPI1046, JNJ460, V10,367, GPI1048, and GPI1485, to name the most prominent compounds.

Several FKBP-sensitive signaling pathways have been found by this approach. Neurotrophic and neuroprotective properties of these compounds convincingly show the involvement of enzyme-catalyzed prolyl isomerizations as a crucial element in neuronal cell signaling under physiological conditions [94–96]. The inhibitory compounds mostly belong to a group of truncated FK506 derivatives. However, the nature of the cellular FKBP substrates displaced from the enzyme by these inhibitors is still unknown. Despite the abundance of FKBP12, its inhibition is unlikely to be functionally significant in neuronal tissues. The FKBP38/CaM/Ca^{2+} complex displays a proapoptotic role, which is abrogated by the potent FKBP38/CaM/Ca^{2+} inhibitor GPI1046, suggesting that secondary messenger-activated FKBP are critical enzymes in neuronal cells [86].

There are several FKBP genes reported to have a defective function in inherited diseases, such as the Williams Beuren syndrome (Williams syndrome) and Leber's congenital amaurosis for FKBP36 (FKBP6) and FKBP44 (AIPL1) mutations, respectively [97–99].

Generally, proteins associated with G_1 regulation have been shown to play a key role in proliferation, differentiation, and oncogenic transformation as well as apoptosis, and represent promising targets for cancer treatment. Reduced cell growth and cyclin D1 levels in fibroblasts from FKBP12-deficient (FKBP12$^{-/-}$) mice corresponds to cell cycle arrest in G_1 phase, and these cells can be rescued by FKBP12 transfection [100]. Both rapamycin and CCI-779, a water-soluble ester analog of rapamycin with improved pharmaceutical properties, have demonstrated impressive activity against a broad range of human cancers growing in tissue culture and in human tumor xenograft [101].

The mechanism suggested for the FKBP12-driven cell cycle is based on the regulation of transforming growth factor β (TGF-β) receptor signaling. The prototypic FKBP12 interacts with the TGF-β receptor [102], acting as a negative regulator of receptor endocytosis [100,103].

Prototypical FKBPs, such as FKBP12 and FKBP12.6, belong to the receptor-associated folding helper enzymes [104]. For example, they copurify and physically interact with intracellular Ca^{2+} release channels, the ryanodine receptors (RyR), and contribute to the regulation of the release of intracellular Ca^{2+} stores [105]. RyR binding affinities have been found in the high nanomolar range [106]. FKBP12.6$^{-/-}$ deficiency in mice causes serious defects including exercise-induced cardiac ventricular arrhythmias. Generally, deficiency of prototypic FKBP and failure of RyR mutants to bind prototypic FKBP have similar phenotypic responses [107].

In many microorganism, multidomain FKBPs of the Mip-type have been detected, characterized by a homodimeric assembly of Mip monomers consisting of a N-terminal dimerization module, a long (65 Å) connecting a-helix, and a C-terminal FKBP domain [108–110]. Mip proteins have been identified as virulence factors of many human pathogens including *Legionella pneumophila, Neis-*

seria meningitidis, Chlamydia trachomatis, Coxiella burnetii, Trypanosoma cruzi, Aeromonas hydrophila, and Salmonella typhimurium. Monoclonal antibodies raised against the active site of L. pneumophila Mip (LpFKBP25) significantly inhibit the early establishment and initiation of an intracellular infection of the bacteria in Acanthamoeba castellanii, the natural host, and in the human U937 macrophages [111].

10.4
Trigger Factor

Trigger factors merit special attention in that they constitute a subfamily of FKBPs. They are present in all species of eubacteria but lack affinity for FK506. Trigger factors are the only ribosome-bound PPIases identified to date [112,113]. Apparently, bacterial cells contain only a single member of this FKBP subfamily. A cellular concentration of 20 μmol L^{-1} promotes the characterization of this CsA and FK506 insensitive PPIase activity in the E. coli cytosol [112]. In principle, the trigger factor is capable of assisting newly synthesized polypeptide chains in folding because its position is in proximity to the tunnel exit site, near ribosomal proteins L23 and L29, located on the back of the 50 S subunit of the E. coli ribosome [114–116]. Consequently, it interacts with both the ribosome and the nascent polypeptide chain in a manner dependent on the functional state of the ribosome. The signal recognition particle (SRP) and trigger factor bind simultaneously to the ribosome with little affinity interference, showing that these factors have separate binding sites on L23 [117,118]. The SRP is an essential component used in the export of polytopic membrane proteins to the cytoplasmic membrane. Interestingly, no eukaryotic cell have to date been shown to harbor a trigger factor homolog. The trigger factor of E. coli has a molecular mass of 48 kDa, and consists of a central FKBP domain, an N-terminal domain comprising the ribosome-binding polypeptide segment of 144 amino acids bearing the well-conserved FRxGxxP consensus motif, and a 184 amino acid residue C-terminal extension. The interaction between vacant ribosomes and trigger factor is characterized by a dissociation constant of about 1 μmol L^{-1} with an average lifetime of the complex of 30 s at 20 °C [119]. The active site of the catalytic domain of the Vibrio cholerae trigger factor is occupied by a loop from the C-terminal domain pointing to its regulatory function for the FKBP-like PPIase activity of trigger factor [120]. Upon binding of specific substrates the insertion loop may dissociate from the active site of the PPIase and may activate catalytic functions. Trigger factor of the thermophilic eubacterium Thermus thermophilus contains Zn^{2+} as a functionally significant constituent in an 1:1 stoichiometry [121].

There is evidence that the trigger factor and the hsp70 chaperone DnaK, a PPIase and a secondary amide peptide bond cis-trans isomerase (APIase) respectively, contribute to the formation of native proteins by apparently overlapping functions with the trigger factor as the primary interaction partner of the emerging polypeptide chain [122–124]. Consequently, synthetic lethality was observed between combined null alleles of trigger factor and DnaK. In contrast, single gene deletions of either DnaK or trigger factor produced viable bacteria under normal growth conditions.

Both proteins are found to associate with empty or translating ribosomes, and to bind to a certain fraction of newly synthesized polypeptides under normal conditions. Biochemically, DnaK resembles trigger factor in that its APIase activity represents an intrinsic property of the protein chain. For oligopeptide substrates to be catalyzed, neither ATP hydrolysis nor accessory protein factors are necessary.

Functionally, DnaK and trigger factor share some common characteristics that arise from the combination of peptide bond isomerizing and polypeptide binding properties. The subsite specificity of the trigger factor at the k_{cat}/K_m level, which is related to the magnitude of transition state binding energy, is virtually unknown for polypeptide substrates despite a considerable body of data describing its ground state affinity for members of a large oligopeptide library [125]. Site-directed mutagenesis of the PPIase catalytic site have been used to uncover the biochemical factors governing assistance of protein folding by trigger factor in the cell. Expression of the apparently inactive trigger factor Phe198Ala variant prevented global protein misfolding and synthetic lethality at temperatures between 20 and 34 °C in $\Delta tig\Delta dnaK$ E. coli cells [126]. On the other hand, overexpression of the trigger factor Phe233Tyr variant, which also displays a greatly reduced PPIase activity in a oligopeptide-based assay, cannot rescue the reduced low-temperature survival in the Δtig background of E. coli cells [127]. Similarly, the PPIase activity of the RopA trigger factor is essential for maturation of a cysteine protease following its secretion from the *Streptococcus pyogenes* bacterial cell [128]. There is reason to believe that functional complementation experiments are sensitive to both subtle differences in the catalytic susceptibility of endogenous substrates involved and the conditional response of the biological assay used. Assistance of trigger factor catalysis by proteinacous cofactors can also be hypothesized but experimental evidence is still lacking. Beside non-native polypeptide chains, GroEL and the proteins L23 and L29 of the large subunit of the E. coli ribosome represent the only proteins with proven trigger factor affinity.

10.5
Parvulins

In 1994, the PPIase family of parvulins was discovered by characterizing the prototypical enzyme Par10 from E. coli [129,130]. This PPIase does not exhibit sequence similarity to either cyclophilins or FKBPs. The mature enzyme, which is 92 amino acids long, is enzymatically active in its monomeric state. A similarly small enzyme has, to the best of our knowledge, not been reported so far. Also, no other organism was reported to possess such a small parvulin. Using a PPIase activity approach, E. coli Par10 was identified at the protein level in E. coli cytosol. Identification at the genetic level suffers from many problems. It is known that during open reading frame (ORF) identification small proteins often escape detection due to their size falling below an arbitrary, researcher-defined minimum cut-off, or the inability to precisely define a promoter or translational start [131]. However, future identification of Par10 homologs can be predicted. Many multidomain parvulins have been characterized in prokaryotes. So far, the existence of only two genes, *par14* and *pin1*, encoding parvulin proteins has been described in the human genome, which is a low frequency when compared with both Cyp and FKBP sequences (Fig. 10.3).

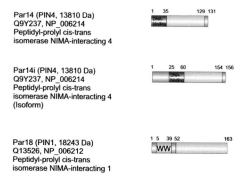

Par14 (PIN4, 13810 Da)
Q9Y237, NP_006214
Peptidyl-prolyl cis-trans
isomerase NIMA-interacting 4

Par14i (PIN4, 13810 Da)
Q9Y237, NP_006214
Peptidyl-prolyl cis-trans
isomerase NIMA-interacting 4
(Isoform)

Par18 (PIN1, 18243 Da)
Q13526, NP_006212
Peptidyl-prolyl cis-trans
isomerase NIMA-interacting 1

Fig. 10.3 Human parvulins. Protein nomenclature is according to Fischer [3]. The gene name and the molecular mass of the unprocessed proteins are shown in the brackets. In the second row, the accession number of the SWISS-PROT/TREMBL database and an example of an alternative name of the protein is given. The amino acid residues that border the protein domains or functional modules are designated according to SWISS-PROT or Pfam databases. The parvulin domain is depicted in yellow. Signal sequence regions are shown as colorless boxes. WW, WW domain.

Among human PPIases, Pin1 provides a rare example of pronounced substrate specificity because polypeptide chains require a pSer(pThr)-Pro- moiety (where p denotes phosphoesterification) if orderly Pin1 catalysis is to occur [132,133]. Pin1 contains an N-terminal WW domain and a C-terminal parvulin-like catalytic domain connected by a flexible linker [132]. Genetic deletion of the single parvulin of *Saccharomyces cerevisiae*, the human Pin1 homolog *ESS1* (also termed PTF1), exhibited a lethal phenotype. Temperature-sensitive mutant yeast strains carrying this mutation arrest at mitosis. There is a functional interchangeability of pSer(pThr)-Pro- specific parvulins between frog, fly, various plants, and humans. In contrast, deletion of the *par10* gene did not lead to a clear phenotype in *E. coli*. Since Pin1 is involved in cell cycle progression its low level expression is associated with proliferation of normal cells. However, Pin1 is strikingly overexpressed in many different human cancers, and inhibition of Pin1 can suppress transformed phenotypes and inhibit tumor growth [134].

Two conserved signature sequences containing histidine motifs (His-Xaa-Val(Ile)-Xaa-Lys and Gly-Xaa-His-Ile(Leu,Val)-Ile), which are separated by a stretch of about 70–85 amino acids containing a few other conserved residues, characterize the catalytic core of all parvulins. In various plants, one member corresponds to the prototypical form of parvulins (it is 117–119 amino acids in length) because it does not show an N-terminal or C-terminal extension when compared with Par10 [135–137]. The difference in chain length to *E. coli* Par10 is due to a single polypeptide segment of 23 amino acids inserted in the N-terminal half of the protein that adopts a loop conformation in the $\alpha 1/\beta 1$ region [138]. In two other eukaryotic parvulins functional domains complement the catalytic domain N-terminally: a DNA binding domain and a (PO_3H_2)Ser/Thr-directed type IV WW domain in Par14 and Pin1 respectively [139–141].

Parvulins showed a high catalytic efficiency comparable to that of Cyp18 toward specific oligopeptide substrates. Unlike other PPIase families the parvulins exhibit an extraordinary pattern of substrate specificity for the amino acid position preceding proline, ranging from a positively charged arginine side-chain to the double-negative phosphoserine (phosphothreonine) side-chain for hPar14 and plant Par13 or hPin1 respectively. Of the active site residues, only those that are necessary for the recognition of characteristic substrates of hPin1 and hPar14 are missing in *E. coli* Par10. The three-dimensional structure of *E. coli* Par10 comprises four β-strands that form a curved β-sheet, enclosed between two α-helices on its convex side and a further α-helix on the concave side, resulting in an $\alpha\beta\alpha$-sandwich structure [142] The loop regions are less well-defined, especially the peptide segments corresponding to Ile39-Gly50, Gln54-Pro59, and Ser67-Pro73, indicating high flexibility regions. The putative substrate binding pocket contains a large aggregation of lipophilic side-chains. In combination with helix 4, the curved β-sheet forms a lipophilic gap on the protein surface, enabling oligopeptides to bind.

The mechanism by which hPin1 exerts its essential role in the cell cycle and other critical physiological events is directly linked to proline-directed phosphotransfer reactions at serine and threonine sites of proteins, as was performed by the specific examples of cyclin-dependent protein kinases, Map kinases, and glycogen synthase 3. In contrast, phosphorylation at tyrosine residues does not lead to substrates especially sensitive to Pin1 catalysis. Approaches using knockdown of enzyme expression along with biochemical tools in vitro have provided evidence for a role of PPIase catalysis in regulation of protein phosphorylation [143,144]. On the other hand, once formed in cells, these phosphoproteins require dephosphorylation in a timed manner. Consequently, hPin1 is involved in the dephosphorylation of pSer(Thr)-Pro sites, as was found by combining in vitro studies on isomer-specific dephosphorylation by protein phosphatase 2a (PP2a) with dephosphorylation studies in *Xenopus* mitotic extracts and rescue experiments in yeast [133,145]. More specifically, the cell cycle regulatory protein phosphatase Cdc25C contains the pThr48-Pro- and pThr67-Pro- moieties that represent critical regulatory phosphorylation sites. The prolyl bond conformation can be either *cis* or *trans* or a mixture of both. For PP2a the inability to dephosphorylate *cis* pThr(Ser)-Pro moieties can be detected in vivo by reciprocal genetic interactions in temperature-sensitive PP2a-deficient *PPH* and the Pin1 homolog-deficient *ESS1/PTF1* mutant strains of budding yeast and in vitro using oligopeptide dephosphorylation experiments. Despite the presence of sufficient activity of PP2a, a fraction of Cdc25C containing *cis* pThr-Pro is left over until it slowly isomerizes to the *trans* isomer. Pin1 markedly accelerated this CTI, thereby altering the rate-limiting step of dephosphorylation. Transgenic expression of point-mutated variants of Pin1 and isolated Pin1 domains in yeast indicates that the enzymatically active PPIase domain is necessary and sufficient to carry out conformational tuning of a PP2a substrate to become dephosphorylated in time. The group IV WW domain alone, although exhibiting affinity to the pThr(pSer)-Pro-containing polypeptide chains, cannot assist in this function. However, depending on the bound phosphopeptide, both domains might be able to communicate in substrate recognition [146].

Many mitotic proteins, transcription factors, and RNA processing proteins, apoptotic proteins, the tumor suppressor p53 and p73 as well as the cytoskeleton protein tau have been shown to be functionally linked to Pin1 catalysis in cells [147].

Par14, on the other hand, has some specificity for arginine preceding the pro-line residue [139]. This enzyme seems to localize to the nucleus and the N-terminal extension of the catalytic domain might mediate interactions with the pre-ribosomal nucleoprotein complex [148].

10.6
Secondary Amide Peptide Bond *Cis-Trans* Isomerases

The dynamics of CTI demonstrate the fundamental similarity between prolyl bonds and secondary amide peptide bonds, making it probable that enzymes exist for the rate acceleration of both types of reactions. The relatively low spontaneous rates indicate the potential importance of CTI of secondary amide peptide bonds as rate-limiting step in protein backbone rearrangements preceding the formation of biologically active proteins.

Whereas peptidyl prolyl *cis-trans* isomerases constitute a well-characterized enzyme class comprising well over 1000 members with small sequence variations in the proteins of different species, the discovery of secondary amide peptide bond *cis-trans* isomerases (APIases) had to await the development of suitable enzyme assays. Fortunately, spectral differences in the UV region between *cis* and *trans* isomers of dipeptides could be exploited to identify and quantify isomerization rate-enhancing factors in biological material [149].

Following fractionation by gel filtration, an Ala-Ala dipeptide-based CTI assay gave two main activity peaks at the 20 and 70 kDa range in *E. coli* lysate, one of which proved to be the bacterial hsp70 homolog DnaK [4]. Consequently, cytosolic fractions prepared from a DnaK knockout strain did not show the activity peak around 70 kDa. Previously, DnaK and other members of the heat-shock protein (hsp70) family were thought to promote protein folding and translocation, both constitutively and in response to stress, by binding to unfolded polypeptide chains. Two functional domains of DnaK have been characterized: a substrate-binding domain binds the polypeptide and an ATPase domain hydrolyzes adenosine triphosphate (ATP), thereby facilitating dissociation of bound polypeptide chains. The APIase function adds an important new component in providing folding assistance since a critical step of chain folding, a slow conformational interconversion, is directly targeted at a specific site.

There has at present been no experimental identification at the amino acid sequence level of an APIase other than members of hsp70 family. However, the formation of the hsp90/*cis* geldanamycin complex was hypothesized to involve hsp90-catalyzed *trans* to *cis* isomerization of geldanamycin [150]. This ansamycin benzoquinone, which contains an *N*-acylated enamine moiety, adopts the *trans* conformation free in solution but has *cis* conformation in the complex. Ab initio quantum chemical calculations have led to a rather high barrier to rotation for the amide moiety of geldanamycin, rendering a catalyzed interconversion useful. In fact, in vitro reconstitution of an hsp90/co-chaperone complex using four co-chaperones including Hsp70, Hsp40, Hop, and p23, which have been shown to be required for constituting a fully functional Hsp90 state in vitro, increased the apparent affinity of a geldanamycin derivative 50-fold when compared with Hsp90 alone [151].

Recombinantly produced DnaK was utilized to characterize its APIase function enzymatically. Co-chaperones, such as DnaJ and GrpE in the presence of ATP might contribute to create the subsite specificity of DnaK. For oligopeptide substrates, the APIase function of DnaK does not require concomitant ATP hydrolysis. A functional overlap of a PPIase and an APIase, trigger factor and DnaK, respectively, could not be observed in an APIase and a standard PPIase assay [127]. Generally, PPIases fail to accelerate CTI of secondary amide peptide bonds in peptide substrates and folding intermediates [152].

Using two different approaches with short peptides, APIase activity of DnaK proved to be stereo- and regioselective. In a series of Ala-Xaa dipeptides DnaK preferentially catalyzes compounds with Xaa = Met, Ala, Ser, Glu, and Leu, whereas CTI of dipeptides with Xaa = Asp, Tyr, Thr, Asn, Phe, Arg, Val, Gln, and Pro could not be accelerated. Interestingly, preferred catalytic subsites are at variance with the hydrophobic core affinity sites concluded from a cellulose-bound library of 13mer oligopeptides [153]. This binding assay reflects ground state affinity of DnaK, yet fails to report susceptibility to catalysis. Only leucine is a preferred residue of both ground state and transition state binding.

To evaluate stereospecificity of catalysis, proton magnetization transfer studies of pentapeptides of the general structure Ala-Xaa-Tyr-Ala-Ala were performed in the presence of DnaK. Since an oligopeptide with an alanine residue at position Xaa was found to be sensitive to DnaK catalysis, the substitution of D-Ala for L-Ala could be used to explore the influence of reversing the N-terminal stereocenter of the targeted peptide bond. Exchange crosspeak intensity on the level of the uncatalyzed CTI of the D-Ala-Tyr bond indicated the importance of proper bond alignment when approaching the transition state of DnaK catalysis [4].

DnaK is also able to catalyze the CTI of a Tyr-Ala bond in a proteinaceous substrate, the native-like folding intermediate of RNase T1 Pro39Ala variant.

The NR peptide (Asn-Arg-Leu-Leu-Leu-Thr-Gly) competitively inhibited the APIase function of DnaK, thereby blocking the DnaK-catalyzed renaturation of chemically denatured luciferase. Thus, it is conceivable that the NR peptide binds to the APIase site of DnaK. Structural investigations already showed that the NR peptide is bound to the peptide binding domain of DnaK in an extended conformation through a loop-defined groove from the β-sandwich subdomain [154].

In conclusion, the biochemical properties of the APIase function of hsp70 molecular chaperones allow kinetic proofreading of folding by rapid scanning of APIase-sensitive sites of unfolded polypeptide chains.

10.7
Catalytic Mechanism of Peptide Bond *Cis-Trans* Isomerases

Despite the amount of data and the simplicity of the chemical reaction catalyzed, the molecular basis of the catalytic mechanism of PPIases and APIases is still only poorly understood [155]. The considerable degree of amino acid sequence dissimilarity between the subgroups of peptide bond *cis-trans* isomerases also raises the challenging question of the mechanistic relatedness among the enzymes. At present there is a lack of detailed mechanistic investigations on APIases and multidomain PPIases. Thus, prototypic PPIases of all three families serve as the bases for unraveling catalytic pathways. One or more potential transition-state structures for enzyme-catalyzed prolyl isomerizations, alone or in combination, are consistent with the acceleration of the spontaneous rate of prolyl isomerization (Fig. 10.4).

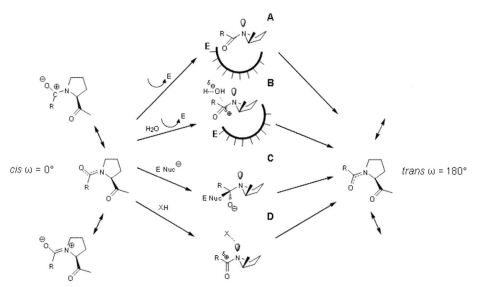

Fig. 10.4 Resonance structures of the prolyl bond and potential transition state structures in the catalytic pathway of PPIases. Noncovalent stabilization of a twisted carbonyl-proline moiety (A). Electrostatic stabilization of the developing positive charge at the amide carbonyl group by a polarized water molecule of the enzyme protein (B). Enzyme-assisted nucleophilic catalysis (C). Enzyme- or substrate-assisted general acid catalysis (D).

One advantage of mechanistic investigations on PPIases is that the accelerated peptide bond rotation is already achieved by globular low-molecular-mass enzymes, making these proteins a perfect subject for structural investigations. Many high-resolution crystal structures of PPIases complexed to oligopeptide substrates, tight binding, reversible inhibitors, and binding proteins have now been collected in the Brookhaven database, and should find great utility in analyses of the catalytic pathway. A major drawback to this approach is that substrate-containing crystals may generate dead-end complexes, making mechanistic conclusions unreliable. It adds to the picture that a PPIase crystal in its solid state has not yet been shown to catalyze CTI of bound substrates. Accordingly, from the structural studies a clear, generally accepted picture of the mechanism of catalysis has not yet emerged. It is also noteworthy that a three-dimensional structure of an oligopeptide substrate complexed with the active site of FKBP or parvulin is still lacking. The location of the active site of these enzymes has been evaluated on the basis of enzyme/inhibitor structures (Fig. 10.5).

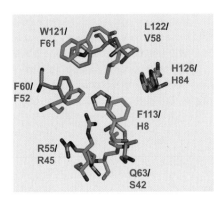

Fig. 10.5 Active site residues of Cyp18 and Par10 were manually superpositioned using their three-dimensional structures in PyMOL representation. Cyp18 residues and Par10 residues were labelled in green and magenta, respectively.

A mechanistic model has been proposed for PPIase catalysis in which a twisted peptide bond, a structure involving substrate strain, is stabilized by noncovalent interaction with the enzyme [156]. However, catalytic antibodies generated to transition state analogs containing twisted carbonyl moieties do not show a PPIase-like catalytic efficiency [157,158]. Consequently, small detergent micelles and phosphatidylcholine membranes are able to catalyze CTI of typical PPIase substrates in a manner reminiscent of that observed for catalytic antibodies [159]. Apparently, sequestration of hydrophobic substrates within the enzyme may account for both a small portion of the catalytic power of FKBP and the acceleration of CTI by catalytic antibodies. Despite overall amino acid sequence dissimilarity the structural features making up the active sites of prototypic enzymes such as Cyp18 and Par10 proved to be similar (Fig. 10.6).

A

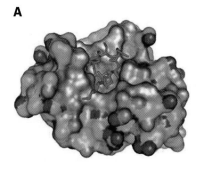

B

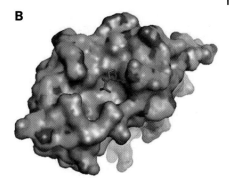

Fig. 10.6 Surfaces of the putative active sites of human FKBP12 (1FKF) and human Pin1 (1PIN1) liganded to the inhibitors FK506 and Ala-Pro, respectively. Inhibitors are depicted as red stick models. Protein surfaces are shown in green, carbon; red, oxygen; blue, nitrogen. Coloration is in pale blue for the putative active sites. Active site residues of human FKBP12 include Arg12, Tyr26, Phe36, Asp37, Phe46, Glu54, Val55, Ile56, Trp59, Tyr82, His87, Ile91, and Phe99 [166]. Active site residues of human Pin1 include His59, Lys63, Arg68,69, Cys113, Leu122, Met130, Phe134, and His157 [132].

Cyclophilins may benefit from general acid catalysis because site-directed muta-genesis of Arg55Ala inactivates the hCyp18 enzyme, and secondary kinetic β-deu-terium effects did not support a tetrahedral intermediate involving a side-chain nucleophile of the enzyme in the catalytic pathway [20]. Enhanced prolyl isomeri-zation rates were already observed by intramolecular general acid catalysis of properly positioned arginine or histidine side-chains in peptides and proteins in aqueous solution [160,161]. Through attractive interactions with a substrate, Arg55, Gln63, and His126, Phe60 and Phe113 may form the active site, which puts Arg55 in a favorable position to perform its catalytic role in the transition state of a Cyp18/substrate complex. In addition, crystal structures revealed a sand-wich-like arrangement of the prolyl ring between the Arg55 side-chain and His126 both in hydrogen bonding distance to the prolyl bond nitrogen of a tetra-peptide substrate that makes this residue a suitable candidate for electrophilic assistance.

The positioning of Arg55 and the geometrical complementarity of the transition state structure compared to that of ground state were taken as criteria to study the role of near attack conformers (NACs) in Cyp18 catalysis [162]. In this model the contention that transition state Cyp18 catalysis is more stable than the ground state has less significance. However, a recent theoretical study using potential of mean force (PMF) simulations with a hybrid QM/MM potential function indi-cated that the population of NACs is not relevant to catalysis [163].

Although the X-ray structure of a Pin1-dipeptide complex implicated covalent catalysis, a lack of nucleophilic assistance in the catalytic pathway can be derived from the covalent modification of parvulins by 5-OH-1,4-naphthoquinone (juglone). A juglone-inactivated *E. coli* Par10 contains two inhibitor molecules that are covalently bound to the side-chains of Cys41 and Cys69 because of a Michael addition of the thiol groups to juglone. However, thiol group modification was shown to proceed 5-fold faster than the rate of enzyme inactivation. Thus, blocking major nucleophils was considered as a necessary but not sufficient condition for inactivation. Partial unfolding of the active site of the juglone-modified parvulins was thought to be the cause of deterioration of PPIase activity [164].

Two types of kinetic experiments provide key information concerning the involvement of a catalytic water molecule in the mechanism of PPIases. First, the physical background of the encounter of substrate with a PPIase molecule has been evaluated. Second, proton inventory studies were done to get reliable data about the kinetic solvent deuterium isotope effects (KSIEs) and to probe the role of exchangeable protons in catalysis. The result of a previous experiment indicated that Cyp18 catalysis does not show a KSIE in k_{cat}/K_m because the magnitude of the effect was in the range of unity. It apparently demonstrated that a proton is not in "flight" during the progression of the reaction from free reactants to the transition state. Now, evidence supporting the hypothesis of an important proton and a common catalytic pathway of cyclophilins and parvulins has been furnished by kinetic studies in solvents of different viscosity as well as by proton inventory data. For both prototypic enzymes the very high catalytic efficiency (k_{cat}/K_m ~10^7 M^{-1} s^{-1}) strongly depends on solvent viscosity, suggesting that the catalyzed CTI of oligopeptide substrates proceeds at diffusion control. If so, these insights could be exploited to determine true KSIEs from the experimentally observed values considering the different viscosities of H_2O and D_2O. For both enzymes, inverse KSIEs resulted, the magnitude of which were found to be remarkably low, in the range of 0.5–0.6. When plotted versus mole fraction of D_2O, the rate data gave a bulging down curve with a large net inverse KSIE, indicating a smaller fractionating factor in the ground state than in the transition state [165]. A catalytic mechanism consistent with the proton inventory data and other information about hCyp18 and *E. coli* Par10 catalysis involves a single proton transfer from Gln63 (Ser42) side-chain to a basic site coordinated in proximity of Gln63 (Ser42), which is believed to occur during transition state formation. The latent iminol anion of the Gln63 side-chain electrostatically stabilizes the developing positive charge at the carbonyl group during progression to the perpendicular transition state configuration of the substrate. Thus, general acid catalysis is assisted in a concerted reaction by electrostatic stabilization of the developing positive charge at the carbonyl carbon during rotation (Fig. 10.7).

Interestingly, FKBP12 does not show diffusion control of k_{cat}/K_m, and proton inventory does not show a bulging down curve. This case underlines the idea that a distinct catalytic mechanism exists for FKBP compared with cyclophilins and parvulins.

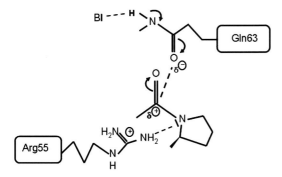

Fig. 10.7 Proposed mechanism for enzymatic catalysis of prolyl *cis-trans* isomerization using two-dimensional representation of the reactant structures for Cyp18. Only those Cyp18 residues whose mutagenesis was highly critical to enzyme activity are shown. Arrows symbolize electron redistribution during approaching the transition state. As the prolyl bond rotates and the carbonyl carbon atom develops a positive charge in the transition state, the weak interaction of the base B with the amide proton of Gln63 becomes strong.

References

1 G. Fischer, H. Bang, C. Mech, *Biomed. Biochim. Acta* **1984**, *43*, 1101–1111.

2 F.X. Schmid, In: *Protein Folding Handbook*, J. Buchner, T. Kiefhaber, Eds. Wiley-VCH, Weinheim, **2005**, 916–939.

3 G. Fischer, *Angew. Chem. Int. Ed.* **1994**, *33*, 1415–1436.

4 C. Schiene-Fischer, J. Habazettl, F. X. Schmid, G. Fischer, *Nat. Struct. Biol.* **2002**, *9*, 419–424.

5 G. Fischer, B. Wittmann Liebold, K. Lang, T. Kiefhaber, F. X. Schmid, *Nature* **1989**, *337*, 476–478.

6 N. Takahashi, T. Hayano, M. Suzuki, *Nature* **1989**, *337*, 473–475.

7 M. W. Harding, A. Galat, D. E. Uehling, S. L. Schreiber, *Nature* **1989**, *341*, 758–760.

8 J. J. Siekierka, S. H. Hung, M. Poe, C. S. Lin, N. H. Sigal, *Nature* **1989**, *341*, 755–757.

9 B. Adams, A. Musiyenko, R. Kumar, S. Barik, *J Biol. Chem.* **2005**, *280*, 24308–2431.

10 J. Friedman, I. Weissman, *Cell* **1991**, *66*, 799–806.

11 J. Liu, J. D. Farmer, Jr., W. S. Lane, J. Friedman, I. Weissman, S. L. Schreiber, *Cell* **1991**, *66*, 807–815.

12 B. Janowski, G. Fischer, *Bioorg. Med. Chem.* **1997**, *5*, 179–186.

13 T. Zarnt, K. Lang, H. Burtscher, G. Fischer, *Biochem. J.* **1995**, *305*, 159–164.

14 J. L. Kofron, P. Kuzmic, V. Kishore, G. Gemmecker, S. W. Fesik, D. H. Rich, *J. Am. Chem. Soc.* **1992**, *114*, 2670–2675.

15 D. Kern, G. Kern, G. Scherer, G. Fischer, T. Drakenberg, *Biochemistry* **1995**, *34*, 13594–13602.

16 J. J. Burbaum, R. T. Raines, W. J. Albery, J. R. Knowles, *Biochemistry* **1989**, *28*, 9293–9305.

17 J. F. Brandts, L.-N. Lin, *Methods Enzymol.* **1986**, *131*, 107–126.

18 G. Fischer, H. Bang, E. Berger, A. Schellenberger, *Biochim. Biophys. Acta* **1984**, *791*, 87–97.

19 C. Schiene, U. Reimer, M. Schutkowski, G. Fischer, *FEBS Lett.* **1998**, *432*, 202–206.

20 R. L. Stein, *Adv. Protein Chem.* **1993**, *44*, 1–24.

21 T. R. Gemmill, X. Wu, S. D. Hanes, *J. Biol. Chem.* **2005**, *280*, 15510–15517.

22 D. A. Bosco, E. Z. Eisenmesser, S. Pochapsky, W. I. Sundquist, D. Kern, *Proc. Natl Acad. Sci. USA* **2002**, *99*, 5247–5252.

23 R. M. Justice Jr., A. D. Kline, J. P. Sluka, W. D. Roeder, G. H. Rodgers, N. Roehm, J. S. Mynderse, *Biochem. Biophys. Res. Commun.* **1990**, *171*, 445–450.

24 A. Galat, *Proteins* **2004**, *56*, 808–820.

25 D. M. Sayah, E. Sokolskaja, L. Berthoux, J. Luban, *Nature* **2004**, *430*, 569–573.

26 A. Galat, *Curr. Top. Med. Chem.* **2003**, *3*, 1315–1347.

27 B. Ryffel, B. M. Foxwell, A. Gee, B. Greiner, G. Woerly, M. J. Mihatsch, *Transplantation* **1988**, *46*, 90S-96S.

28 F. M. Goldner, J. W. Patrick, *J. Comp. Neurol.* **1996**, *372*, 283–293.

29 L. Arckens, E. Van der Gucht, G. Van den Bergh, A. Massie, I. Leysen, E. Vandenbussche, U. T. Eysel, R. Huybrechts, F. Vandesande, *Eur. J. Neurosci.* **2003**, *18*, 61–75.

30 J. Colgan, M. Asmal, B. Yu, J. Luban *J. Immunol.* **2005**, *174*, 6030–6038.

31 R. H. Porter, P. W. Burnet, S. L. Eastwood, P. J. Harrison, *Brain Res.* **1996**, *710*, 97–102.

32 K. Sykes, M. J. Gething, J. Sambrook, *Proc. Natl Acad. Sci. USA* **1993**, *90*, 5853–5857.

33 G. Küllertz, A. Liebau, P. Rücknagel, W. Schierhorn, B. Diettrich, G. Fischer, M. Luckner, *Planta* **1999**, *208*, 599–605.

34 L. Andreeva, R. Heads, C. J. Green, *Int. J. Exp. Pathol.* **1999**, *80*, 305–31.

35 F. Vuadens, D. Crettaz, C. Scelatta, C. Servis, M. Quadroni, W. V. Bienvenut, P. Schneider, P. Hohlfeld, L. A. Applegate, J. D. Tissot, *Electrophoresis* **2003**, *24*, 1281–1291.

36 F. Boraldi, L. Bini, S. Liberatori, A. Armini, V. Pallini, R. Tiozzo, I. Pasquali-Ronchetti, D. Quaglino, *Proteomics* **2003**, *3*, 917–929.

37 D. Roy, D. Ghosh, S. Gupta-Bhattacharya, *Biochem. Biophys. Res. Commun.* **2003**, *307*, 422–429.

38 K. Dolinski, S. Muir, M. Cardenas, J. Heitman, *Proc. Natl Acad. Sci. USA* **1997**, *94*, 13093–13098.

39 P. Wang, M. E. Cardenas, C. M. Cox, J. R. Perfect, J. Heitman, *EMBO Rep.* **2001**, *2*, 511–518 .

40 M. J. Bossard, P. L. Koser, M. Brandt, D. J. Bergsma, M. A. Levy, *Biochem. Biophys. Res. Commun.* **1991**, *176*, 1142–1148.

41 D. J. Bergsma, C. Eder, M. Gross, H. Kersten, D. Sylvester, E. Appelbaum, D. Cusimano, G. P. Livi, M. M. McLaughlin, K. Kasyan, T. G. Porter, C. Silverman, D. Dunnington, A. Hand, W. P. Pritchett, M. J. Bossard, M. Brandt, M. A. Levy, *J. Biol. Chem.* **1991**, *266*, 23204–23214.

42 F. X. Schmid, C. Frech, C. Scholz, S. Walter, *Biol. Chem.* **1996**, *377*, 417–424.

43 M. J. W. Thies, J. Mayer, J. G. Augustine, C. A. Frederick, H. Lilie, J. Buchner, *J. Mol. Biol.* **1999**, *293*, 67–79.

44 J. W. Montague, F. M. Hughes, J. A. Cidlowski, *J. Biol. Chem.* **1997**, *272*, 6677–6684.

45 R. G. Nicieza, J. Huergo, B. A. Connolly, J. Sanchez, *J. Biol. Chem.* **1999**, *274*, 20366–20375.

46 B. Schmidt, T. Tradler, J. U. Rahfeld, B. Ludwig, B. Jain, K. Mann, K. P. Rücknagel, B. Janowski, A. Schierhorn, G. Küllertz, J. Hacker, Fischer, G. *Mol. Microbiol.* **1996**, *21*, 1147–1160.

47 A. Manteca, J. Sanchez, *J. Bactiol.* **2004**, *186*, 6325–6326.

48 S. Arai, G. B. Vogelsang, *Blood Rev.* **2000**, *14*, 190–204.

49 E. Gremese, G. F. Ferraccioli *Clin. Exp. Rheumatol.* **2004**, *22*, S101–S107.

50 M. A. Navia *Curr. Opin. Struct. Biol.* **1996**, *6*, 838–847.

51 M. H. Schreier G. Baumann G. Zenke *Transplant Proc.* **1993**, *25*, 502–507.

52 L. Jin, S. C. Harrison, *Proc. Natl Acad. Sci. USA* **2002**, *99*, 13522–13526.

53 Q. Huai, H. Y. Kim, Y. D. Liu, Y. D. Zhao, A. Mondragon, J. O. Liu, H. M. Ke, *Proc. Natl Acad. Sci. USA* **2002**, *99*, 12037–12042.

54 P. Truffa-Bachi, I. Lefkovits, J. R. Frey, *Mol. Immunol.* **2000**, *37*, 21–28.

55 R. Baumgrass, Y. Zhang, F. Erdmann, A. Thiel, A. Radbruch, M. Weiwad,

G. Fischer, *J. Biol. Chem* **2004**, *279*, 2470–2479.

56 Y. X. Zhang, R. Baumgrass, M. Schutkowski, G. Fischer, *Chembiochem* **2004**, *5*, 1006–1009.

57 L. H Zhang, H. D. Youn, J. O. Liu, *J. Biol. Chem.* **2001**, *276*, 43534–43540.

58 G. Zenke, U. Strittmatter, S. Fuchs, V. F. Quesniaux, V. Brinkmann, W. Schuler, M. Zurini, A. Enz, A. Billich, J. J. Sanglier, T. Fehr, *J. Immunol.* **2001**, *166*, 7165–7171.

59 T. D. Batiuk, F. Pazderka, P. F. Halloran, *Transplant. Proc.* **1995**, *27*, 840–841.

60 T. D. Batiuk, L. N. Kung, P. F. Halloran, *J. Clin. Invest.* **1997**, *100*, 1894–1901.

61 L. Wei, J. P. Steiner, G. S. Hamilton, Y. Q. Wu, *Bioorg. Med. Chem. Lett.* **2004**, *14*, 4549–4551.

62 S.R. Bartz, E. Hohenwalter, M. K. Hu, D. H. Rich, and M. Malkovsky, *Proc. Natl Acad. Sci. USA* **1995**, *92*, 5381–5385.

63 P. C. Waldmeier, J. J. Feldtrauer, T. Qian, J. J. Lemasters, *Mol. Pharmacol.* **2002**, *62*, 22–29.

64 L. Demange, M. Moutiez, C. Dugave, *J. Med. Chem.* **2002**, *45*, 3928–3933.

65 G. Kontopidis, P. Taylor, M. D. Walkinshaw, *Acta Crystallogr. D Biol. Crystallogr.* **2004**, *60*, 479–485.

66 F. Edlich, G. Fischer, *Handbook of Experimental Pharmacology-Molecular Chaperones in Health and Disease.* Springer-Verlag, Berlin, in press, **2006**, pp. 359–404.

67 J. P. Lee, H. C. Palfrey, V. P. Bindokas, G. D. Ghadge, L. Ma, R. J. Miller, R. P. Roos, *Proc. Natl Acad. Sci. USA* **1999**, *96*, 3251–3256.

68 K. N. Brazin, R. J. Mallis, D. B. Fulton, A. H. Andreotti, *Proc. Natl Acad. Sci. USA* **2002**, *99*, 1899–1904.

69 A. H. Andreotti, *Biochemistry* **2003**, *42*, 9515–9524.

70 J. Colgan, M. Asmal, M. Neagu, B. Yu, J. Schneidkraut, Y. Lee, E. Sokolskaja, A. Andreotti, J. Luban, *Immunity* **2005**, *21*, 189–201.

71 Z. G. Jin, M. G. Melaragno, D. F. Liao, C. Yan, J. Haendeler, Y. A. Suh, J. D. Lambeth, B. C. Berk, *Circ. Res.* **2000**, *87*, 789–796.

72 J. Alberti, J. G. Valenzuela, V. B. Carruthers, S. Hieny, J. Andersen, H. Charest, C. R. E. Sousa, A. Fairlamb, J. M. Ribeiro, A. Sher, *Nat. Immunol.* **2004**, *4*, 485–490.

73 G. L. Coaker, A. Falick, B. Staskawicz, *Science* **2005**, *308*, 448–450.

74 M. T. Decenzo, S. T. Park, B. P. Jarrett, R. A. Aldape, O. Futer, M. A. Murcko, D. J. Livingston, *Protein Eng.* **1996**, *9*, 173–180.

75 T. Tradler, G. Stoller, K. P. Rucknagel, A. Schierhorn, J. U. Rahfeld, G. Fischer, *FEBS Lett.* **1997**, *407*, 184–190.

76 C. Scholz, T. Zarnt, G. Kern, K. Lang, H. Burtscher, G. Fischer, F. X. Schmid, *J. Biol. Chem.* **1996**, *271*, 12703–12707.

77 S. Veeraraghavan, T. F. Holzman, Nall, B. T. *Biochemistry* **1996**, *35*, 10601–10607.

78 T. Maruyama, M. Furutani, *Front. Biosci.* **2000**, *5*, D821–D836.

79 A. Galat, *Comput. Biol. Chem.* **2004**, *28*, 129–140.

80 S. Wawra, A.-K. Paschke, G. Fischer, submitted **2006**.

81 L. Van Melderen, S. Gottesman, *Proc. Natl Acad. Sci. USA* **1999**, *96*, 6064–6071.

82 W. B. Pratt, D. O. Toft, *Endocrine Rev.* **1997**, *18*, 306–360.

83 T. H. Davies, E. R. Sanchez, *Int. J. Biochem. Cell Biol.* **2005**, *37*, 42–47.

84 E. Lam, M. Martin, G. Wiederrecht, *Gene* **1995**, *160*, 297–302.

85 S. Fong, L. Mounkes, Y. Liu, M. Maibaum, E. Alonzo, P. Y. Desprez, A. D. Thor, M. Kashani-Sabet, R. J. Debs, *Proc. Natl Acad Sci. USA* **2003**, *100*, 14253–14258.

86 F. Edlich, M. Weiwad, F. Erdmann, J. Fanghänel, F. Jarczowski, J.-U. Rahfeld, G. Fischer, *EMBO J.* **2005**, *24*, 2688–2698.

87 F. Edlich, M. Weiwad, S. Kilka, F. Jarczowski, M. Dorn, M.-C. Mouty, G. Fischer, submitted **2006**.

88 X. Xu, B. Su, R. J. Barndt, H. Chen, H. Xin, G. Yan, L. Chen, D. Cheng, J. Heitman, Y. Zhuang, S. Fleischer, W. Shou, *Transplantation* **2002**, *73*, 1835–1838.

89 A. M. Marsland, C. E. M. Griffiths, *Eur. J. Dermatol.* **2002**, *12*, 618–621.

90 J. E. Dancey, *Expert Opin. Investigat. Drugs* **2005**, *14*, 313–328.

91 E. K. Rowinsky, *Curr. Opin. Oncol.* **2004**, *16*, 564–575.

92 B. D. Kahan, *Transplant. Proc.* **2004**, *36*, 71–75.

93 L. A. Banaszynski, C. W. Liu, T. J. Wandless, *J. Am. Chem. Soc.* **2005**, *127*, 4715–4721.

94 C. Christner, T. Herdegen, G. Fischer, *Mini Rev. Med. Chem.* **2001**, *1*, 377–397.

95 B. G. Gold, E. Udina, D. Bourdette, X. Navarro, *Neurol. Res.* **2004**, *26*, 371–380.

96 R. B. Birge, S. Wadsworth, R. Akakura, H. Abeysinghe, R. Kanojia, M. MacIelag, J. Desbarats, M. Escalante, K. Singh, S. Sundarababu, K. Parris, G. Childs, A. August, J. Siekierka, D. E. Weinstein, *Neuroscience* **2004**, *124*, 351–366.

97 X. Meng, X. Lu, C. A. Morris, M. T. Keating, *Genomics* **1998**, *52*, 130–137.

98 V. Ramamurthy, M. Roberts, F. van den Akker, G. Niemi, T. A. Reh, J. B. Hurley, *Proc. Natl Acad. Sci. USA* **2003**, *100*, 12630–12635.

99 M. M. Sohocki, S. J. Bowne, L. S. Sullivan, S. Blackshaw, C. L. Cepko, A. M. Payne, S. S. Bhattacharya, S. Khaliq, S. Qasim Mehdi, D. G. Birch, W. R. Harrison, F. F. Elder, J. R. Heckenlively, S.P. Daiger, *Nat. Genet.* **2000**, *24*, 79–83.

100 B. Aghdasi, K. Q. Ye, A. Resnick, A. Huang, H. C. Ha, X. Guo, T. M. Dawson, V. L. Dawson, S. H. Snyder, *Proc. Natl Acad. Sci. USA* **2001**, *98*, 2425–2430.

101 M. Hidalgo, E. K. Rowinsky, *Oncogene* **2000**, *19*, 6680–6686.

102 T. W. Wang, P. K. Donahoe, *Front. Biosci.* **2004**, *9*, 619–631.

103 D. Y. Yao, J. J. E. Dore, E. B. Leof, *J. Biol. Chem.* **2000**, *275*, 13149–13154.

104 C. Schiene-Fischer, C. Yu, *FEBS Lett.* **2001**, *495*, 1–6.

105 T. Jayaraman, A. M. Brillantes, A. P. Timerman, S. Fleischer, H. Erdjument-Bromage, P. Tempst, A. R. Marks, *J. Biol. Chem.* **1992**, *267*, 9474–9477.

106 L.H. Jeyakumar, L. Ballester, D. S. Cheng, J. O. McIntyre, P. Chang, H. E. Olivey, L. Rollins-Smith, J. V. Barnett, K. Murray, H. B. Xin, S. Fleischer, *Biochem. Biophys. Res. Commun.* **2001**, *281*, 979–986.

107 N. Tiso, M. Salamon, A. Bagattin, G. A. Danieli, F. Argenton, M. Bortolussi, *Biochem. Biophys. Res. Commun.* **2002**, *299*, 594–598.

108 A. Riboldi-Tunnicliffe, B. Konig, S. Jessen, M. S. Weiss, J. Rahfeld, J. Hacker, G. Fischer, R. Hilgenfeld, *Nat. Struct. Biol.* **2001**, *8*, 779–783.

109 F. A. Saul, J. P. Arie, B. V. L. Normand, R. Kahn, J. M. Betton, G. A. Bentley, *J. Mol. Biol.* **2004**, *335*, 595–608.

110 P. J. B. Pereira, M. C. Vega, E. Gonzalez-Rey, R. Fernandez-Carazo, S. Macedo-Ribeiro, F. X. Gomis-Ruth, A. Gonzalez, M. Coll, *EMBO Rep.* **2002**, *3*, 88–94.

111 J. H. Helbig, B. Konig, H. Knospe, B. Bubert, C. Yu, C. P. Luck, A. Riboldi-Tunnicliffe, R. Hilgenfeld, E. Jacobs, J. Hacker, G. Fischer, *Biol. Chem.* **2003**, *384*, 125–137.

112 G. Stoller, K. P. Rucknagel, K. H. Nierhaus, F. X. Schmid, G. Fischer, J. U. Rahfeld, *EMBO J.* **1995**, *14*, 4939–4948.

113 T. Hesterkamp, S. Hauser, H. Lutcke, B. Bukau, *Proc. Natl Acad. Sci. USA* **1996**, *93*, 4437–4441.

114 G. Kramer, T. Rauch, W. Rist, S. Vorderwulbecke, H. Patzelt, A. Schulze-Specking, N. Ban, E. Deuerling, and B. Bukau, *Nature* **2002**, *419*, 171–174.

115 L. Ferbitz, T. Maier, H. Patzelt, B. Bukau, E. Deuerling, N. Ban, *Nature* **2004**, *431*, 590–596.

116 G. Blaha, D. N. Wilson, G. Stoller, G. Fischer, R. Willumeit, K. H. Nierhaus, *J. Mol. Biol.* **2003**, *326*, 887–897.

117 A. Raine, N. Ivanova, J. E. S. Wikberg, M. Ehrenberg, *Biochimie* **2004**, *86*, 495–500.

118 I. Buskiewicz, E. Deuerling, S. Q. Gu, J. Jockel, M. V. Rodnina, B. Bukau, W. Wintermeyer, *Proc. Natl Acad. Sci. USA* **2004**, *101*, 7902–7906.

119 R. Maier, C. Scholz, F. X. Schmid, *J. Mol. Biol.* **2001**, *314*, 1181–1190.

120 A. V. Ludlam, B. A. Moore, Z. H. Xu, *Proc. Natl Acad. Sci. USA* **2004**, *101*, 13436–13441.

121 R. Suno, H. Taguchi, R. Masui,
M. Odaka, M. Yoshida, *J. Biol. Chem.*
2004, *279*, 6380–6384.

122 K. Beck, L. F. Wu, J. Brunner, M. Muller,
EMBO J. **2000**, *19*, 134–143.

123 S. A. Teter, W. A. Houry, D. Ang,
T. Tradler, D. Rockabrand, G. Fischer,
P. Blum, C. Georgopoulos, F. U. Hartl,
Cell **1999**, *97*, 755–765.

124 E. Deuerling, A. Schulze-Specking,
T. Tomoyasu, A. Mogk, B. Bukau, (1999)
Trigger factor and DnaK cooperate in
folding of newly synthesized proteins.
Nature **1999**, *400*, 693–696.

125 H. Patzelt, S. Rudiger, D. Brehmer,
G. Kramer, S. Vorderwulbecke,
E. Schaffitzel, A. Waitz, T. Hesterkamp,
L. Dong, J. Schneider-Mergener,
B. Bukau, *Proc. Natl Acad. Sci. USA*
2001, *98*, 14244–14249.

126 G. Kramer, H. Patzelt, T. Rauch,
T. A. Kurz, S. Vorderwulbecke, B. Bukau,
E. Deuerling, *J. Biol. Chem.* **2004**, *279*,
14165–14170.

127 C. Schiene-Fischer, J. Habazettl,
T. Tradler, G. Fischer, *Biol. Chem.* **2002**,
383, 1865–1873.

128 W. R. Lyon, M. G. Caparon, *J. Bacteriol.*
2003, *185*, 3661–3667.

129 J. U. Rahfeld, K. P. Rucknagel,
B. Schelbert, B. Ludwig, J. Hacker,
K. Mann, G. Fischer, *FEBS Lett.* **1994**,
352, 180–184.

130 J. U. Rahfeld, A. Schierhorn, K. Mann,
G. Fischer, *FEBS Lett.* **1994**, *343*, 65–69.

131 W. C. Ray, R. S Munson, C. J. Daniels,
Bioinformatics **2001**, *17*, 1105–1112.

132 R. Ranganathan, K. P. Lu, T. Hunter,
J. P. Noel, *Cell* **1997**, *89*, 875–886.

133 X. Z. Zhou, O. Kops, A. Werner, P. J. Lu,
M. Shen, G. Stoller, G. Kullertz,
M. Stark, G. Fischer, K. P. Lu, *Mol. Cell*
2000, *6*, 873–883.

134 L. Bao, A. Kimzey, G. Sauter,
J. M. Sowadski, K. P. Lu, D. G. Wang,
Am. J. Pathol. **2004**, *164*, 1727–1737.

135 J. L. Yao, O. Kops, P. J. Lu, K. P. Lu,
J. Biol. Chem. **2001**, *276*, 13517–13523.

136 M. Metzner, G. Stoller, K. P. Rucknagel,
K. P. Lu, G. Fischer, M. Luckner,
G. Küllertz, *J. Biol. Chem.* **2001**, *276*,
13524–13529.

137 I. Landrieu, L. De Veylder, J. S. Fruchart,
B. Odaert, P. Casteels, D. Portetelle,
M. Van Montagu, D. Inze, G. Lippens,
(2000). The *Arabidopsis thaliana* PIN1At
gene encodes a single-domain phospho-
rylation-dependent peptidyl prolyl *cis/
trans* isomerase. *J. Biol. Chem.* **2000**,
275, 10577–10581.

138 I. Landrieu, J. M. Wieruszeski,
R. Wintjens, D. Inze, G. Lippens, *J. Mol.
Biol.* **2002**, *320*, 321–332.

139 T. Uchida, F. Fujimori, T. Tradler,
G. Fischer, J. U. Rahfeld, *FEBS Lett.*
1999, *446*, 278–282.

140 T. A. Surmacz, E. Bayer, J. U. Rahfeld,
G. Fischer, P. Bayer, *J. Mol. Biol.* **2002**,
321, 235–247.

141 M. A. Verdecia, M. E. Bowman, K. P. Lu,
T. Hunter, J. P. Noel, *Nat. Struct. Biol.*
2000, *7*, 639–643.

142 A. Kuhlewein, G. Voll, B. Hernandez-
Alvarez, H. Kessler, G. Fischer, Rahfeld,
J.-U. G. Gemmecker, *Protein Sci.* **2004**,
13, 2378–2387.

143 M. Weiwad, A. Werner, P. Rucknagel,
A. Schierhorn, G. Kullertz, G. Fischer,
J. Mol. Biol. **2004**, *339*, 635–646.

144 Y. Obata, K. Yamamoto, M. Miyazaki,
K. Shimotohno, S. Kohno,
T. Mastuyama, *J. Biol. Chem.* **2005**, *280*,
18355–18360.

145 K. P. Lu, Y. C. Liou, X. Z. Zhou, *Trends
Cell Biol.* **2002**, *12*,164–172.

146 D. M. Jacobs, K. Saxena, M. Vogtherr,
P. Bernado, M. Pons, K. M. Fiebig,
J.Biol. Chem. **2003**, *278*, 26174–26182.

147 G. Wulf, G. Finn, F. Suizu, K. P. Lu,
Nat. Cell Biol. **2005**, *7*, 435–441.

148 S. Fujiyama, M. Yanagida, T. Hayano,
Y. Miura, T. Isobe, N. Takahashi, *J. Biol.
Chem.* **2002**, *277*, 23773–23780.

149 C. Schiene-Fischer, G. Fischer, *J. Am.
Chem. Soc.* **2001**, *123*, 6227–6231.

150 Y. S. Lee, M. G. Marcu, L. Neckers,
Chem. Biol. **2004**, *11*, 991–998.

151 A. Kamal, L. Thao, J. Sensintaffar,
L. Zhang, M. F. Boehm, L. C. Fritz,
F. J. Burrows, *Nature* **2003**, *425*,
407–410.

152 C. Scholz, G. Scherer, L. M. Mayr,
T. Schindler, G. Fischer, F. X. Schmid,
Biol. Chem. **1998**, *379*, 361–365.

153 S. Rudiger, L. Germeroth,
J. Schneidermergener, B. Bukau, *EMBO
J.* **1997**, *16*, 1501–1507.

154 X. T. Zhu, X. Zhao, W. F. Burkholder,
A. Gragerov, C. M. Ogata,
M. E. Gottesman, W. A. Hendrickson,
Science **1996**, *272*, 1606–1614.

155 J. Fanghanel, G. Fischer, *Front. Biosci.*
2004, *9*, 3453–3478.

156 M. Orozco, J. Tiradorives,
W.L. Jorgensen, *Biochemistry* **1993**, *32*,
12864–12874.

157 J. T. Yli-Kauhaluoma, J. A. Ashley,
C. H. L. Lo, J. Coakley, P. Wirsching,
K. D. Janda, (1994) Catalytic antibodies
with peptidyl prolyl *cis/trans* isomerase
activity. *J. Am. Chem. Soc.* **1996**, *118*,
5496–5497.

158 L. F. Ma, L. C. Hsieh-Wilson,
P. G. Schultz, *Proc. Natl Acad. Sci. USA*
1998, *95*, 7251–7256.

159 M. L. Kramer, G. Fischer, *Biopolymers*
1997, *42*, 49–60.

160 U. Reimer, N. Elmokdad,
M. Schutkowski, G. Fischer, *Biochemistry* **1997**, *36*, 13802–13808.

161 F. L. Texter, D.B. Spencer, R. Rosenstein,
C. R. Matthews, *Biochemistry* **1992**, *31*,
5687–5691.

162 S. Hur, T. C. Bruice, *J. Am. Chem. Soc.*
2002, *124*, 7303–7313.

163 P. K. Agarwal, A. Geist, A. Gorin, *Biochemistry* **2004**, *43*, 10605–10618.

164 L. Hennig, C. Christner, M. Kipping,
B. Schelbert, K. P. Rucknagel,
S. Grabley, G. Kullertz, G. Fischer, *Biochemistry* **1998**, *37*, 5953–5960.

165 J. Fanghanel, G. Fischer, **2006**, in preparation.

166 G. D. Van Duyne, R. F. Standaert,
P. A. Karplus, S. L. Schreiber, J. Clardy,
Science **1991**, *252*, 839–842.

11
Tailoring the *Cis-Trans* Isomerization of Amides

Luis Moroder, Christian Renner (11.1, 11.2)
John J. Lopez, Manfred Mutter, and Gabriele Tuchscherer (11.3)

11.1
Introduction

Proline plays a key role in structure modulation and induction of specific conformations in peptides and proteins. The fine-tuning of the impact on *cis-trans* isomerization (CTI) and onset of particular backbone angles is achieved by introducing substituents in the cyclic structure of proline.

In this chapter we give an overview of the large number of naturally or chemically substituted proline derivatives, their conformational analysis and role in native polypeptides, with an emphasis on their impact on collagen-like structures as well as mimetics for the induction of various turn conformations.

Then we focus on the synthesis and chemical and structural properties of pseudoprolines as solubilizing, structure-inducing and -disrupting building blocks in peptides, particularly on the potential of tailoring the *cis* to *trans* ratio at will, depending on the ring substitution. This paves the way for a variety of biological applications as exemplified by the study of structure–activity relationships and bioactive conformations of therapeutically relevant targets.

11.2
Substituted Prolines

Among the large diversity of proline analogs identified as constituents of natural products, including those with ring restriction and ring expansion, and the related alkylated, dehydro-, mono-, and bis-hydroxylated or oxo derivatives, (2S,3S)-hydroxyproline ((3S)-Hyp or 3-*trans*-Hyp) and (2S,4R)-hydroxyproline ((4R)-Hyp or 4-*trans*-Hyp) are the most common proteinogenic proline derivatives [1–3]. These hydroxylated derivatives result from stereoselective posttranslational hydroxylation, particularly of all types of collagens, by the enzymes prolyl-3-hydroxylase and prolyl-4-hydroxylase, respectively [4,5].

cis-trans Isomerization in Biochemistry. Edited by Christophe Dugave
Copyright © 2006 WILEY-VCH Verlag GmbH & Co. KGaA, Weinheim
ISBN: 3-527-31304-4

11.2.1
Hydroxyprolines

The occurrence of (3S)-Hyp is much less frequent than that of (4R)-Hyp in the total amino acid content of collagens and in contrast to (4R)-Hyp, which exclusively occupies the Yaa position of the (Xaa-Yaa-Gly) repetitive collagen triplets, (3S)-Hyp is found only in the Xaa sequence position, with (4R)-Hyp in the Yaa position [6–9]. Since structural analysis of native collagens is difficult to perform, most information about the stabilizing effects of the collagen triple helix has been derived from synthetic model peptides. Comparative analysis of such model peptides has shown that (Pro-(4R)-Hyp-Gly)$_{10}$ forms a triple helix that is characterized by significantly higher thermal stability than (Pro-Pro-Gly)$_{10}$ [10,11]. Since neither (Pro-(4S)-Hyp-Gly)$_{10}$ [12] nor ((4S)-Hyp-Pro-Gly)$_{10}$ [12] or ((4R)-Hyp-Pro-Gly)$_{10}$ [13] self-associate into triple-helical trimers, it was concluded that (4R)-Hyp in the Yaa position exerts a strong stabilizing effect on the triple helix of collagen-like peptides [11,14,15] and that it increases their folding rates [16]. These results also suggested that (4R)-Hyp in the Xaa sequence position prevents triple helix formation, a fact that would agree with the exclusive location of this proline analog in the Yaa position of collagens. However, more recently it was observed that (Gly-Xaa-Yaa)$_{10}$ peptides with both the Xaa and Yaa occupied by (4R)-Hyp do form triple helices that are even more stable than those with the naturally occurring (Pro-(4R)-Hyp-Gly) repeats [17–19]. Rather unexpected were the findings that the collagenous peptide Ac-(Gly-(3S)-Hyp-(4R)-Hyp)$_{10}$-NH$_2$ that contains the only combination of the two hydroxyprolines occurring in natural collagens, does not fold into a triple helix. Conversely, the identical inability of Ac-(Gly-Pro-(3S)-Hyp)$_{10}$-NH$_2$ to fold into a triple helix was expected from the unnatural position of the 3-hydroxyproline [20]. However, when the triplet (Gly-(3S)-Hyp-(4R)-Hyp) was incorporated into a strongly structuring host–guest peptide such as Ac-(Gly-Pro-(4R)-Hyp)$_3$-Gly-Xaa-Yaa-(Gly-Pro-(4R)-Hyp)$_4$-Gly-Gly-NH$_2$ a destabilizing effect could not be detected [20], while in the less stable peptide (Pro-(4R)-Hyp-Gly)$_3$-Xaa-Yaa-Gly-(Pro-(4R)-Hyp-Gly)$_3$ a slightly reduced triple-helix stability was recorded [21]. This fact may well suggest a fine-tuning of local stabilities in natural collagen structures as required for their interaction with other components of the extracellular matrix [20,21].

In contrast to the previous hypothesis that hydroxyprolines exert their stabilizing effects mainly via a network of bridging water molecules [22], high resolution X-ray analyses of collagen model peptides [23–27] and computed structural models [28–31] combined with above experimental data provided convincing evidence that two factors are mainly responsible for stabilizing the collagen triple helix: (1) the repetitive network of interchain Gly NH/Xaa Co hydrogen bonds and (2) the φ dihedral angle limitation by the pyrrolidine ring structure. Detailed comparative analyses of simple derivatives of the amino acids Pro (**1**), (4R)-Hyp (**2**), and (4S)-Hyp (**3**) as well as (3S)-Hyp (**4**) and (3R)-Hyp (**5**) (Fig. 11.1) clearly revealed the gauche effect of (4R)-Hyp as responsible in the Yaa position for favoring the C$^\gamma$-*exo* pucker of the pyrrolidine ring, while in the Xaa position this residue would disfavor the C$^\gamma$-endo pucker (Fig. 11.2). Concurrently, the stereoinductive effects of 4R-hydoxylation stabilize the *trans* Xaa-Hyp peptide bond conformation and strengthen the interchain NH/CO hydrogen bond [15,32–38]. Similarly, main chain dihedral angles of (3S)-Hyp are favorable for the Xaa position, while in Yaa position this residue would be destabilizing. Moreover, compared with a proline residue the effect of (3S)-Hyp on the peptide bond isomerization is negligible [21].

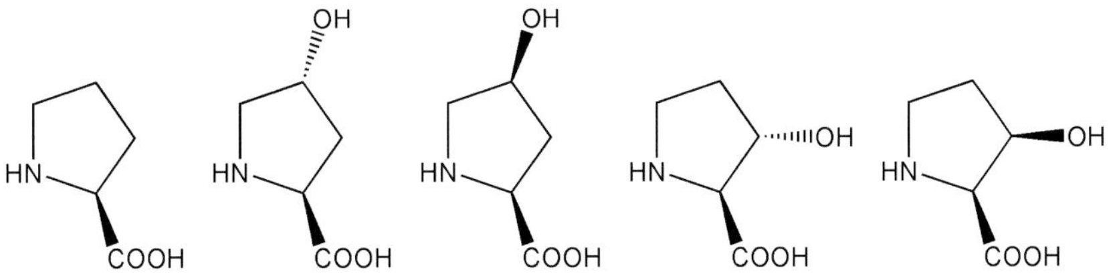

Fig. 11.1 Structures of mono- and dihydroxylated prolines: (**1**) (2S)-proline (Pro), (**2**) (2S,4R)-4-hydroxyproline ((4R)-Hyp), (**3**) (2S,4S)-4-hydroxyproline ((4S)-Hyp), (**4**) (2S,3S)-3-hydroxyproline ((3S)-Hyp), (**5**) (2S,3R)-3-hydroxyproline ((3R)-Hyp), (**6**) 2,3-*trans*-3,4-*cis*-3,4-dihydroxyproline ((3R,4S)-Dhp), and (**7**) 2,3-*cis*-3,4-*trans*-3,4-hydroxyproline ((3S,4S)-Dhp).

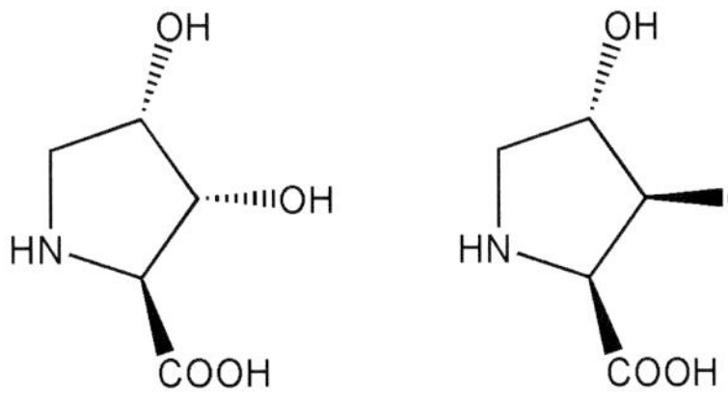

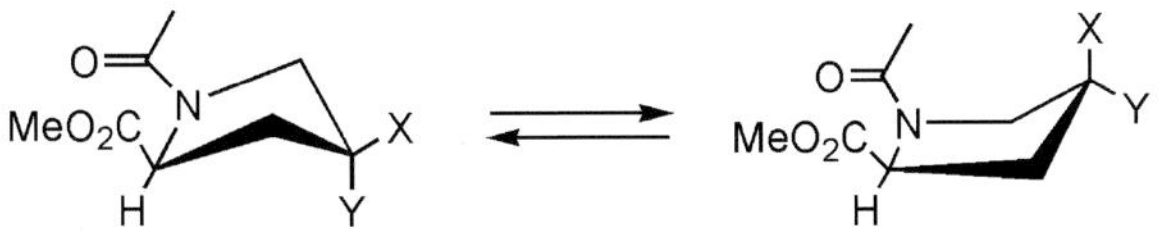

Fig. 11.2 Ring puckers in 4-substituted Ac-Pro-OMe. The C$^\gamma$-endo pucker (right) is favored when X = H, OH, or F and Y = H. The C$^\gamma$-exo pucker (left) is favored when X = H and Y = OH or F (for discussion on the fluorine derivatives see Sections 11.2.1 and 11.2.3).

These conclusions were drawn from the X-ray structures of some of the hydroxylated prolines and related *N*-acetyl methyl ester derivatives [39,40] as well as from the thermodynamic parameters obtained by NMR conformational analysis of the model compounds Ac-Pro*-OMe, with Pro* corresponding to 1–4 [15,21,32], and Ac-Phe-Pro*-NHMe, with Pro* being 1–5 (Tables 11.1 and 11.2) [41]. While *O*-acetylation or *O*-trifluoroacetylation of the 4-hydroxyproline increases the inductive effects [40], and thus stabilizes the C$^\gamma$-*exo* pucker and *trans*-conformation of aminoacyl-Hyp bonds, unfavorable steric interactions reduce these effects in terms of triple-helix induction and stability [40]. In addition, Table 11.2 reports the thermodynamic parameters of (3*R*,4*S*)-Dhp (**6**) which displays a similar behavior

Table 11.1 Thermodynamic parameters of Ac-Pro*-OMe with Pro* = **1–4**.

Compound	$K_{t/c}$ (D$_2$O, 25 °C)	ΔG at 25 °C[a] (kJ mol^{-1})	ΔH[b] (kJ mol^{-1})	ΔS[b] (J mol^{-1} K^{-1})
Ac-Pro-OMe	4.6	−3.8	−4.2	−2.1
Ac-(4*R*)-Hyp-OMe	6.1	−4.6	−8.0	−10.9
Ac-(4*S*)-Hyp-OMe	2.4	−2.1	−	−
Ac-(3*S*)-Hyp-OMe	4.9	−3.8	−	−

a Calculated by using $\Delta G = -RT \ln K$.
b Extracted from Van't Hoff plots using $\Delta G = \Delta H - T\Delta S$.

Table 11.2 Thermodynamic Parameters of Ac-Phe-Pro*-MHMe with Pro* = **1–7**.

Compound	$K_{t/c}$[a] D$_2$O, 25 °C	ΔG at 25 °C[b] (kJ mol^{-1})	$\Delta\delta H$[c] at 25 °C (ppm)	ΔH[d] (kJ mol^{-1})	ΔS[d] (J mol^{-1} K^{-1})
Ac-Phe-Pro-NHMe	2.1	−1.8	−0.73	+1.1	+9.6
Ac-Phe-(4*R*)-Hyp-NHMe	5.0	−4.0	−0.53	−1.0	+9.2
Ac-Phe-(4*S*)-Hyp-NHMe	1.9	−1.6	−0.15	+5.1	+22.6
Ac-Phe-(3*S*)-Hyp-NHMe	2.4	−2.2	−0.44	+2.7	+16.3
Ac-Phe-(3*R*)-Hyp-NHMe	2.2	−2.0	−0.81	+2.0	+13.4
Ac-Phe-(3*R*,4*S*)-Dhp-NHMe	5.6	−4.3	−0.27	−1.7	+8.4
Ac-Phe-(3*S*,4*S*)-Dhp-NHMe	2.5	−2.3	−0.63	+3.4	+19.3

a Determined by integration of well-resolved signals in the ^{1}H-NMR spectrum.
b Calculated by using $\Delta G = -RT \ln K$.
c Chemical shift difference δH_a (*trans*) − δH_a (*cis*).
d Extracted from Van't Hoff plots using $\Delta G = \Delta H - T\Delta S$.

to that of (4R)-Hyp, with a stronger preference for the C^γ-*exo* pucker because of the superior alignment of backbone dihedral angles to stabilize the *trans* peptide bond and thus enhance the *trans/cis* ratio [41]. The additional gauche interaction and the extra electron-withdrawing capabilities of the 3-hydroxyl group account for the similar but enhanced behavior of this proline analog relative to (4R)-Hyp. As expected, the opposite effects were observed for (3S,4S)-Dhp (7).

In the context of the collagen triple helix it can be concluded that the C^γ-*exo* pucker of the proline in Yaa position is favored by stereoelectronic effects of a (4R)-OH substituent which also stabilizes the *trans*-Xaa-(4R)-Hyp peptide bond, thus preorganizing this residue in a conformation that best befits a triple helix. Conversely, in the Xaa position the Pro residue is preferred in the C^γ-*endo* pucker; the related dihedral angles are reported in Table 11.3.

Table 11.3 Values of φ and ψ dihedral angles for the *trans*-amide conformers of Ac-Pro*-OMe with Pro* = Pro, (4R)-Hyp, (4R)-FPro and (4S)-FPro [34,39].

Pro*	Ring pucker	φ (deg)	ψ (deg)
Pro	C^γ-*exo*	−58.6	143.0
Pro	C^γ-*endo*	−70.0	152.1
(4R)-Hyp[a]	C^γ-*exo*	−62.0	156.4
(4R)-Hyp[a]	C^γ-*exo*	−50.9	145.2
(4R)-FPro	C^γ-*exo*	−59.2	140.8
(4S)-FPro	C^γ-*endo*	−76.4	169.0

a Ac-(4R)-Hyp-OMe contains two forms in the crystal unit cell [39].

11.2.2
Mercaptoproline

The (4R)- and (4S)-mercaptoproline diastereoisomers, as chalcogen analogs of the related hydroxyprolines, were synthesized [42–44] and used sporadically for side-chain cyclizations of peptides via thioether or disulfide bridges to restrict the conformational space of peptidic macrocycles [45–47]. However, a possible effect of the thiol group on the conformational preferences of the pyrrolidine ring has not been investigated to date. Compared with the hydroxyl group, the effect of a thiol group is expected to be significantly reduced.

11.2.3
Halogenated Prolines

The occurrence of halogen-substituted prolines in natural products has not been reported, but 4-fluoro-, 4-chloro-, and 4-bromoprolines are all incorporated into natural compounds by microorganisms when applied to the culture medium [1]. Among halogenated prolines, fluoroproline (4-FPro) has attracted early attention for its direct incoporation into procollagens by matrix-free connective tissue cells [48,49]. Although incorporation was successful, the use of (4S)-FPro prevented the expressed collagen chains from folding into stable triple helices, a fact in full agreement with the present-day knowledge in the field (*vide infra*).

A comparative analysis of the fluorine-substituents at C4 or C3 of the pyrrolidine ring as *cis* and *trans* isomers (**8, 9, 11**, and **12** in Fig. 11.3) [15,32,50] as well as of the difluoro derivative **10** in model compounds such as Ac-FPro-OMe [50–52] confirmed the expected effect of a stronger electron withdrawal both in terms of *γ-exo* and *γ-endo* puckering of the 5-membered ring and of *cis/trans* isomerization of the acyl-FPro imide bond, as shown in Tables 11.3 and 11.4.

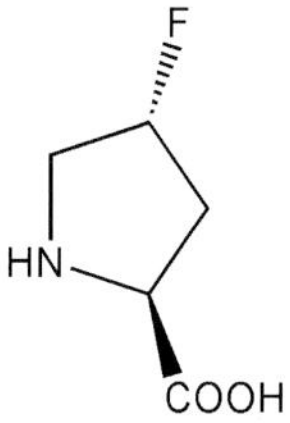

(**8**) 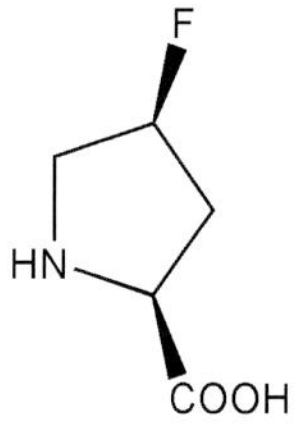   (**9**) 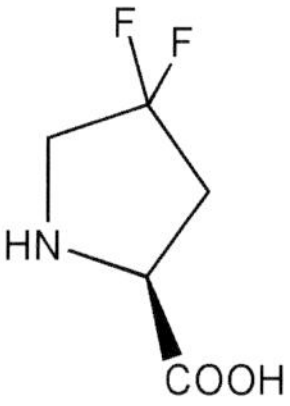   (**10**)

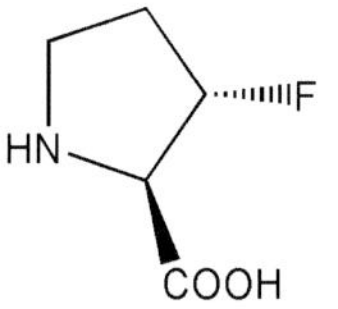

(**11**) 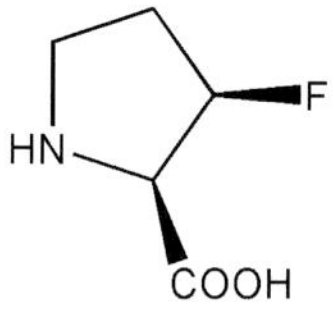   (**12**)

Fig. 11.3 Structures of fluoroprolines: (**8**) (2S,4R)-4-fluoroproline ((4R)-FPro), (**9**) (2S,4S)-4-fluoroproline ((4S)-FPro), (**10**) (2S)-4,4-difluoroproline (4,4-F₂Pro), (**11**) (2R,3S)-3-fluoroproline ((3S)-FPro), and (**12**) (2R,3R)-3-fluoroproline ((3R)-FPro).

Table 11.4 Activation enthalpy ($\Delta H^{\neq}$) and entropy ($\Delta S^{\neq}$) as derived from Eyring plots. Error limits obtained from the residuals of the linear least-squares fitting were 1–2 % for entry 1, 3, and 4, and 2–3% for entry 2. Additionally, the free energy of activation ΔG^* at 27 °C is given.

Compounds	trans→cis			cis→trans			$K_{t/c}$
	$\Delta H^{\neq}$ (kJ mol^{-1})	$\Delta S^{\neq}$ (J mol^{-1} K^{-1})	$\Delta G^{\neq}$ (kJ mol^{-1})	$\Delta H^{\neq}$ (kJ mol^{-1})	$\Delta S^{\neq}$ (J mol^{-1} K^{-1})	$\Delta G^{\neq}$ (kJ mol^{-1})	
Ac-Pro-OMe	92.3	12.8	88.4	87.2	9.00	84.5	4.7
Ac-(4R)FPro-OMe	91.8	14.8	87.4	84.2	5.37	82.6	6.8
Ac-(4S)-FPro-OMe	87.5	4.17	86.3	84.7	2.42	84.0	2.5
Ac-4,4-F$_2$Pro-OMe	103.7	66.38	83.8	98.5	59.11	80.8	3.3

Introduction of fluorine at C4 as a *trans* isomer strongly favors the C$^{\gamma}$-*exo* pucker by gauche effect and decreases the energy barrier to isomerization by affecting the pyramidalization of the pyrrolidine nitrogen (Fig. 11.2). As a consequence, (4R)-FPro stabilizes the *trans* acyl-FPro conformation, while (4S)-FPro favors the *cis* conformation. These properties are fully reflected by the hyperstability of the triple helical structure of collagenous peptides consisting of (Pro-(4R)-FPro-Gly) triplets [35,36], while the opposite effect is exerted by replacing (4R)-FPro with (4S)-FPro in this Yaa position [32]. Since fluorine can only form very weak hydrogen bonds [53], with this result the hypothesis of collagen stabilization by a water-bridged hydrogen bonding network was fully contradicted [35,36], a fact that agrees with the previously observed stability of triple helices in alcohols (i.e. in the absence of water) [11,54]. By favoring the C$^{\gamma}$-*exo* pucker of the pyrrolidine ring in the Yaa position and the ω main-chain dihedral angle for the *trans* conformation ($\omega = 180°$) with (4R)-FPro [15,32,34,50], the entropic penalty of triple-helix formation is decreased to larger extents than with (4R)-Hyp. In the latter case some entropy loss may additionally derive by ordering of water molecules around the 4-hydroxyl group [55]. As expected from the results with (4S)-Hyp in Yaa position, even with (4S)-FPro collagenous peptides do not associate into homotrimeric triple helices [32]. Conversely, (4S)-FPro was expected to stabilize the triple helix by favoring the C$^{\gamma}$-*endo* pucker without steric clashes in the Xaa position [28]. Indeed, peptides of the type ((4S)-FPro-Pro-Gly)$_7$ [56] and ((4S)-FPro-Pro-Gly)$_{10}$ [55,57] assume a collagen structure of significant thermal stability [56], while in combination with (4R)-Hyp in the Yaa position (i.e. in a homotrimeric peptide consisting of ((4S)-FPro-(4R)-Hyp-Gly)$_5$ which was even C-terminally crosslinked by a cystine knot to reduce the entropy loss of self-assembly into homotrimers) such a fold was not detected even at low temperatures (4 °C) [58]. This would suggest that interpretation of experimental data of the proline analogs in collagens may not be simply reduced to a favored puckering of the pyrrolidine rings and *trans* conformation of the aminoacyl-Pro bonds, but has to account for multifactorial aspects including the water shell, despite its fast exchange process [59].

In contrast to the multiple substitutions of proline residues in the repeating sequences of collagens, single replacements of proline residues in proteins may be more informative in terms of *cis/trans* peptide bond isomerization affecting the overall fold. In fact, by replacing the proline-46 in the barstar variant Cys40Ala/Cys82Ala/Pro27Ala, which is in *cis* conformation, by (4S)-FPro or (4R)-FPro, the thermal stability was markedly enhanced (T_m = 69.4 °C) and decreased (T_m = 61.5 °C), respectively, if compared with the parent protein (T_m = 64.3 °C) [50]. As in the case of barstar, the cysteine-rich N-terminal domain of minicollagen contains a type VI β-turn with *cis*-Ala-Pro as central residues [60]. By replacing this Pro residue with (4S)-FPro or (4R)-FPro significantly enhanced and reduced oxidative folding rates were observed as a result of the lowered and increased activation energy for the *trans*-to-*cis* isomerization of Ala-(4S)-FPro and Ala-(4R)-FPro, respectively [61]. In addition, catalysis of CTI of Suc-Ala-Ser-Xaa-Phe-*p*NA (Xaa = (4R)-, (4S)-FPro and 4,4-F$_2$Pro) as well as of the quadruple variants of barstar with Pro48/(4R)-FPro, Pro48/(4S)-FPro, and Pro48/4,4-F$_2$Pro replacements, when used as substrates of PPIases, showed stereospecificity for monofluoro-substitution at the 4-position of the pyrrolidine ring. These replacements did not impair productive interactions with the majority of PPIases analyzed, although clearly revealing considerable differences for the activation of the catalytic machinery among members of a particular family of PPIases [62]. Similar studies with the substrates Suc-Ala-Ala-Xaa-Phe-*p*NA with Xaa = 3- and 4-fluoroprolines revealed that these substitutions led to weaker substrates for cyclophilin (hCyp-18) by one order of magnitude relative to the Pro-containing peptide except for the (4R)-FPro analog [51,52].

11.2.4
Other Proline Analogs

Among other proline analogs such as **13–17** shown in Fig. 11.4, NMR analysis of the *N*-acetyl derivatives of **13–16** in neutral solutions did not reveal any significant difference in the *trans/cis* ratio compared with *N*-acetyl-proline except for 5-oxo-proline (**15**) which exhibits an exclusive *trans*-amide conformation [63]. However, when replacing proline in Ala-Pro-*p*NA with Aze (**13**) and Pip (**14**) the *cis*-amide content was 20% and 13%, respectively, compared with 6% for the Pro-peptide. Rates of CTI were accelerated 14.5- and 49-fold relative to the Pro-peptide [64].

Because of the similar size of amino and hydroxyl groups, but the higher basicity of an amino group (4R)-Amp (**17**) has been synthesized and used for analyzing comparatively in pH dependency the triple-helix stability of the collagen model peptides (Pro-Yaa-Gly)$_7$ with Yaa = (4R)-Hyp and (4R)-Amp [65]. Upon end-capping of the peptides to avoid endgroup effects [66], the (4R)-Amp-peptide formed a triple helix which at both pH 3.0 and 12.0 was more stable than the related (4R)-Hyp peptide. In the protonated state the 4-amino group remarkably enhances the thermal stability when compared with the deprotonated form. The stereoelectro-

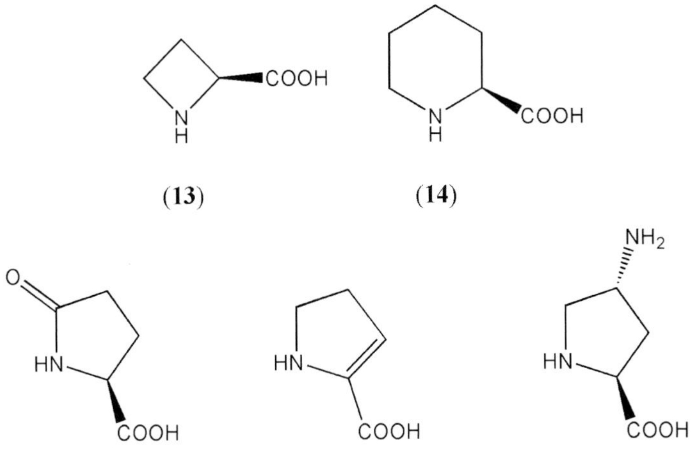

Fig. 11.4 Structures of (2S)-azetidine-2-carboxylic acis (Aze, **13**), (2S)-piperidine-2-carboxylic acid (Pip, **14**), (2S)-5-oxoproline (**15**), 2,3-dehydroproline (**16**), (2S,4R)-4-aminoproline ((4R)-Amp, **17**).

nic effects of this proline derivative on the peptide backbone dihedral angles and the role of electrostatic interactions have not been analyzed in detail.

11.2.5
Alkylated Proline Analogs

A great variety of alkylated proline derivatives are found in natural products from microorganisms and others were prepared synthetically [1,2]. Among these the cyclic imino acids **18–22** (Fig. 11.5) are the most interesting in terms of preferential *trans* or *cis* conformation of the aminoacyl-proline bonds. While for **22** as N-acetyl N-methylamide derivative only a *trans* amide bond was detected by NMR in aqueous solution, both the *anti*- (**18**) and *syn*-3-methylproline (**19**) showed only slightly differing *cis/trans* ratios compared with the related proline derivative [67]. Conversely, the Tbp (**20**) [68,69] was found to increase the *cis*-amide bond population in the N-acetyl N-methylamide derivative and particularly in Ac-Xaa-Tbp-NHMe dipeptides when enhanced bulkiness of the Xaa side-chain leads to steric interactions with the *tert*-butyl group (e.g. for Xaa = Ala 79% and for Xaa = Phe 90% *cis*-amide isomer were determined) [70–73]. The bulky 5-*tert*-butyl substituent lowers the barrier for amide isomerization by 15.5 kJ mol^{-1} compared with N-acetyl-Pro-NHMe. The acceleration of amide isomerization was attributed in part to ground-state destabilization resulting from the bulky substituent skewing the amide bond away from planarity [74]. This property has been successfully exploited to stabilize β-turns of the type VIa and VIb in bioactive peptides [75]. In similar manner, incorporation of the bulky *tert*-butyl group at C6 of pipecolic acid augmented the population of the *cis*-amide isomer and lowered the energy barrier for isomerization of aminoacyl-pipecolic acid amide, but to lower extents than in the case of the related proline derivative [76].

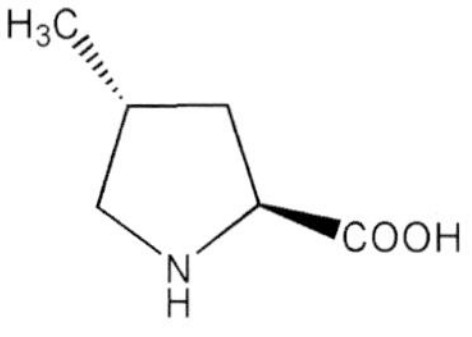

Fig. 11.5 Structures of alkylated proline derivatives:
(**18**) (2S,4R)-4-methylproline, (**19**) (2S,4S)-4-methylproline,
(**20**) (2S,5R)-5-*tert*-butylproline (Tbp), (**21**) (2S)-5,5-dimethyl-
proline (Dmp), and (**22**) (2S)-2-methylproline.

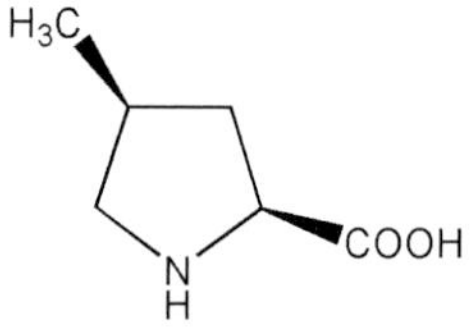

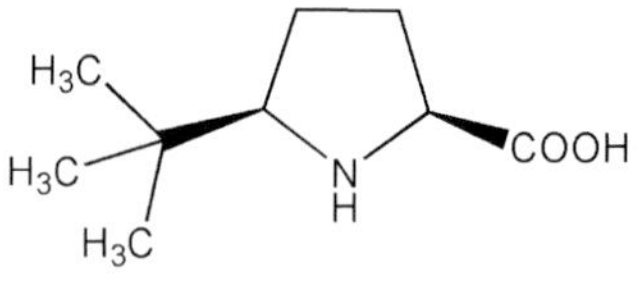

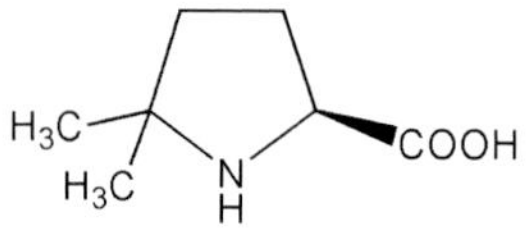

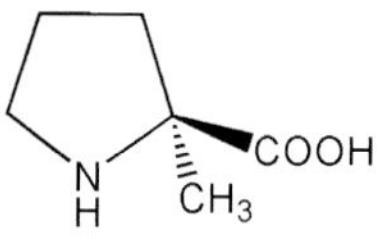

Based on early reports that 2,2-dimethylthiazolidine carboxylic acid gives rise exclusively to the *cis*-amide bond isomer in a model peptide [77], the (2S)-5,5-dimethylproline (Dmp, **21**) was synthesized and analyzed for its propensity to stabilize the *cis*-isomeric state [78,79]. For Boc-Phe-Dmp-OMe and Ac-Tyr-Dmp-Asn-OH only a *cis* conformer was detected, while for Ac-Asn-Dmp-Tyr-OH the high content of *cis*-amide bond was found to slightly decrease in function of temperature from 90.3% at 6 °C to 78.9% at 80 °C. This allowed the extraction of the thermodynamic parameters for the CTI (Table 11.5). In contrast to (2S)-5,5-dimethyl-proline, the (2S)-3,3-dimethylproline leads only to a slight increase of the *cis* isomer population of the related N-acetyl N-methylamide derivative compared with proline. However, the rates of *cis*-to-*trans* and *trans*-to-*cis* isomerization were strongly reduced, a fact that was attributed to steric interactions between the 3-methyl groups and the C-terminal amide that restrict the values away from $\varphi \approx 0°$ [80].

Table 11.5 Thermodynamic parameters determined for the CTI of a Dmp-tripeptide at 25 °C and the related activation energy [79]. For comparison the parameters of Ac-Pro-OMe are reported [50].

Compound	$\Delta G°$ (kJ mol^{-1})	$\Delta H°$ (kJ mol^{-1})	$\Delta S°$ (J mol^{-1} deg^{-1})	$\Delta G^{\neq}_{ct}$ (kJ mol^{-1})
Ac-Asn-Dmp-Tyr-OH	4.82 ± 0.04	10.1 ± 0.04	16.8 ± 0.1	54 ± 4
Ac-Pro-OMe	3.90 ± 0.04	5.03 ± 0.05	3.81 ± 0.04	85 ± 1

The different propensities of the alkylated proline analogs Tbp and Dmp for the *cis* conformation were recently exploited to modulate the activation of 5-hydroxy-tryptamine type 3 receptor upon site-specific incorporation of these residues at the place of a critical proline residue. A direct comparison of Pro, Pip, Aze, Tbp, and Dmp at a defined position of the loop between the second and third trans-membrane helices allowed to correlate directly the intrinsic *cis*-to-*trans* energy gap of these residues with the activation of the channel [81].

11.2.6
Bridged Bicyclic Proline Analogs

Among the naturally occurring and synthetic bridged bicyclic proline analogs **22–28** [82–95], compounds **24** [87,96] and **25** [86,97,98] have been analyzed for their effects on the amide bond conformation (Fig. 11.6). For example, in model Ac-2,4-MePro-NHMe or Ac-Tyr-2,4-MePro-NHMe the *trans* conformation of the peptide bond preceding the bridged proline analog is strongly stabilized (≥95%) over the *cis* conformation, making this proline mimetic well suited for a selective stabilization of *trans* peptide bonds. This property has been exploited in the design of TRH [99], bradykinin, and angiotensin II analogs [100]. Besides the 4,5-methano-proline diastereoisomers **27–28** [89], 2,3- and 3,4-methanoprolines were also synthesized [101–103], but so far these proline analogs have not been investigated in their conformational preferences. Of the 3,4-methanoproline a related 3,4-(ami-nomethano)proline derivative has been synthesized in enantiomerically pure form and suitably protected for incorporation into peptides as mimic of a lysine residue [104]. An additional interesting proline analog is the 3,4-fulleroproline (**30**) which shows a preference for the *trans*-amide state in organic solvent as *N*-acetyl *tert*-butyl ester, when compared with Ac-Pro-OtBu, and a reduced activation energy by about 30 kJ mol^{-1}. This is due to the electron withdrawing effect of the fullerene moiety [105].

More recently, 5-hydroxy and 5-fluoro derivatives of the 3,5-methanoproline (**26**) were analyzed comparatively as *N*-acetyl methyl esters (Fig. 11.7) by NMR and in the crystalline state for their preferred conformations [90]. In 3,5-methanoproline both pyrrolidine ring puckers are incorporated into the same framework (Fig. 11.7), which allows the dissection of the relative contributions of ring pucker and inductive effects to the conformation of the proline compounds. The hydroxy

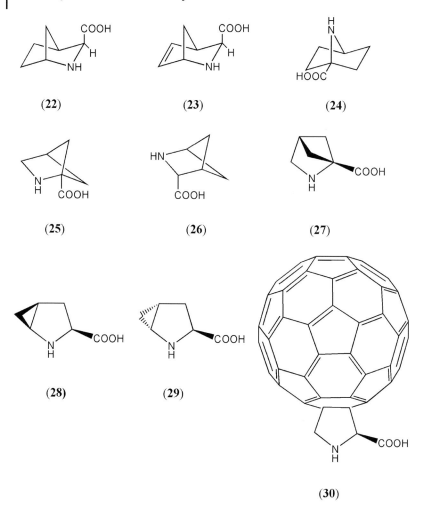

Fig. 11.6 Structures of bridged bicyclic proline mimetics: (**22**) 2-aza-bicyclo[2.2.1]heptane-3-carboxylic acid, (**23**) 2-aza-bicyclo[2.2.1]hept-5-ene-3-carboxylic acid, (**24**) 7-aza-bicyclo[2.2.1]heptane-1-carboxylic acid, (**25**) 2-aza-bicyclo[2.2.1]hexane-1 carboxylic acid, (**26**) 2-aza-bicyclo[2.2.1]hexane-3arboxylic acid, (**27**) (1 S,4S)-2-aza-bicyclo[2.1.1]hexane-1-carboxylic acid (2,4-methanoproline), (**28**) (1 S,3 S,5 S)-2-aza-bicyclo[3.1.0]hexane-3-carboxylic acid, (**29**) (1 R,3 S,5 R)-2-aza-bicyclo[3.1.0]hexane-3-carboxylic acid; (**30**) 3,4-fulleroproline.

and fluoro derivatives of 3,5-methanoproline are analogs of (4S)-hydroxyproline and (4S)-fluoroproline, respectively, with substituents in an orientation that disfavors the C$^\gamma$-exo pucker, as shown in Fig. 11.2 and discussed in Sections 11.1.2 and 11.1.3. The *cis/trans* ratios of compounds **31** (3.5), **32** (3.6), and **33** (3.5) are very similar, and in water somewhat higher than in organic solvents [90]. These values are intermediate between those of Ac-Pro-OMe and Ac-(4S)-Hyp-OMe and

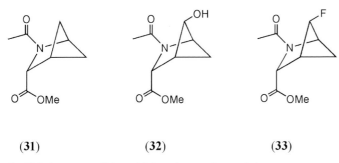

(31)   (32)   (33)

Fig. 11.7 Structures of N-acetyl-3,5-methanoproline methyl ester (**31**) and related 4-hydroxy (**32**) and 4-fluoro (**33**) derivatives of defined stereochemistry.

(4S)-FPro-OMe (Table 11.1) and demonstrate that rigidifying the pyrrolidine ring with a methano bridge abolishes any inductive effect exerted by the fluoro and hydroxy substituent on the *cis/trans* ratio of its peptide bond [15,32,50].

11.2.7
Locked Proline Mimetics

Loops, and in particular β-turns, play an important role in protein folding and recognition. Therefore considerable effort has been focussed on developing conformationally restricted mimics of the backbone geometry, intramolecular hydrogen bonding, and side-chain orientations exhibited by these secondary structure motifs [106,107]. Proline is often encountered in chain reversals of polypeptides with formation of β- or γ-turn structural motifs as shown in Fig. 11.8, but also capping of helices by a proline residue occurs frequently. The type of turn depends upon the sequence position of the proline as well as on the *trans* or *cis* conformation of the aminoacyl-Pro peptide bond. Because of the conformational restriction of the φ dihedral angle of proline, this residue frequently occupies the position i+1 of type I and II β-turns and the position i+2 of type VIa and VIb β-turns in peptides and proteins of ribosomal origin, while in peptides resulting from enzymatic synthesis proline is often replaced by N-methylated amino acids [108,109]. The type VI β-turn is a unique secondary structure element as it features an amide *cis*-isomer N-terminal to the proline at the i+2 position of the peptide bend [108,110]. In the type VIa the proline ψ dihedral angle is near 0° and an intramolecular hydrogen bond exists between the CO of the i and NH of i+3 residue. In type VIb the proline ψ angle is around 150° and thus an intramolecular hydrogen bond cannot be formed. While in peptides, particularly cyclic peptides from natural sources, biologically relevant CTIs are taking place, loops in proteins are generally conformationally homogeneous and therefore a higher tendency for the type VIa β-turn is observed [108,110].

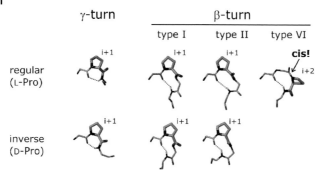

Fig. 11.8 Regular turns containing proline. The dotted line indicates the stabilizing and characteristic hydrogen bonds (i,i+2 for γ-turns; i,i+3 for β-turns). Note the *cis*-prolyl configuration required for the type VI β-turn.

With the use of modified prolines such as those discussed in the preceding sections, modulation of such structural motifs has often been successful.

In first approaches to peptidomimetic structures capable of inducing and stabilizing the various types of turns, rigid functional spacers were proposed that simulate a pair of residues preferring i+1 and i+2 positions of reverse-turn peptides [111–113]. In subsequent studies attention has been paid to retaining the correct peptide raster framework of repeating N–C_a–CO units with defined angular and torsional parameters for isosteric replacement of dipeptide units in addition to the important 10-membered intramolecular hydrogen bond that orients the two peptide extensions from each end. The main design concept relies on the covalent rigidification of the amino acid side-chains on the dipeptide backbone, which leads to an 1-azaoxobicycloalkane skeleton as a general template (Fig. 11.9). From this general framework that incorporates a complete dipeptide unit with one carboxyl "end," a central amide group, and an "amino" end, a variety of structures have been developed synthetically, which often raised serious problems of diastereoselectivity at the chiral centers of the ring-fusion center Z as well as at the amino or carboxyl terminus, respectively [106]. All these azabicycloalkane amino acids, which may contain additional heteroatoms (X and Y) and differ in the ring sizes of the fused bicyclic framework, can be considered as surrogates of Xaa-Pro as *trans*-amide isomer (Fig. 11.10).

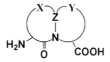

Fig. 11.9 General framework of dipeptide mimics with a *trans*-amide conformation; X and Y represent possible heteroatoms in the fused rings which can vary in the size and bear additional substitutents to mimic the amino acid side-chains. A stereocontrol of the ring-fusion center Z is difficult to achieve.

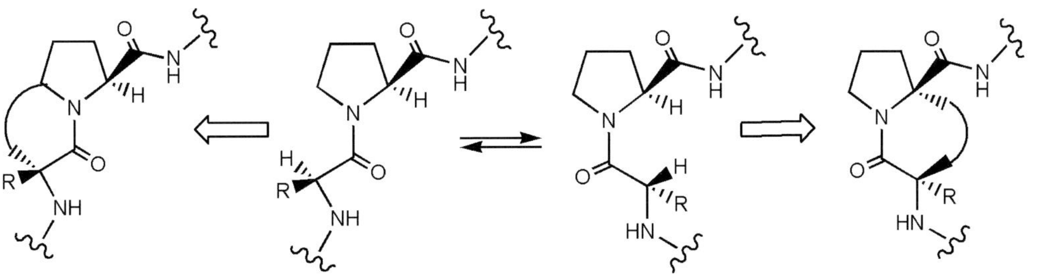

Fig. 11.10 Mimetics of the *trans* Xaa-Pro (left) and *cis* Xaa-Pro (right).

Despite the difficulty in controlling the stereochemistry of the ring-fusion center as well as the low yields of the multistep synthesis, which lead to serious drawbacks of these strategies, a large variety of such bicyclic compounds could be synthesized that mimic the central dipeptide of β-turns in general and in particular the Xaa-Pro dipeptide as *trans* isomer [106,114–118]. A few examples of such dipeptide mimetics are shown in Fig. 11.11.

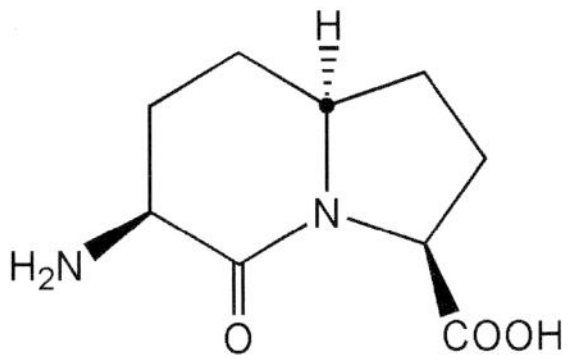

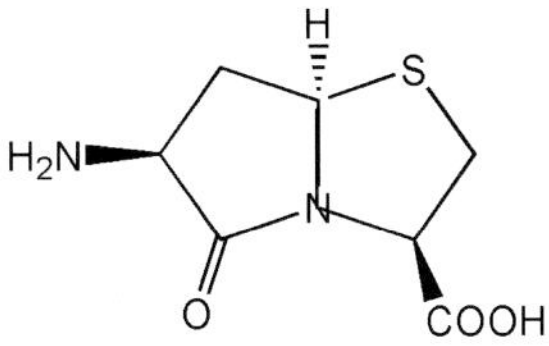

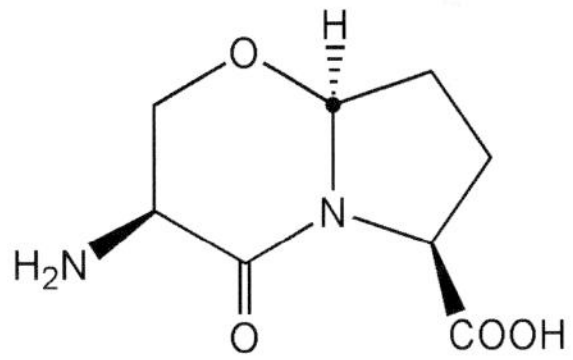

Fig. 11.11 Mimetic structures of Xaa-Pro with a *trans* amide bond (**34** [119], **35** [120], **36** [121], **37** [122], **38** [123]) and of Pro-Xaa (**39** [124]).

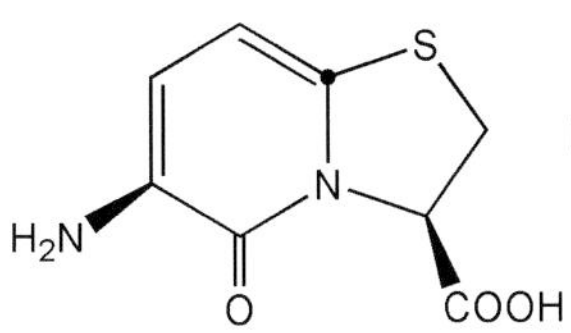

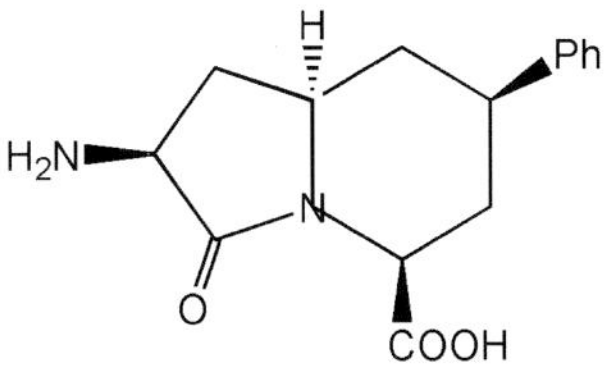

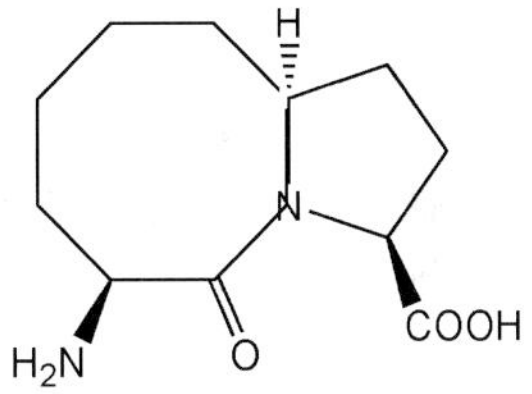

In the case of the type VI β-turn competent replacements for the backbone geometry of the central residues with a *cis*-amide bond are obtained by tethering the C_a of the N-terminal amino acid residue to the proline C_a in a dipeptide lactame (Fig. 11.10). Again azabicycloalkane amino acids are obtained which may contain additional heteroatoms and differ in the ring sizes of the fused bicyclic framework. Selected examples are shown in Fig. 11.12.

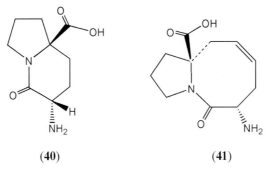

(40)   (41)

Fig. 11.12 Two examples of *cis* Xaa-Pro mimetics: **(40)** (6S,8aR)-6-amino-5-oxo-octahydroindolizine-8a-carboxylic acid was obtained in in good yields and enantiomerically homogeneous form [125–129]. The compound (S)-1-((S,Z)-2-aminohept-4-enoyl)pyrrolidine-carboxylic acid **(41)** was derived from MD simulations and synthesized as enantiomerically pure compound by Grubb's ring-closing olefin metathesis [130].

11.3
Pseudoprolines in Chemical Synthesis and Biology

11.3.1
From Proline to Pseudoproline

The cyclic nature of the proline residue confers unique conformational properties to the peptide or protein backbone when compared with the common proteinogenic amino acids. For instance, on the one hand, the proline ring serves to intrinsically restrict its Φ dihedral angle around $-60° \pm 15$, while on the other hand, the imidic bond formed with the preceding residue (Xaa-Pro) is readily subject to CTI. The CTI of the Xaa-Pro tertiary bond is characterized by relaxation times from ten to hundreds of seconds at room temperature and is inherently accompanied by a dramatic change of the backbone direction by 180° [131]. Accordingly, proline residues are usually encountered in loop or turn motifs with the utmost preference at the i+1 position for β-turn type II when the Xaa-Pro imide bond is *trans* ($\omega_i = 180°$) or at the i+2 position of turn type VI ($\omega_{i+1} = 0$) in the *cis* form [132–135]. In addition, the characteristic low activation barrier for isomerization combined with the small free energy difference between the two Xaa-Pro peptide bond isomers provides a rationale for the putative role of proline in the limiting steps of the protein-folding pathway [136]. The prevalence of proline residues in biological processes such as protein folding and protein recognition has led to the development of numerous mimetics and substituted proline analogs intended to constrain and control the peptide backbone in reverse turn motifs or to alter the imide *cis-trans* ratio [137].

For example, substitution with β-alkyl proline results in a Ψ angle constrained by the *syn-β* substituent around 0° while (S)-α-methyl substituted proline induces preferentially the *trans* form of the Xaa-Pro bond due to unfavorable steric interactions between the ethyl group and the α-proton of Xaa in the *cis* form, whereas bulky substituents at C5 of the cyclic Pro system result in *cis*-imide bond formation [138–142]. However, the chemical synthesis and incorporation into peptide or protein backbones of these Pro-surrogates hampers this strategy to become a routinely applied tool in biomimetic chemistry. During our extensive work on the study of the relationship between conformational preference of a growing peptide chain and its impact upon its physical and chemical properties [143–147], the particular role of Pro in disrupting potentially secondary structure forming peptides has been noticed very early [148]. In particular, the finding of its tendency to prevent β-sheet conformations and aggregation, paralleled by a pronounced solubilizing effect [149,150], had a strong impact upon solid-phase synthesis. Most notably, our systematic studies in collaboration with Toniolo's group on the onset and disruption of β-sheet structures of oligopeptides set the stage for taking conformational effects upon solubility, solvation, and reaction kinetics in the synthesis of hydrophobic, self-associating peptides (later termed "difficult sequence" [151,152]) as major elements in designing strategies for peptide synthesis. Based on these observations in our ongoing work on peptidomimetics for tuning secondary structure formation, we have noticed that the temporary transformation of serine (Ser), threonine (Thr), and cysteine (Cys) residues into cyclic structures by established procedures [153–155] results in proline-like compounds, termed pseudoprolines, (ΨPro, Fig. 11.13) [156–161].

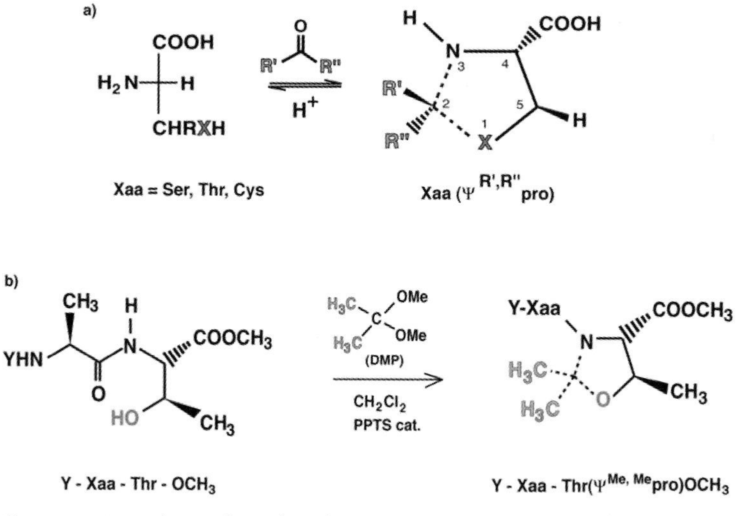

Fig. 11.13 (A) Synthesis of pseudoprolines (ΨPro) by condensation reaction of aldehydes or ketones with Xaa (Ser, Thr, or Cys). (B) Direct insertion ("post-insertion") of ΨPro systems into dipeptide derivatives for use in peptide synthesis [153–161].

The physical, chemical, and conformational properties of the resulting oxazolidine or thiazolidine systems prove to be strongly dependent upon the character of the C2 substituents of the cyclic ring. For example, as shown by a chymotrypsin-coupled assay and NMR studies using ΨPro-containing model peptides, the *cis*-content of Xaa$_{i-1}$-Pro$_i$ imide bonds can be tailored by varying the aldehyde or ketone component (i.e. R^1, R^2) in the cyclization reaction (Fig. 11.14) [162]. As the use of bulky substituents at C2 (e.g. dialkyl groups) induces up to 100% *cis* conformation, this technique served to target bioactive conformations in Pro-containing peptide ligands (see below). The CTI rate is remarkably increased compared to Pro, which is probably due to the increased length of the amide bond between Xaa and the pseudoproline. X-ray data of the C2 dimethylated thiazolidine Fmoc-Ala($\Psi^{Me,Me}$Pro)-OH revealed a distance of 1.39 Å for the tertiary amide bond $\omega 1$ between Ala and ΨPro, slightly longer than the tertiary amide bond in Ac-Ala-Pro-NHMe (1.345 Å) [163,164]. The twisted amide bond ($\omega_1 = -9.7°$) indicates steric complications due to the C2 methyl groups. This leads to a decrease of amide resonance due to out-of-plane distortion and can be an explanation for the observed increase of isomerization rate. Similarly, the chemical stability of ΨPro systems strongly depends on the electronic effects of the C2 substituents, ranging from highly acid labile (e.g. electron-donating substituents in oxazolidines) to completely acid stable (e.g. unsubstituted thiazolidines) derivatives (Fig. 11.2). C2 monoarylated ΨPro dipeptides (R_1 = Ph, R_2 = H, X = O) were found to be differential *cis*-inducers, depending on the stereochemistry at C2. The C2(S) stereoisomers exhibit 40–60% *cis* in organic solvents, while their C2(R) analogs are mainly in the *trans* conformation (~95%) [156–161,165–167]. In summary, modulation of the C2 substitution allows the *cis:trans* ratio to be tailored in Xaa-ΨPro units as well as its interconversion at will, offering a novel tool for studying peptide structure and function.

11.3.2
Synthesis of Pseudoprolines

The facile synthetic accessibility of ΨPro building blocks exhibiting differential chemical stabilities and *cis:trans* ratios allows us to mimic and even to enhance the unique properties of Pro in natural peptides and proteins, opening a broad palette of applications in bioorganic chemistry.

Pseudoprolines (ΨPro) are obtained by reacting the free amino acids Ser, Thr, or Cys with aldehydes or ketones [153–161] (Fig. 11.13).

The coupling of amino acid derivatives to a growing peptide chain containing N-terminal ΨPro generally results in low yields because of the sterically hindered nature of the oxazolidine (thiazolidine) ring system and the decreased nucleophilicity of the nitrogen atom. Consequently, the preformation of suitably protected dipeptide derivatives of the type Y-Xaa-ΨPro-OH is preferable for use in peptide synthesis [168–172]. Two conceptually different approaches have been elaborated for preparing the oxazolidine and thiazolidine ring-containing dipeptide derivatives: (1) the acylation of Ser-, Thr-, (Cys)-derived oxazolidines (thiazolidines)

a)

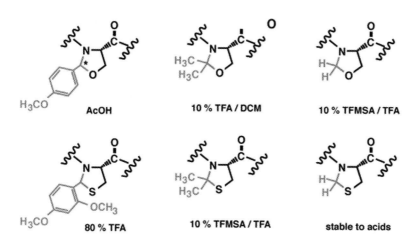

b)

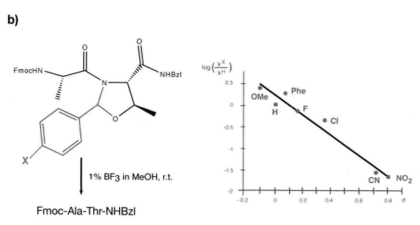

Fmoc-Ala-Thr-NHBzl

Fig. 11.14 (A) Stability of various ΨPro systems towards acids as used for the selective ring cleavage in peptide synthesis [153.161]. (B) Differential influence of paraphenyl substituents upon the cleavage kinetics by Lewis acids of C2-monosubstituted ΨPro systems in dipeptide Fmoc-Ala-Thr(ΨPro). The Hammett plot reveals a strong correlation between acid lability and inductive effect of substituent X [161].

using acid fluorides or N-carboxyanhydrides (UNCA); and (2) the direct conversion of Ser, Thr, Cys into the corresponding 5-membered ring systems by intraresidual N,O- or N,S-acetalization applying catalytic amounts of paratoluene-sulfonic acid (PTSA) [156–161].

In general, the incorporation of ΨPro systems into a growing peptide chain in solid-phase peptide synthesis (SPPS) proceeds preferentially via their preformed dipeptide derivatives of the type Fmoc-Xaa-ΨPro-OH. These commercially available ΨPro-containing building blocks [171] can be coupled without racemization according to standard procedures [168–172]. Oxazolidine- or thiazolidine-containing dipeptides present a significantly greater polar character than conventionally protected Ser, Thr, Cys derivatives due to the formation of hydrogen bonds to the solvent-exposed ring heteroatom O or S, similar to cyclic ethers. Consequently, ΨPro act as a polar protecting technique, contributing to higher solvation of protected peptides. In addition, ΨPro derivatives show remarkable thermodynamic stability and are stable over a long period of time at room temperature. ΨPro dipeptides are generally isolated as crystalline compounds and prove readily soluble in solvents used in peptide synthesis.

11.3.3
Pseudoprolines for the Synthesis of Difficult Sequences

So far, pseudoprolines have found applications in providing solubilizing, secondary structure-disrupting building blocks for the synthesis of difficult peptide sequences and for the reversible induction of *cis*-amide bonds into peptide backbones (Fig. 11.15).

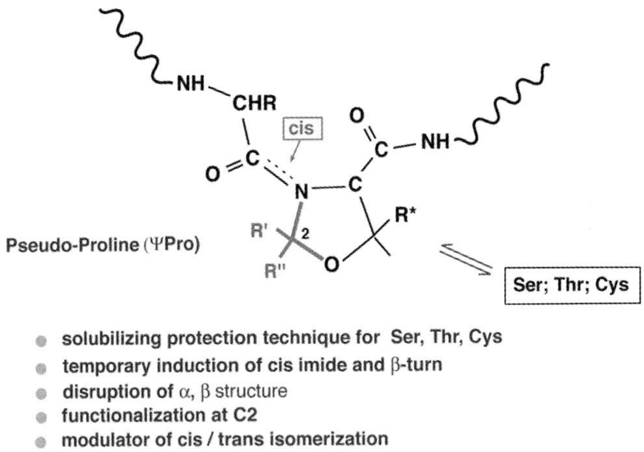

- solubilizing protection technique for Ser, Thr, Cys
- temporary induction of cis imide and β-turn
- disruption of α, β structure
- functionalization at C2
- modulator of cis / trans isomerization

Fig. 11.15 The concept of pseudoprolines (ΨPro): The temporary insertion of a Ser-, Thr-, or Cys-derived ΨPro system results in a "kink-conformation," giving access to a number of applications in peptide synthesis and bioorganic chemistry.

Here, the use of ΨPro systems as a temporary protection technique for Ser-, Thr-, or Cys-containing peptides proved to strongly improve the yields and purity of otherwise difficult sequences as shown by a steadily increasing number of comparative studies [156–162,168–181]. Most notably, the use of C2 dimethyl substituted ΨPro building blocks [168–172] results in the formation of a *cis*-amide bond in the regular peptide backbone ("kink conformation"), thus preventing β-sheet formation and self-aggregation during peptide synthesis. As most native peptides contain one or several Ser, Thr, or Cys residues, the ΨPro strategy appears particularly appealing for the chemical synthesis of long peptides. In standard Fmoc strategy [178], the final cleavage of the target peptide from an acid labile resin results in a completely deprotected peptide including ΨPro ring opening (i.e. transformation of the ΨPro units to the corresponding "native" Ser, Thr, or Cys). For preserving the solubilizing, structure-disrupting effect of ΨPro, alternative protection schemes have to be applied. The use of orthogonal protection techniques (e.g. allyl/aloc side-chain protection, super acid labile linkers) gives access to ΨPro-containing, side-chain-unprotected (water-soluble) peptides, thus strongly facilitating the chemical synthesis of long peptides or even proteins by modern chemoslective ligation techniques [179–181].

11.3.4
Pseudoprolines in Bioactive Peptides

More recently, the ΨPro concept was extended to targeting *cis*-amide bonds in biologically relevant recognition processes (Fig. 11.16).

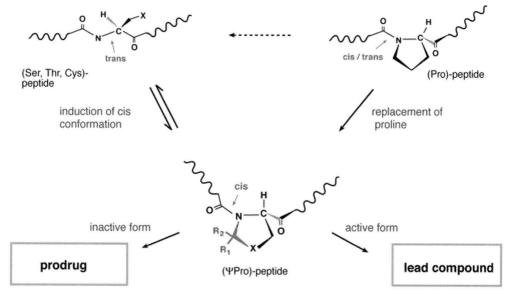

Fig. 11.16 The ΨProline concept in molecular recognition and drug design.
(A) Direct insertion of ΨPro units into biologically active, all-*trans* peptide sequences containing Ser, Thr, or Cys allows modulation of the physicochemical, biological, and pharmacokinetic properties (e.g. as applied in prodrug design).
(B) In Pro-containing peptides the dynamic process of CTI can be shifted towards *cis*, in replacing Pro by ΨPro, thus targeting bioactive conformations for use in lead finding [160].

As Nature uses proline for switching the shape of a potentially bioactive molecule, thus modulating its affinity to a given receptor, or, alternatively, directing the pathway of protein folding, the CTI of Xaa-Pro imidic bonds is one of the fundamental principles of the mechanism of cellular communication. Direct experimental determination of the bioactive, receptor-bound conformation of such peptides is yet not feasible and conformational studies on linear peptide ligands in solution suffer from the inherent high flexibility. Furthermore, the CTI around the peptide bond preceding a Pro residue in a peptide is a rapid dynamic process, thus preventing the isolation of the two isomers at room temperature. Due to the small energy difference between *cis* and *trans* conformers [131] a peptide with a high *trans:cis* ratio in solution might nevertheless adopt a *cis* conformation in the receptor-bound state where the relatively higher energy is compensated by favorable ligand–receptor interactions.

As a representative example to address the question of the bioactive conformation applying the ΨPro concept, the opioid peptide agonist morphiceptin has been chosen, a tetrapeptide (H-Tyr-Pro-Phe-Pro-NH$_2$) with CTI propensity around the Tyr^1-Pro^2 bond [182]. To exclude the *trans* isomer as the receptor-bound species, the proline residue was substituted by C2 dimethylated pseudoprolines capable of forcing the Tyr-ΨPro amide bond into the *cis* conformation. In a series of thiazolidine- and oxazolidine-containing morphiceptin analogs, all dimethylated compounds retained full μ agonist potency in guinea-pig ileum assay and high μ receptor selectivity, presenting strong evidence that morphiceptin has a *cis* conformation around the Tyr-Pro peptide bond as bioactive conformation.

Similarly, the introduction of the *cis*-prolyl mimic 2,2-dimethyl thiazolidine in position 7 of the peptide hormone oxytocin resulted in a retained high binding affinity for the oxytocin receptor, but in a 10-fold reduction in agonistic activity compared with oxytocin [183]. This is consistent with the reduction of the *trans* conformation from 90% for oxytocin to 5–8% for the dimethylated thiazolidine analog as determined by NMR, strongly supporting the hypothesis that a *cis/trans* conformational change plays an important role in oxytocin receptor binding and activation.

11.3.5
Pseudoprolines for Enhancing Peptide Cyclization and Turn Induction

Due to its capacity for modulating the *cis:trans* ratio in linear peptides, the ΨPro concept has been applied for enhancing the well-known tendency of Pro for peptide cyclization [184,185] and the induction of turn conformations [186–188]. As a most challenging target, the linear tripeptide H-Pro-ΨPro-Pro-OH containing a preformed *cis*-Pro Thr($\Psi^{Me,Me}$pro) tertiary amide bond cyclizes instantaneously and free of formation of oligomeric structures to cyclo-[Pro-Thr($\Psi^{Me,Me}$pro)-Pro] with all peptide bonds in the *cis* conformation [189]. These results indicate the enhanced cyclization tendency of *cis*-amide bond-containing peptides of short chain length.

The incorporation of turn-inducing elements (e.g. Gly, Pro, N-alkylated, or D-amino acids) into a linear peptide sequence to enhance cyclization rates and yields in head-to-tail cyclizations is well known [178]. However, to cyclize small peptides that do not contain turn inducers is often difficult. In general, cyclization depends on the propensity of the linear precursor peptide to adopt a conformation similar to the transition state required for cyclization, resulting in a conformationally controlled reaction [184,185]. Therefore, CTI was shown to have a pronounced impact on the cyclization reactions of short peptides [190]. Here, the use of pseudoprolines as removable turn inducers offers an elegant way to take advantage of the temporary induction of a *cis* conformation by forming a type VI β-turn structure to facilitate cyclization and subsequent cleavage of the pseudoproline ring system to yield the target cyclic peptide (Fig. 11.15).

Similarly, the importance of type VI β-turns in molecular recognition has mobilized great efforts to stabilize the *cis*-amide bonds and to mimic type VI β-turns in peptide chains. There is a broad palette of peptidomimetics that are used to induce a *cis*-conformation, most importantly, mimetics carrying sterically bulky groups on the C_δ-atom of the proline residue [191–193]. Introducing a sterically bulky group on the C_δ-atom of the proline residue changes the interactions of the atoms adjacent to the amide bond. Because the substituent(s) on the C_δ-atom is larger than that of the C_α-atom, it favors the *cis*-amide bond preceding the proline residue. Despite the stabilizing effect, spatially demanding substituents alter the steric geometry of the turn structure and modify the hydrogen-bonding characteristics, which can severely affect the recognition specificity. To reduce unfavorable steric features that could compromise receptor interactions, aza-amino acids, in particular aza-proline have been proposed as mimetics of type VI β-turns. The amino acid analog azaproline (azPro) contains a nitrogen atom in place of the C_α of proline and has unique conformational properties due to its diacylhydrazide backbone (e.g. favoring the cis-amide bond preceding the azPro residue as a result of the unfavorable lone-pair/lone-pair repulsion in the *trans*-amide conformation). Peptides containing azPro were shown to stabilize the *cis*-amide conformer for the Xaa-azPro bond and prefer type VI β-turns both in crystals and in organic solvents by NMR [194]. As the azPro derivatives generally stabilize the *cis*-amide bond representing a mimic of type VI β-turn without the incorporation of additional sterically bulky groups, this chemical modification of the peptide backbone offers some attractive features for the introduction of conformational constraints in biologically active peptides.

11.3.6
Pseudoprolines for Modulating Polyproline Helices

The use of pseudoprolines to induce and lock in oligopeptides in an all-*cis* conformation was demonstrated on the example of a polyproline I helix [167]. Depending on the experimental conditions, the formation and oligomerization of C2 monoarylated pseudoproline dipeptide building units can be controlled with respect to their configuration and conformation by aromatic stacking interactions

[165]. Electrophilically induced cyclic acetyl formation of *O*-benzyl dipeptide esters (e.g. Fmoc-Pro-Thr-OBn) leads predominantly to the (*R*) diastereoisomers at the C2 position of the resulting substituted 1,3-oxazolidine (ΨPro) unit, while upon acetalization of the corresponding methyl ester the C2 (*S*) epimer is formed due to the lack of a transannular stacking relay along the reaction coordinate. Thus, the assembly into peptide oligomers of alternating Pro-ΨPro (ΨPro as the C2 (*R*) epimer) constitution should direct, based on a related Pro-aromatic stacking effect, the *cis/trans* equilibrium along the substituted amide bonds into a homogeneous *cis* (polyproline I, PPI) conformation. For example, the circular dichroism (CD) spectrum of a nonamer of type Ac-[Pro-ΨPro]₄-Pro-OH is almost solvent independent and adopts the mirror image of an all-*trans*-polyproline helix as reported in literature [167]. In contrast to oligoproline sequences, no transition to a more extended, backbone-exposed polyproline II helix (all-*trans*) could be enforced by solvent exchange due to the stabilization of the more compact PPI-like helix by intramolecular aromatic stacking interactions. Most notably, the stabilization of secondary structure by applying the concept of pseudoprolines can be further extended to investigate polyproline I/II conformational transitions as a prerequisite for molecular recognition and cellular communication.

Protein–protein interactions are often mediated by binding of proline-rich ligands of the consensus sequence Pro-Xaa-Xaa-Pro that adopt a left-handed polyproline II helical conformation with unique geometric properties as a structural requirement for effective binding (e.g. to SH3 domains). Pseudoproline building blocks with enhanced inherent properties of natural ʟ-Pro can play a dual role in targeting molecular recognition (Fig. 11.17) [195].

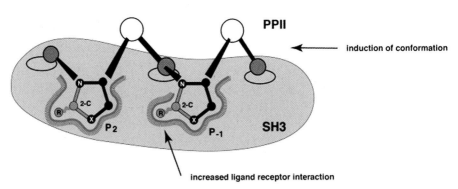

Fig. 11.17 The dual function of pseudoprolines is demonstrated for the example of proline-rich peptides as ligands for Src homology domains (SH3). ΨPro building blocks induce the relevant PPII conformation of the ligand and allow in addition the modulation of affinity and specificity by tuning van der Waals contacts and hydrogen bonding interactions of the substituents R at C2 of ΨPro to the receptor molecule [195].

In particular, the nature of the substituents at the C2 position of ΨPro permits the CTI equilibrium to occur and thus the polyproline transitions (PPI/PPII) to be tailored at will. Besides its potential for adopting a PPII helical conformation, the incorporation of substituents of variable size and polarity at the C2 position of ΨPro offers a powerful tool to modulate ligand–receptor interactions. As depicted in Fig. 11.17, ΨPro units exert a dual functionality in (1) inducing and stabilizing the relevant PPII conformation and (2) increasing and optimizing the van der Waals contacts and the formation of hydrogen bonds to the receptor. Most notably, a defined stereochemistry at the C2 position as well as the generation of a library of different substituents at C2 allows the factors contributing to affinity and specificity in protein–protein interactions to be explored, and to further elucidate ligand recognition at a molecular level [195].

11.3.7
Pseudoprolines for Modulating Structure and Function of Cyclosporins

Numerous chemical methods for the modulation or optimization of the physico-chemical, pharmacological, and biological properties of natural peptides or lead compounds have been successfully applied in drug development. Pseudoprolines are a versatile alternative to existing strategies for chemical modification of bioactive peptides. So far, the incorporation of Pro systems into peptide backbones has been achieved by performing the corresponding ΨPro building blocks and subsequent coupling to the peptide chain. A next step would be the direct incorporation of pseudoprolines into complex peptide structures (e.g. cyclosporin analogs), aiming at efficient preparation of a large variety of derivatives with differential structural and functional properties (Fig. 11.18).

Cyclosporins are hydrophobic, cyclic undecapeptides with a remarkable variety of biological functions ranging from immunosuppression to blocking of HIV-1 replication [196,197]. In the search of drugs specifically addressing one of the two pathways, research on cyclosporin derivatives has intensified recently. Upon binding of cyclosporin A (CsA) to the peptidylprolyl *cis/trans*-isomerase cyclophilin (Cyp) and a second protein, the cellular phosphatase calcineurin (Cn), a tenary complex (CsA-Cyp)-Cn is formed, being responsible for immunosuppression by inhibiting the transcription of cytokine genes (e.g. interleukin-2) [131]. Similarly, CypA is involved in HIV-1 replication by binding capsid protein Gag, Pr55^{gag} and additionally, its incorporation into the HIV-1 virion [198,199]. As the presence of cyclosporin prevents these interactions, strong ligands for CypA are of utmost interest to interrupt the interaction between Cyp and Pr55^{gag}. Starting from the Thr^2-containing analog CsC, we succeeded in the one-step insertion of ΨPro moieties featuring a variety of C2 substituents by treating the 2-threonine hydroxyl group and the preceding amide nitrogen with a number of both arylated and nonarylated dimethyl acetals (Fig. 11.18) [200]. These monosubstituted derivatives were obtained in good yield, with high stereo- and regioselectivity and showed characteristic enhanced conformational backbone rigidity. Despite the drastic conformational constraints within the cyclic peptide backbone induced by the ΨPro system in the direct vicinity of the receptor (CypA) binding site, most of the ΨPro-containing analogs retained their binding capacity to CypA, indicating that bioactive conformations are induced and stabilized. The choice of *para*-substituted aryl

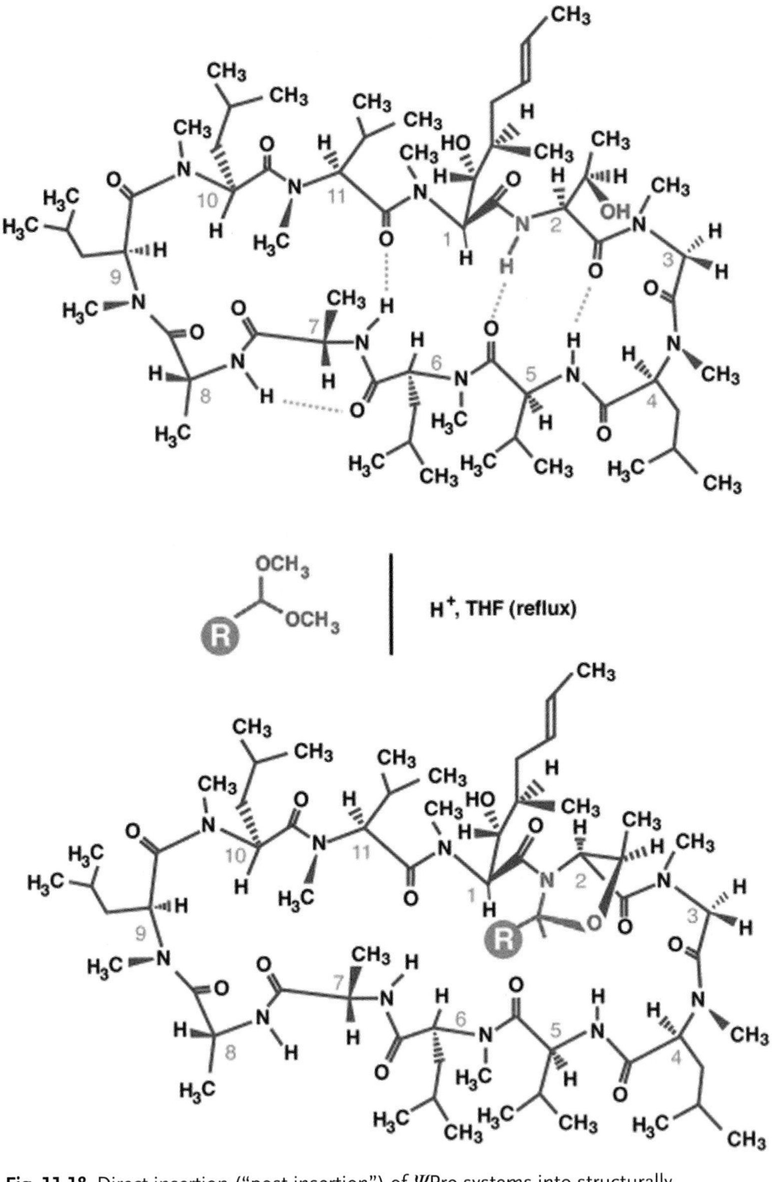

Fig. 11.18 Direct insertion ("post-insertion") of ΨPro systems into structurally complex molecules (i.e. cyclosporin C [165,200]). Different C2′ substituents (R) allow for the modulation of the pharmacokinetic and biological profiles.

dimethyl acetals allows the inhibitory properties of the corresponding derivatives to be modulated to either prodrugs or moderately strongly binding CsC derivatives of differential bioavailability. Consequently, the direct insertion of ΨPro into Cs analogs [200,201] offers a novel access to active-site inhibitors with tailored pharmacokinetic properties.

On a molecular level, the amino acid residues 3–8 of CsA are involved in Cn binding whereas residues at positions 10, 11, and 13 interact with Cyp, and inhibition of the HIV-1 replication process by CsA occurs independently of calcineurin inhibition [202,203]. To address these two modes of action of CsA separately, increased research has been devoted to the synthesis of derivatives that bind with high selectivity to CypA but have drastically reduced affinity for Cn (i.e. to develop CsA-derived drugs with antiviral but devoid of immunosuppressive acitivity). For this purpose, the structural and biological impact of ΨPro at position 5 of CsA was of particular interest and required a novel CsA analog, Thr^{5}CsA, for the selective insertion of ΨPro building blocks [204]. To test the hypothesis of a *cis*-amide bond between residues 4 and 5 of CsA, Val at position 5 was replaced by a Ser or Thr following a synthesis strategy based on ring opening, Edman degradation, and introduction of the ΨPro unit. Detailed conformational analysis and receptor binding studies revealed that e.g. C2 dimethyl ΨPro-containing Thr^{5}CsA derivatives maintain binding to Cyp and Cn and feature a *cis*-amide bond between residues 5 and 6 with all remaining amide bonds *trans*. In summary, novel synthetic routes for generating ΨPro containing cyclosporin derivatives pave the way for extended structure–activity studies aiming at the design of pharmacologically active compounds with a selective activity profile.

11.3.8
Pseudoprolines for Targeting *Cis* Bonds in Peptides and Proteins

CTIs are often the on/off switch for biological activity or, when located on the protein surface, crucial for launching a biochemical chain reaction. As demonstrated on opioid peptides, such as morphiceptin, a *cis*-peptide bond can strongly be enhanced by the introduction of a pseudoproline residue solely depending on its C2 substituents. In particular, C2 dimethylated thiazolidine derivatives can induce up to 100% *cis* conformation in the Xaa$_{i-1}$–ΨPro$_i$ peptide bond, making ΨPro-containing peptides ideal candidates for mimicking biologically relevant *cis*-prolyl conformations. As a challenging target to test this hypothesis, we have chosen the V3 loop of the envelope protein gp120 of HIV-1, being the major HIV-1 neutralizing epitope [205] (Fig. 11.19).

A generally conserved tetrapeptide motif Gly-Pro-Gly-Arg is situated at the tip of the V3 loop, forming a type II β-turn. It is suggested that a CTI towards a type VI β-turn with a *cis*-proline peptide bond is a necessary conformational change for the onset of viral infection. Targeting or blocking this key event by a monoclonal antibody therefore represents a potential tool for interfering with this pathway. For probing the infection-active site *cis*-prolyl loop tip conformation, ΨPro building blocks have been introduced in V3 loop analogs to constrain the peptide in a *cis* conformation. A cyclic, ΨPro-containing, *cis*-constrained V3 loop mimetic has been used as immunogen for the preparation of polyclonal and monoclonal antibodies; the results of immunological studies demonstrate that the antibodies selectively distinguish between the *cis* and *trans* conformation of Xaa-Pro imide bonds in cyclic peptides as well as their linear analogs, pointing to the versatility

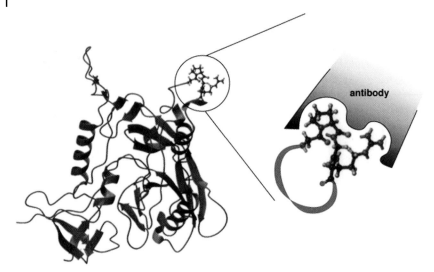

Fig. 11.19 NMR-derived structure of the envelope protein gp120 of HIV-1; the V3 loop corresponds to the major neutralizing epitope (left, encircled). A highly conserved tetrapeptide adopting a type II β-turn is considered to isomerize towards a type VI β-turn with a *cis*-proline peptide bond in the infection-active state. For targeting the relevant *cis* conformation, ΨPro-containing, *cis*-constrained V3 loop analogs were prepared and used as immunogens to raise antibodies that selectively recognize *cis*-amide bonds (right, close-up) [205].

of Ψ-prolines for detecting conformational changes relevant in biological processes.

11.4
Conclusions and Perspectives

Since its introduction in the early 1990s, the concept of pseudoprolines is presently becoming recognized as versatile strategy for the synthesis of difficult peptides. The commercial availability of Fmoc-protected building blocks of the type Fmoc-Xaa-ΨPro facilitates the introduction of this solubilizing, structure-disrupting dipeptide as a temporary protection technique for Ser, Thr, and Cys. Comparative studies have documented the superiority of the ΨPro strategy in comparison to other existing solubilizing techniques in SPPS [156–161,168–172,174–177].

In enhancing and tailoring the intrinsic conformational effects of proline, notably its *cis:trans* ratio, the ΨPro concept opens new avenues for biological applications. For example, bioactive conformations in β-turn or loop structures of peptide ligands could be delineated, and the temporary transformation of Ser-, Thr-, or Cys-containing bioactive peptides to drugs and prodrugs of altered pharmacological profile has been successfully achieved.

Due to its pronounced structure-disrupting effect, its use as building block in β-sheet breaking peptides in the potential treatment of amyloid β (Aβ) fibrillogenesis has been explored [206]. The interesting finding that the insertion of a ΨPro-induced *cis*-amide bond – resulting in a "kink" conformation – did not significantly improve the β-breaking effect of Aβ-derived peptides compared with Soto's original β-breaking lead compound (Ac-Leu-Pro-Phe-Phe-Asp-NH₂) stimulated us to design molecules exhibiting two distinct conformational states termed "switch-peptides" [207,208] (Fig. 11.20). These are capable of favorable intermolecular interactions with a β-template (e.g. Aβ fibrils, recognition state, S_{off}) and switching by in situ induction of acyl migration to a functional β-sheet and fibril-disrupting state (S_{on}), as depicted schematically in Fig. 11.20.

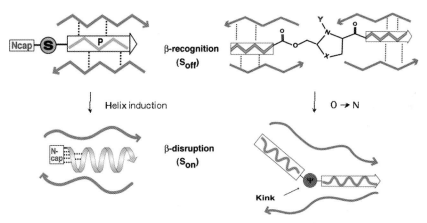

Fig. 11.20 The concept of switch-peptides [207,208]: in situ induction of conformational changes by enzyme-triggered acyl migration results in the nucleation of a helix (left) or a *cis*-amide bond-induced kink conformation (right) for the breaking of β-sheet templates (e.g. amyloid fibrils).

In addition to their use in the in situ induction of conformational transitions, switch-peptides (*O*-acyl isopeptides) may serve as alternative solubilizing technique in peptide synthesis [207–214]. The elaboration of these concepts, notably the rational design of dynamic pharmacophores as new type of therapeutically useful compounds, is the subject of our current research.

References

1 Mauger, A. B. **(1977)** *The Chemistry of Cyclic Alpha-imino Acids in Chemistry and Biochemistry of Amino Acids, Peptides and Proteins*, B. Weinstein, B., Ed. Marcel Dekker, New York, 179–240.

2 Wagner, I., Musso, H. **(1983)** *Angew. Chem. Int. Ed. Engl.* 22, 816–828.

3 Kuttan, R., Radhakrishnan, A. N. **(1973)** *Adv. Enzymol.* 37, 273–347.

4 Cardinale, G. J., Udenfriend, S. **(1974)** *Adv. Enzymol.* 41, 245–300.

5 Schomburg, D., Stephan, D. **(1994)** Oxidoreductases. In *Enzyme Handbook*, Vol. 8. Schomburg, D., Stephan, D., Eds. Springer-Verlag, Berlin, 1.13–1.97.

6 Gaill, F., Mann, K., Wiedemann, H., Engel, J., Timpl, R. **(1995)** *J. Mol. Biol.* 246, 284–294.

7 Gryder, R. M., Lamon, M., Adams, E. **(1975)** *J. Biol. Chem.* 250, 2470–2474.

8 Rexrodt, F. W., Hopper, K. E., Fietzek, P. P., Kuhn, K. **(1973)** *J. Biol. Chem.* 38, 384–395.

9 Baum, J., Brodsky, B. **(2000)** Folding of the collagen triple helix and its naturally occurring mutants. In *Mechanisms of Protein Folding*, 2nd edn. Pain, R. H., Ed. Oxford University Press, Oxford, 330–351.

10 Sakakibara, S., Inouye, K., Shudo, K., Kishida, Y., Kobayashi,Y., Prockop, D. J. **(1973)** *Biochim. Biophys. Acta* 303, 198–202.

11 Engel, J., Chen, H. T., Prockop, D. J., Klump, H. **(1977)** *Biopolymers* 16, 601–622.

12 Inouye, K., Sakakibara, S., Prockop, D. J. **(1976)** *Biochim. Biophys. Acta* 420, 133–141.

13 Inouye, K., Kobayashi, Y., Kyogoku, Y., Kishida, Y., Sakakibara, S., Prockop, D. J. **(1982)** *Arch. Biochem. Biophys.* 219, 198–203.

14 Privalov, P. L. **(1982)** *Adv. Protein Chem.* 35, 1–104.

15 Eberhardt, E. S., Panasik, N., Raines, R. T. **(1996)** *J. Am. Chem. Soc.* 118, 12261–12266.

16 Boudko, S., Frank, S., Kammerer, R. A., Stetefeld, J., Schulthess, T., Landwehr,

R., Lustig, A., Bächinger, H. P., Engel, J. **(2002)** *J. Mol. Biol.* 317, 459–470.

17 Mizuno, K., Hayashi, T., Peyton, D. H., Bächinger, H. P. **(2004)** *J. Biol. Chem.* 279, 38072–38078.

18 Berisio, R., Granata, V., Vitagliano, L., Zagari, A. **(2004)** *J. Am. Chem. Soc.* 126, 11402–11403.

19 Doi, M., Nishi, Y., Uchiyama, S., Nishiuchi, Y., Nishio, H., Nakazawa, T., Ohkubo, T., Kobayashi, Y. **(2005)** *J. Peptide Sci.* 11, 609–616.

20 Mizuno, K., Hayashi, T., Peyton, D. H., Bächinger, H. P. **(2004)** *J. Biol. Chem.* 279, 282–287.

21 Jenkins, C. L., Bretscher, L. E., Guzei, I. A., Raines, R. T. **(2003)** *J. Am. Chem. Soc.* 125, 6422–6427.

22 Bella, J., Brodsky, B., Berman, H. M. **(1995)** *Structure* 3, 893–906.

23 Berisio, R., Vitagliano, L., Mazzarella, L., Zagari, A. **(2002)** *Protein Sci.* 11, 262–270.

24 Hongo, C., Nagarajan, V., Noguchi, K., Kamitori, S., Okuyama, K., Tanaka, Y., Nishino, N. **(2001)** *Polymer J.* 33, 812–818.

25 Okuyama, K., Nagarajan, V., Kamitori, S. **(1999)** *Proc. Indian. Acad. Sci. Chem. Sci.* 111, 19–34.

26 Bella, J., Eaton, M., Brodsky, B., Berman, H. M. **(1994)** *Science* 266, 75–81.

27 Okuyama, K., Tanaka, N., Ashida, T., Kakudo, M., Sakakibara, S., Kishida, Y. **(1972)** *J. Mol. Biol.* 72, 571–576.

28 Improta, R., Mele, F., Crescenzi, O., Benzi, C., Barone, V. **(2002)** *J. Am. Chem. Soc.* 124, 7857–7865.

29 Improta, R., Benzi, C., Barone, V. **(2001)** *J. Am. Chem. Soc.* 123, 12568–12577.

30 Vitagliano, L., Berisio, R., Mastrangelo, A., Mazzarella, L., Zagari, A. **(2001)** *Protein Sci.* 10, 2627–2632.

31 Vitagliano, L., Berisio, R., Mazzarella, L., Zagari, A. **(2001)** *Biopolymers* 59, 459–464.

32 Bretscher, L. E., Jenkins, C. L., Taylor, K. M., DeRider, M. L., Raines, R. T. **(2001)** *J. Am. Chem. Soc.* 123, 777–778.

33 DeRider, M. L., Wilkens, S. J., Dzakula, Z., Raines, R. T., Markley, J. L. (**1998**) *Abstracts of Papers of the American Chemical Society* 216, U718–U718.

34 DeRider, M. L., Wilkens, S. J., Waddell, M. J., Bretscher, L. E., Weinhold, F., Raines, R. T., Markley, J. L. (**2002**) *J. Am. Chem. Soc.* 124, 2497–2505.

35 Holmgren, S. K., Bretscher, L. E., Taylor, K. M., Raines, R. T. (**1999**) *Chem. Biol.* 6, 63–70.

36 Holmgren, S. K., Taylor, K. M., Bretscher, L. E., Raines, R. T. (**1998**) *Nature* 392, 666–667.

37 Jenkins, C. L., Bretscher, L. E., Raines, R. T. (**2001**) *Biochemistry* 40, 8658–8658.

38 Kersteen, E. A., Raines, R. T. (**2001**) *Biopolymers* 59, 24–28.

39 Panasik, N., Eberhardt, E. S., Edison, A. S., Powell, D. R., Raines, R. T. (**1994**) *Int. J. Peptide Prot. Res.* 44, 262–269.

40 Jenkins, C. L., McCloskey, A. I., Guzei, I. A., Eberhardt, E. S., Raines, R. T. (**2005**) *Biopolymers* 80, 1–8.

41 Taylor, C. M., Hardre, R., Edwards, P. J. B. (**2005**) *J. Org. Chem.* 70, 1306–1315.

42 Kemp, D. S., Rothman, J. H. (**1995**) *Tetrahedron Lett.* 36, 4019–4022.

43 Eswarakrishnan, V., Field, L. (**1981**) *J. Org. Chem.* 46, 4182–4187.

44 Verbiscar, A. J; Witkop, B. (**1970**) *J. Org. Chem.* 35, 1924–1927.

45 Nikiforovich, G. V., Kao, J. L. F., Plucinska, K., Zhang, W. J., Marshall, G. R. (**1994**) *Biochemistry* 33, 3591–3598.

46 Plucinska, K., Kataoka, T., Yodo, M., Cody, W. L., He, J. X., Humblet, C., Lu, G. H., Lunney, E., Major, T. C., Panek, R. L., Schelkun, P., Skeean, R., Marshall, G. R. (**1993**) *J. Med. Chem.* 36, 1902–1913.

47 Kataoka, T., Beusen, D. D., Clark, J. D., Yodo, M., Marshall, G. R. (**1992**) *Biopolymers* 32, 1519–1533.

48 Uitto, J., Hoffmann, H. P., Prockop, D. J. (**1976**) *Arch. Biochem. Biophys.* 173, 187–200.

49 Uitto, J., Prockop, D. J. (**1977**) *Arch. Biochem. Biophys.* 181, 293–299.

50 Renner, C., Alefelder, S., Bae, J. H., Budisa, N., Huber, R., Moroder, L. (**2001**) *Angew. Chem. Int. Ed.* 40, 923–925.

51 Demange, L. (**2001**). PhD dissertation, University Paris-Sud.

52 Dugave, C., Demange, L. (**2003**) *Chem. Res.* 103, 2475–2532.

53 Dunitz, J. D., Taylor, R. (**1997**) *Chem. Eur. J.* 3, 89–98.

54 Engel, J., Prockop, D. J. (**1998**) *Matrix Biol.* 17, 679–680.

55 Nishi, Y., Uchiyama, S., Doi, M., Nishiuchi, Y., Nakazawa, T., Ohkubo, T., Kobayashi, Y. (**2005**) *Biochemistry* 44, 6034–6042.

56 Hodges, J. A., Raines, R. T. (**2003**) *J. Am. Chem. Soc.* 125, 9262–9263.

57 Doi, M., Nishi, Y., Uchiyama, S., Nishiuchi, Y., Nakazawa, T., Ohkubo, T., Kobayashi, Y. (**2003**) *J. Am. Chem. Soc.* 125, 9922–9923.

58 Barth, D., Milbradt, A. G., Renner, C., Moroder, L. (**2004**) *ChemBiochem* 5, 79–86.

59 Melacini, G., Bonvin, A., Goodman, M., Boelens, R., Kaptein, R. (**2000**) *J. Mol. Biol.* 300, 1041–1048.

60 Milbradt, A. G., Boulegue, C., Moroder, L., Renner, C. (**2005**) *J. Mol. Biol.* 354, 591–600.

61 Boulègue, C., Milbradt, A. G., Renner, C., Moroder, L. (**2006**) Unpublished results.

62 Golbik, R., Yu, C., Weyher-Stingl, E., Huber, R., Moroder, L., Budisa, N., Schiene-Fischer, C. (**2005**) *Biochemistry* 44, 16026–16034.

63 Galardy, R. E., Alger, J. R., Liakopoulou-Kyriakides, M. (**1982**) *Int. J. Peptide Protein Res.* 19, 123–132.

64 Kern, D., Schutkowski, M., Drakenberg, T. (**1997**) *J. Am. Chem. Soc.* 119, 8403–8408.

65 Babu, I. R., Ganesh, K. N. (**2001**) *J. Am. Chem. Soc.* 123, **2079–2080**.

66 Venugopal, M. G., Ramshaw, J. A. M., Braswell, E., Zhu, D., Brodsky, B. (**1994**) *Biochemistry* 33, 7948–7956.

67 Delaney, N. G., Madison, V. (**1982**) *J. Am. Chem. Soc.* 104, 6635–6641.

68 Halab, L., Bélec, L., Lubell, W. D. (**2001**) *Tetrahedron* 57, 6439–6446.

69 Beausoleil, E., Larcheveque, B., Bélec, L., Atfani, M., Lubell, W. D. (**1996**) *J. Org. Chem.* 61, 9447–9454.

70 Beausoleil, E., Lubell, W. D. (**1996**) *J. Am. Chem. Soc.* 118, 12902–12908.

71 Halab, L., Lubell, W. D. (**1999**) *J. Org. Chem.* 64, 3312–3321.

72 Halab, L., Lubell, W. D. (**2001**) *J. Peptide Sci.* 7, 92–104.

73 Halab, L., Lubell, W. D. (**2002**) *J. Am. Chem. Soc.* 124, 2474–2484.

74 Beausoleil, E., Lubell, W. D. (**2000**) *Biopolymers* 53, 249–256.

75 Bélec, L., Slaninova, J., Lubell, W. D. (**2000**) *J. Med. Chem.* 43, 1448–1455.

76 Swarbrick, M. E., Gosselin, F., Lubell, W. D. (**1999**) *J. Org. Chem.* 64, **1993–2002**.

77 Savdra, J. (**1976**) *Cis-trans* isomerism of N-acyl derivatives of proline and its analogs: linear peptides with *cis* peptide bonds. In *Peptides*, Loffet, A., Ed. University of Brussels, Brussels, 653–656.

78 Magaard, V. W., Sanchez, R. M., Bean, J. W., Moore, M. L. (**1993**) *Tetrahedron Lett.* 34, 381–384.

79 An, S. S. A., Lester, C. C., Peng, J. L., Li, Y. J., Rothwarf, D. M., Welker, E., Thannhauser, T. W., Zhang, L. S., Tam, J. P., Scheraga, H. A. (**1999**) *J. Am. Chem. Soc.* 121, 11558–11566.

80 Beausoleil, E., Sharma, R., Michnick, S. W., Lubell, W. D. (**1998**) *J. Org. Chem.* 63, 6572–6578.

81 Lummis, S. C. R., Beene, D. L., Lee, L. W., Lester, H. A., Broadhurst, R. W., Dougherty, D. A. (**2005**) *Nature* 438, 248–252.

82 Hughes, P., Martin, M., Clardy, J. (**1980**) *Tetrahedron Lett.* 21, 4579–4580.

83 Bell, E. A., Qureshi, M. Y., Pryce, R. J., Janzen, D. H., Lemke, P., Clardy, J. (**1980**) *J. Am. Chem. Soc.* 102, 1409–1412.

84 Jung, M. E., Shishido, K., Light, L., Davis, L. (**1981**) *Tetrahedron Lett.* 22, 4607–4610.

85 Gaitanopoulos, D. E., Weinstock, J. (**1985**) *J. Heterocyclic Chem.* 22, 957–959.

86 Montelione, G. T., Hughes, P., Clardy, J., Scheraga, H. A. (**1986**) *J. Am. Chem. Soc.* 108, 6765–6773.

87 Avenoza, A., Busto, J. H., Peregrina, J. M., Rodriguez, F. (**2002**) *J. Org. Chem.* 67, 4241–4249.

88 Bunuel, E., Gil, A. M., Diaz-de-Villegas, M. D., Cativiela, C. (**2001**) *Tetrahedron* 57, 6417–6427.

89 Hanessian, S., Reinhold, U., Gentile, G. (**1997**) *Angew. Chem. Int. Ed. Engl.* 36, 1881–1884.

90 Jenkins, C. L., Lin, G. L., Duo, J. Q., Rapolu, D., Guzei, I. A., Raines, R. T., Krow, G. R. (**2004**) *J. Org. Chem.* 69, 8565–8573.

91 Hughes, P., Clardy, J. (**1988**) *J. Org. Chem.* 53, 4793–4796.

92 Rammeloo, T., Stevens, C. V., De Kimpe, N. (**2002**) *J. Org. Chem.* 67, 6509–6513.

93 Rammeloo, T., Stevens, C. V. (**2002**) *Chem. Commun.* 250–251.

94 Avenoza, A., Cativiela, C., Busto, J. H., Fernandez-Recio, M. A., Peregrina, J. M., Rodriguez, F. (**2001**) *Tetrahedron* 57, 545–548.

95 Campbell, J. A., Rapoport, H. (**1996**) *J. Org. Chem.* 61, 6313–6325.

96 Han, W., Pelletier, J. C., Hodge, C. N. (**1998**) *Bioorg. Med. Chem. Lett.* 8, 3615–3620.

97 Talluri, S., Montelione, G. T., Vanduyne, G., Piela, L., Clardy, J., Scheraga, H. A. (**1987**) *J. Am. Chem. Soc.* 109, 4473–4477.

98 Piela, L., Nemethy, G., Scheraga, H. A. (**1987**) *J. Am. Chem. Soc.* 109, 4477–4485.

99 Mapelli, C., Vanhalbeek, H., Stammer, C. H. (**1990**) *Biopolymers* 29, 407–422.

100 Juvvadi, P., Dooley, D. J., Humblet, C. C., Lu, G. H., Lunney, E. A., Panek, R. L., Skeean, R., Marshall, G. R. (**1992**) *Int. J. Peptide Protein Res.* 40, 163–170.

101 Hercouet, A., Bessières, B., LeCorre, M. (**1996**) *Tetrahedron Asymm.* 7, 1267–1268.

102 Switzer, F. L., Vanhalbeek, H., Holt, E. M., Stammer, C. H., Saltveit, M. E. (**1989**) *Tetrahedron* 45, 6091–6100.

103 Fujimoto, Y., Irreverre, F., Karle, J. M., Karle, I. L., Witkop, B. (**1971**) *J. Am. Chem. Soc.* 93, 3471–3477.

104 Brackmann, F., Schill, H., de Meijere, A. (**2005**) *Chem. Eur. J.* 11, 6593–6600.

105 Bianco, A., Lucchini, V., Maggini, M., Prato, M., Scorrano, G., Toniolo, C. (**1998**) *J. Peptide Sci.* 4, 364–368.

106 Hanessian, S., McNaughton-Smith, G., Lombart, H. G., Lubell, W. D. (**1997**) *Tetrahedron* 53, 12789–12854.

107 Gillespie, P., Cicariello, J., Olson, G. L. (**1997**) *Biopolymers* 43, 191–217.

108 Müller, G., Gurrath, M., Kurz, M., Kessler, H. (**1993**) *Proteins Struct. Funct. Genet.* 15, 235–251.

109 Rose, G. D., Gierasch, L. M., Smith, J. A. (**1985**) *Adv. Protein Chem.* 37, 1–109.

110 Wilmot, C. M., Thornton, J. M. (**1988**) *J. Mol. Biol.* 203, 221–232.

111 Sato, K., Nagai, U. (**1986**) *J. Chem. Soc. Perkin Trans. 1*, 1231–1234.

112 Feigel, M. (**1986**) *J. Am. Chem. Soc.* 108, 181–182.

113 Nagai, U., Sato, K. (**1985**) *Tetrahedron Lett.* 26, 647–650.

114 Kahn, M., Eguchi, M. (**2002**) *Houben-Weyl, Methods of Organic Chemistry. Synthesis of Peptides and Peptidomimetics,* Vol. E22c, Goodman, M., Felix, A., Moroder, L., Toniolo, C., Eds. Georg Thieme Verlag, Stuttgart, 695–740.

115 Kahn, M., Eguchi, M. (**2002**) *Houben-Weyl, Methods of Organic Chemistry, Synthesis of Peptides and Peptidomimetics,* Vol. E22c, Goodman, M., Felix, A., Moroder, L., Toniolo, C., Eds. Georg Thieme Verlag, Stuttgart, 741–758.

116 Stigers, K. D., Soth, M. J., Nowick, J. S. (**1999**) *Curr. Opin. Chem. Biol.* 3, 714–723.

117 Marraud, M., Aubry, A. (**1996**) *Biopolymers* 40, 45–83.

118 Rizo, J., Gierasch, L. M. (**1992**) *Annu. Rev. Biochem.* 61, 387–418.

119 Lombart, H. G., Lubell, W. D. (**1996**) *J. Org. Chem.* 61, 9437–9446.

120 Baldwin, J. E., Freeman, R. T., Schofield, C. (**1989**) *Tetrahedron Lett.* 30, 4019–4020.

121 Baldwin, J. E., Hulme, C., Schofield, C. J., Edwards, A. J. (**1993**) *J. Chem. Soc. Chem. Commun.* 935–936.

122 Tremmel, P., Geyer, A. (**2005**) *Eur. J. Org. Chem.* 3475–3481.

123 Cluzeau, J., Lubell, W. D. (**2004**) *J. Org. Chem.* 69, 1504–1512.

124 Colombo, L., Digiacomo, M., Scolastico, C., Manzoni, L., Belvisi, L., Molteni, V. (**1995**) *Tetrahedron Lett.* 36, 625–628.

125 Kim, K.: Germanas, J. P. (**1997**) *J. Org. Chem.* 62, 2847–2852.

126 Kim, K., Germanas, J. P. (**1997**). *J. Org. Chem.* 62, 2853–2860.

127 Kim, K. H., Dumas, J. P., Germanas, J. P. (**1996**) *J. Org. Chem.* 61, 3138–3144.

128 Gramberg, D., Weber, C., Beeli, R., Inglis, J., Bruns, C., Robinson, J. A. (**1995**) *Helv. Chim. Acta* 78, 1588–1606.

129 Gramberg, D., Robinson, J. A. (**1994**) *Tetrahedron Lett.* 35, 861–864.

130 Hoffmann, T., Lanig, H., Waibel, R., Gmeiner, P. (**2001**) *Angew. Chem. Int. Ed.* 40, 3361–3364.

131 Fischer, G. (**2000**) *Chem. Soc. Rev.* 29, 119–127.

132 Rose, G.D., Gierasch, L.M., Smith, J.A. (**1985**) *Adv. Protein Chem.* 37, 1–109.

133 Müller, G., Gurrath, M., Kurz, M., Kessler, H. (**1993**) *Struct. Funct. Genet.* 15, 235–251.

134 Richardson, J.S. (**1981**) *Adv. Protein Chem.* 34, 116–339.

135 Smith, J.A., Pease, L.G. (**1980**) *CRC Crit. Rev. Biochem.* 8, 315–399.

136 Kim, P.S., Baldwin, R.L. (**1982**) *Annu. Rev. Biochem.* 51, 459–489; Creighton, T.E. (**1993**) *Proteins, Structures and Molecular Properties,* 2nd edn. W.H. Freeman and Co., New York.

137 Dugave, C. (**2002**) *Curr. Org. Chem.* 6, 1397–1431.

138 Delaney, N.G., Madison, V. (**1982**) *J. Am. Chem. Soc.* 104, 6635–6641.

139 Zhang, R., Bronewell, F., Madalengoitia, J.S. (**1998**) *J. Am. Chem. Soc.* 120, 3894–3902.

140 An, S.S.A., Lester, C.C., Peng, J.-L., Li, Y.-J., Rothwarf, D.M., Welker, E., Thannhauser, T.W., Zhang, L.S., Tam, J.P., Scheraga, H.A. (**1999**) *J. Am. Chem. Soc.* 121, 11558–11566.

141 Beausoleil, E., Lubell, W.D. (**2000**) *Biopolymers* 53, 249–256.

142 Koskinen, A.M.P., Helaja, J., Kumpulainen, E.T.T., Koivisto, J., Mansikkamaki, H., Rissanen, K. (**2005**) *J. Org. Chem.* 70, 6447–6453.

143 Pillai, V.N.R., Mutter, M. (**1981**) *Acc. Chem. Res.* 14, 122–130.

144 Mutter, M., Pillai, V.N.R., Anzinger, H., Bayer, E., Toniolo, C. (**1981**) *Procedings*

of the 16th European Peptide Symposium, 660–665.

145 Toniolo, C.,Bonora, G.M., Mutter, M., Maser, F. (**1983**) *J. Am. Chem. Soc., Chem. Commun.* 1298–1299.

146 Mutter, M. (**1985**) *Angew. Chem. Int. Ed.* 24, 639–653.

147 Mutter, M., Vuilleumier, S. (**1989**) *Angew. Chem. Int. Ed.* 28, 535–554.

148 Abd El Rahman, S., Anzinger, H., Mutter, M. (**1980**) *Biopolymers* 19, 173–187.

149 Toniolo, C., Bonora, G.M., Mutter, M., Pillai, V.N.R. (**1981**) *Makromol. Chem.* 182, **1997–2005**.

150 Toniolo, C., Bonora, G.M., Mutter, M., Pillai, V.N.R. (**1981**) *Makromol. Chem.* 182, **2007–2014**.

151 Mutter, M., Altmann, K.H., Bellof, D., Floersheimer, A., Herbert, J., Huber, M., Klein, B., Strauch, L., Vorherr, T., Gremlich, H.U. (**1985**) *Proceedings of the 9th American Peptide Symposium* 397–405.

152 Kent, S.B.H. (**1988**) *Annu. Rev. Biochem.* 57, 957–989.

153 Goodman, M., Niu, G.C.C., Su, K.C. (**1970**) *J. Am. Chem. Soc.* 92, 5219–5220.

154 Seebach, D., Sommerfeld, T.L., Jiang, Q., Venanzi, L.M. (**1994**) *Helv. Chim. Acta* 77, 1313–1330.

155 Seebach, D., Thaler, A., Beck, A.K. (**1989**) *Helv. Chim. Acta* 72, 857–867.

156 Haack, T., Mutter, M. (**1992**) *Tetrahedron Lett.* 33, 1589–1592.

157 Wöhr, T., Wahl, F., Nefzi, A., Rohwedder, B., Sato, T., Sun, X., Mutter, M. (**1996**) *J. Am. Chem. Soc.* 118, 9218–9227.

158 Wöhr, T., Mutter, M. (**1995**) *Tetrahedron Lett.* 36, 3847–3848.

159 Dumy, P., Keller, M., Ryan, D.E., Rohwedder, B., Wöhr, T., Mutter, M. (**1997**) *J. Am. Chem. Soc.* 119, 918–925.

160 Tuchscherer, G., Mutter, M. (**2001**) *Chimia* 55, 306–313.

161 Wöhr, T. (**1997**) PhD thesis, University of Lausanne.

162 Keller, M., Sager, C., Dumy, P., Schutkowski, M., Fischer, G.S., Mutter, M. (**1998**) *J. Am. Chem. Soc.* 120, 2714–2720.

163 Nefzi, A., Schenk, K., Mutter, M. (**1994**) *Protein Pept. Lett.* 1, 66–69.

164 Kang, Y.K. (**2002**) *J. Phys. Chem.* B106, 2074–2082.

165 Keller, M., Lehmann, C., Mutter, M. (**1999**) *Tetrahedron* 55, 413–422.

166 Keller, M., Mutter, M., Lehmann, C. (**1999**) *Synlett* S1, 935–939.

167 Mutter, M., Wöhr, T., Gioria, S., Keller, M. (**1999**) *Biopolymers (Pept. Sci.)* 51, 121–128.

168 White, P., Bloomberg, G., Munns, M. (**1998**) *Proceedings of the 25th European Peptide Symposium*, Bajusz, S., Hudecz, F. Eds. Akadémiai Kiado, Budapest, 120–121.

169 He, Y., Mountzouris, J., Wu, C. (**2005**) *Proceedings of the 19th American Peptide Symposium*, San Diego, in press.

170 Keller, M., Miller, A.D. (**2001**) *Bioorg. Med. Chem. Lett.* 11, 857–859.

171 Novabiochem (**2000**) *Innovations* 5/00.

172 Novabiochem (**2004**) *Innovations* 1/04.

173 Mutter, M., Nefzi, A., Sato, T.,Sun, X., Wahl, F., Wöhr, T. (**1995**) *Pept. Res.* 8, 145–153.

174 Sampson, W.R., Patsiouras, H., Ede, N.J. (**1999**) *J. Pept, Sci.* 5, 403–409.

175 Abedini, A., Raleigh, D.P. (**2005**) *Org. Lett.* 7, 693–696.

176 Chierici, S., Jourdan, M., Figuet, M., Dumy, P. (**2004**) *Org. Biomol. Chem.* 2, 2437–2441.

177 White, P., Keyte, J.W., Bailey, K., Bloomberg G. (**2004**) *J.Pept. Sci.*10, 18–26.

178 Goodmann, M. (Ed.) (**2003**) *Houben-Weyl, Methods of Organic Chemistry*, Vol E22. Thieme, Stuttgart.

179 Tam, J.P., Xu, J., Dong Eom, K. (**2001**) *Biopolymers (Pept. Sci.)* 60, 194–205.

180 Miao, Z., Tam, J.P. (**2000**) *J. Am. Chem. Soc.* 122, 4253–4260.

181 Von Eggelkraut-Gottanka, R., Machova, Z., Grouzmann, E., Beck-Sickinger, A. (**2003**) *ChemBioChem* 4, 425–433.

182 Keller, M., Boissard, C., Patiny, L., Chung, N.N., Lemieux, C., Mutter, M., Schiller, P.W. (**2001**) *J. Med. Chem.* 44, 3896–3903.

183 Wittelsberger,A., Patiny, L., Slaninova, J., Barberis, C., Mutter, M. (**2005**) *J. Med. Chem.* 48, 6553–6562.

184 Mutter, M. (**1977**) *J. Am. Chem. Soc.* 99, 8307–8314.

185 Kessler, H., Haase, B. (**1992**) *Int. J. Pept. Protein Res.* 39, 36–40.

186 Lambert, J.N., Mitchell, J.P., Roberts, K.D. (**2001**) *Perkin Trans. 1*, 471–484.

187 Davies, J.S. (**2003**) *J. Pept. Sci.* 9, 471–501.

188 Skropeta, D., Jollife, K.A., Turner, P. (**2004**) *J. Org. Chem.* 69, 8804–8809.

189 Rückle, T., de Lavallaz, P., Keller, M., Dumy, P., Mutter, M. (**1999**) *Tetrahedron* 55, 11281–11288.

190 Sager, C., Mutter, M., Dumy, P. (**1999**) *Tetrahedron Lett.* 40, 7987–7991.

191 Che, Y., Marshall, g.R. (**2004**) *J. Org. Chem.* 69, 9030–9042.

192 Lecoq, A., Boussard, G., Marraud, M., Aubry, A. (**1993**) *Biopolymers* 33, 1051–1059.

193 Lopez-Areiza, J.J., Rückle, T., Soto-Jara, C. PCT Int. Appl. **2004**, 52pp. CODEN: PIXXD2 WO 2004050689.

194 Bac, A., Rivoal, K., Cung, M.T., Boussard, G., Marraud, M., Soudan, B., Tetaert, D., Degand, P. (**1997**) *Lett. Pept. Sci.* 4, 251–258.

195 Tuchscherer, G., Grell, D., Tatsu, Y., Durieux, P., Fernandez-Carneado, J., Hengst, B., Kardinal, C., Feller, S. (**2001**) *Angew. Chem. Int. Ed.* 40, 2844–2848.

196 Wenger, R.M. (**1990**) *Transplant. Proc.* 22, 1104–1108.

197 Luban, J., Bossolt, K.L., Franke, E.K., Kaplana, G.V., Goff, S.P. (**1993**) *Cell* 73, 1067–1078.

198 Borvak, J., Chou, C.-S., Van Dyke, G., Rosenwirth, B., Vitetta, E.S., Ramilo, O. (**1996**) *J. Infect. Dis.* 174, 850–853.

199 Gamble, T.R., Vajdos, F.F., Yoo, S.H., Worthylake, D.K., Houseweart, M., Sundquist, W.I., Hill, C.P. (**1996**) *Cell* 87, 1285–1294.

200 Keller, M., Wöhr, T., Dumy, P., Patiny, L., Mutter, M. (**2000**) *Chem. Eur. J.* 6, 4358–4363.

201 Guichou, J.-F., Patiny, L., Mutter, M. (**2002**) *Tetrahedron Lett.* 43, 4389–4390.

202 Papageorgiou, C., Florineth, A., Mikol, V. (**1994**) *J. Med. Chem.* 37, 3674–3676.

203 Hubler, F., Rückle, T., Patiny, L., Muamba, T., Guichou, J.-F., Mutter, M., Wenger, R. (**2000**) *Tetrahedron Lett.* 41, 7193–7196.

204 Patiny, L., Guichou, J.-F., Keller, M., Turpin, O., Rückle, T., Lhote, P., Buetler, T.M., Ruegg, U.T., Wenger, R.M., Mutter, M. (**2003**) *Tetrahedron* 59, 5241–5249.

205 Wittelsberger, A., Keller, M., Scarpellino, L., Patiny, L., Acha-Orbea, H., Mutter, M. (**2000**) *Angew. Chem. Int. Ed.* 39, 1111–1115.

206 Adessi, C., Frossard, M.-J., Boissard, C., Fraga, S., Bieler, S., Rückle, T., Vilbois, F., Robinson, S.M., Mutter, M. (**2003**) *J. Biol. Chem.* 278, 13905–13911.

207 Mutter, M., Chandravarkar, A., Boyat, C., Lopez, J., Dos Santos, S., Mandal, B., Mimna, R., Murat, K., Patiny, L., Saucède, L., Tuchscherer, G. (**2004**) *Angew. Chem. Int. Ed.* 43, 4172–4178.

208 Dos Santos, S., Chandravarkar, A., Mandal, B., Mimna, R., Murat, K., Saucède, L., Tella, P., Tuchscherer, G., Mutter, M. (**2005**) *J. Am. Chem. Soc.* 127, 11888–11889.

209 Mutter, M., Hersperger, R. (**1990**) *Angew. Chem. Int. Ed.* 29, 185–187.

210 Mutter, M., Gassmann, R., Buttkus, U., Altmann, K.H. (**1991**) *Angew. Chem. Int. Ed.* 30, 1514–1516.

211 Hamada, Y., Ohtake, J., Sohma, Y., Kimura, T., Hayashi, Y., Kiso, Y. (**2002**) *Bioorg. Med. Chem.* 10, 4155–4167.

212 Carpino, L.A., Krause, E., Sferdean, C.D., Schühmann, M., Fabian, H., Bienert, M., Beyermann, M. (**2004**) *Tetrahedron Lett.* 45, 7519–7523.

213 Sohma, Y., Hayashi, Y., Kimura, M., Chiyomori, Y., Taniguchi, A., Sasaki, M., Kimura, T., Kiso, Y. (**2005**) *J. Pept. Sci.* 11, 441–451.

12

Peptidyl Prolyl Isomerases: New Targets for Novel Therapeutics?

Christophe Dugave

12.1
Introduction

Peptidyl prolyl isomerases (PPIases or rotamases) are ubiquitous proteins that are involved in a large number of biological processes and are implicated in numerous diseases. Consequently, they should be interesting targets for the development of novel therapeutics. For a long time PPIases were only related to immunosuppression and this prompted scientists to call some of them "immunophilins." Therefore, they were not considered as reliable therapeutic targets until the end of the twentieth century. Several reasons may account for this. First, the real implication of PPIases in several diseases was only highlighted in the late 1990s and, in many cases, is still misunderstood. Second, the overexpression of PPIases in such diseases was not clearly demonstrated until recently. Third, immunophilins form a large family of proteins [1,2], sometimes with few differences in terms of substrate specificity, which makes the development of selective inhibitors difficult. In many cases, PPIases were found to be accessory proteins that display a chaperone activity [3] and the fact that PPIases are individually and collectively dispensable for yeast viability [4] and are not essential for mammalian cell viability [5] was rather discouraging. All these reasons meant that immunophilins were for some time excluded from the search for PPIase-directed therapeutics, except in the case of the FK506 binding proteins (FKBP) subfamily which was the object of intensive research for the development of antineurodegenerative agents [6,7].

Since the end of the 1990s, PPIases have been identified as therapeutic targets and novel disease markers due to their important function as helper enzymes, in particular in steroid receptor signaling, coupled gating of the ryanodine receptor, transforming growth factor β (TGF-β) receptor activity, and increased levels in diseased areas [3].

In the very beginning of the search for new therapeutics targeting PPIases, especially hFKBPs, the development of ligands and inhibitors was mainly based on mimicry, with natural high-affinity ligands such as FK506, rapamycin and ascomycin. The relative tolerance of hFKBPs to ligand modifications led to the de-

cis-trans Isomerization in Biochemistry. Edited by Christophe Dugave
Copyright © 2006 WILEY-VCH Verlag GmbH & Co. KGaA, Weinheim
ISBN: 3-527-31304-4

velopment of a large set of small organic compounds and the definition of reliable pharmacophores [8].

Structural resolution of PPIase:ligand complexes by X-ray crystallography and NMR [2,8–11] has improved our understanding of the mode of interaction between PPIases and their substrates or inhibitors. This is particularly useful for the ab initio design and screening of inhibitors with simplified chemical structures and, although their affinities remain somewhat low (in the micromolar range), such compounds might provide valuable leads for the emergence of tomorrow's therapeutics directed towards PPIases [8]. Presently, there are more than 130 structures deposited in the Brookhaven Protein Data Bank with good resolution and, in most cases, ligand–protein interactions have been delineated with sufficient accuracy. Structure–activity relationship (SAR) studies have also yielded important information about the molecular interactions of PPIases and their effectors and about the molecular determinants of their catalytic activity [2,9,12].

In this chapter, we will tackle the involvement of PPIases in biological processes and disorders that may cause diseases. A growing number of natural and synthetic compounds have been used as PPIase inhibitors and, in many cases, their action is well understood and will be briefly described herein. This has opened the way to the design of novel molecules that might be employed as drugs in the coming years. In order to avoid overwhelming the reader with references, we will preferentially cite reviews which summarize previous work rather than the original publications themselves, except for papers published since 2000.

12.2
Implication of PPIases in Biological Processes and Diseases

12.2.1
PPIases and Protein Folding and Trafficking

Since 1984, when Fischer and coworkers isolated a PPIase activity from porcine kidney extracts, the *cis-trans* isomerase activity of PPIases has been related to nascent protein folding. PPIases seem to operate as chaperones in protein folding via multiple pathways that are disconnected from the isomerase activity, particularly in preventing protein aggregation [13–16]. This new function may explain the variety of cyclophilins and FKBPs (see Chapter 10) which have additional domains on top of their catalytic module. This particular function was discussed in depth in Fischer's excellent 2003 review [2]. However, this does not exclude a central role of amide *cis-trans* isomerization (CTI) in protein folding. Moreover, many members of the cyclophilin and FKBP subfamilies are located in the endoplasmic reticulum (ER) and are likely to play a part in the folding of nascent proteins [2] and in their trafficking [17]. The multiplicity of members of the PPIase families in mammals complicates discrimination between the enzyme's catalytic PPIase activity and the associations mediated by peripheral protein regions [1,16]. In fact, the situation is much easier to understand with single-domain PPIases such as hFKBP12 and hFKBP12.6, which interact with receptor protein kinases [2,3] and ion channels [3,18,19].

hFKBP36 deficiency has been reported in humans suffering from Williams syndrome, a developmental disorder affecting multiple organ systems characterized by cardiovascular and renal problems, hypercalcemia, dysmorphic facial features, and mental retardation. However, this congenital disease results from a haploinsufficiency of several genes and the exact role of hFKBP36 gene deletion is not yet clear [20].

PPIases are also associated with motor proteins: the first PPIase domain of hFKBP52 (and to a lesser extent hCyp40) binds to the microtubule-associated motor protein dynein [21] and some authors propose that the amino acyl proline CTI plays an important part in the protein transconformation that directs relative molecular motion in contractile muscle fibers [22]. However, there is presently no evidence that PPIases are directly implicated in muscle diseases.

12.2.2
Immunosuppressive Pathways Through Formation of PPIase:Ligand Complexes

Cyclosporin A (CsA), a cyclic undecapeptide isolated from *Trichoderma polysporum* (Fig. 12.1), has long been known as a potent immunosuppressant and has been successfully employed to prevent allograft rejection [23]. In 1989, two independent research teams identified cyclophilin as the specific target of CsA. At the same time, Schreiber and coworkers, as well as members of Vertex Pharmaceuticals, reported that another cytosolic protein called FKBP12 was the receptor of the immunosuppressant FK506 (tacrolimus), a macrolide extracted from *Streptomyces hygroscopicus*. The macrolides rapamycin and ascomycin have also been related to the FKBP12 immunosuppressive pathway, though rapamycin affects distinct phases of T cell activation [10,23,24] (Fig. 12.1).

The immunosuppressive activity of CsA- and FK506-related compounds has been thoroughly investigated and, though they have distinct chemical structures, both immunosuppressants clearly suppress the transcription of interleukin-2 (IL-2) and other cytokines required for T cell activation and proliferation via a common pathway. In particular, hCyp18 regulates IL-2 tyrosine kinase (Itk) activity, which controls T cell activation, by switching the Asn286-Pro287 motif in the SH2-SH3 domain from *cis* (inactive Itk) to *trans* (active Itk) [25,26]. In fact, hCyp18 (the major cytosolic cyclophilin in human) and hFKBP12 cannot display immunosuppressive activity alone and must interact with CsA and FK506 respectively to form a complex which is able to inhibit calcineurin, a calcium-dependent phosphatase controlling the activity of NF-AT, the nuclear factor of activated T cells [2,6]. The high specificity of CsA and FK506 for the immune system might be explained by the low calcineurin concentrations in T cells, which contrast with higher levels expressed in other cells such as nerves and cardiomyocytes. Pretreatment of T cells with CsA also impairs the transcriptional activity of AP-1 and NF-κB, suggesting the existence of another immunosuppressive pathway. Immunophilin:immunosuppressant drug complexes block the MAPK cascade, which

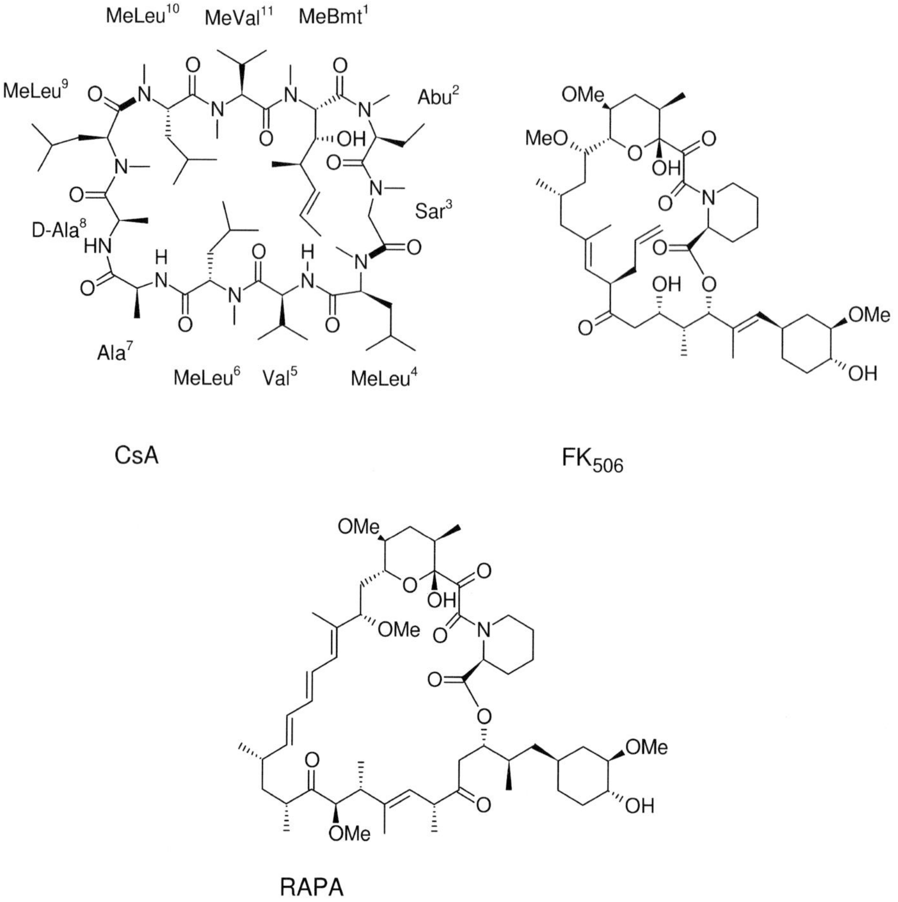

Fig. 12.1 Structures of cyclosporin A (CsA), FK506, and rapamycin (RAPA).

controls the transcription of various transcription factors including AP-1 and protein kinases involved in stress responses like inflammation and apoptosis [24].

Curiously, the immunosuppressive activity of rapamycin (RAPA) does not involve the inhibition of IL-2 synthesis even though RAPA shares some structural similarities with FK506. The hFKBP12:RAPA complex interacts with a protein (RAFT in rat brain and FRAP in human) which controls the mRNA translation promoting the IL-2-stimulated G_1 to S phase transition in T cells [6].

Other compounds, in particular sanglifehrin A (SFA) [27,28] and cyclopeptides such as cyclolinopeptides [29,30], display interesting immunosuppressive activities. Although the interactions between SFA and hCyp18 have been delineated [31], the molecular target of the SFA:hCyp18 complex is not yet identified. It seems that SFA blocks IL-dependent proliferation and cytokine production of T cells via a novel mode of action relative to CsA, FK506, and RAPA [27,28].

In summary, though they act via distinct pathways, immunophilins may be considered as presenter proteins whose activity is conditioned by the nature of the effector interacting at their active site. However, CsA- and FK506-related compounds have additional properties such as neurotrophic, cardioprotective, and antimitotic effects [32] which are not yet completely understood.

12.2.3
Modulation of Ion Channels by PPIases

The ryanodine receptor (Ryr), the major calcium release channel in the ER of skeletal and cardiac muscle, is required for excitation–contraction coupling. The Ryr protein forms a tetrameric channel in specific regions of the ER and includes four associated hFKBP proteins. hFKBP12 is associated with Ryr1 channels found in skeletal muscle whereas hFKBP12.6 binds specifically to cardiac muscle Ryr2 channels. Ryr molecules are organized in regular arrays in which channels are in close contact with each other. When hFKBP12.6 is bound to the channels, these open and close simultaneously as a single Ca^{2+} release entity, whereas absence of PPIases does not disorganize the channel cluster but suppresses the full conductance that results from the simultaneous opening and closing [19,33].

In Ryr1, hFKBP12 putatively catalyzes CTI on the Val2461-Pro2462 moiety. Val to Gly substitution abolishes binding of the channel to FKBPs while mutation from Val to Ile (the corresponding residue in Ryr2) confers selective binding to the channel on FKBP12.6. The Ryr2:hFKBP12.6 interaction is abolished by hyperphosphorylation of a serine residue of the channel as observed in patients suffering from severe heart failure. This situation results in depressed and prolonged Ca^{2+} transients that alter muscle performance and may cause fatal cardiac arrhythmias.

HFKBP12 also binds to the inositol triphosphate receptor IP3R, the other major calcium-release channel in the ER which is found in neurons and cardiac tissue. The interaction takes place at a Val-Pro moiety and is inhibited by FK506, suggesting that the active site of hFKBP12 is involved in the binding [3].

12.2.4
Chaperone Activity of Immunophilins in Steroid Receptor Signaling

Multidomain PPIases such as hCyp40 (CypD), hFKBP51, and hFKBP52 possess several tetratricopeptide repeat (TPR) domains in addition to the catalytic PPIase domain which contain essential recognition motifs, in particular for the hsp90 protein, a constituent of the steroid receptor heterocomplex (see Chapter 10). Upon binding of the steroid hormone, the mature (PPIase-associated) receptor migrates to the nucleus and the heterocomplex components dissociate. The steroid receptor protein is therefore available to form homodimers which act as transcription factors able to bind to DNA. Many PPIases, including hCyp40, hFKBP51, and hFKBP52, may associate with the steroid receptor and the nature of the PPIase in the mature stage might direct the functional diversity of the receptor [3,12,34]. For example, the progesterone receptor preferentially associates with hFKBP51 (despite a 5-fold relative abundance in hFKBP52). HFKBP51 is also preferred over hFKBP52 and hCyp40 in the assembly of the glucocorticoid receptor in vitro. Conversely, hCyp40 predominates in estrogen receptor complexes in bovine myometrial cytosol while hFKBP52 is found as the exclusive PPIase in the estrogen-responsive MCF-7 breast cancer cell line [35,36].

The exact role of PPIases is not yet fully understood. Their catalytic activity does not seem to be implicated in the assembly and composition of heterocomplexes, since these are insensitive to treatment with PPIase inhibitors. It has been proposed that PPIases might control the nuclear targeting of the receptor, a role consistent with the nuclear localization of hFKBP52 and hCyp40. It is noteworthy that the relative abundance of multidomain immunophilins may change in response to varied hormonal stimuli that upregulate the incorporation of TPR immunophilins into steroid receptor complexes [3,34].

Although PPIases seem to have an accessory role as disease biomarkers, they are likely to be important determinants of cellular dysfunction related to steroid hormone signaling as well as reliable targets for drug development [37].

12.2.5
Immunophilins and Neurodegenerative Disorders

Over the past decade, considerable evidence has accrued suggesting that immunophilin ligands display neurotrophic effects in various in vitro and in vivo systems, and that PPIase activity might be involved in neurodegenerative processes [38,39]. Use of [^{3}H]FK506 showed that hFKBP12 levels change in neuronal diseases since they decline in dying neurons and are elevated in less severely injured neurons. In a rat model, the FKBP12 level declines markedly in the hippocampus and increases in the area of cerebral infarction following focal ischemia [40]. In comparison with normal human brain, FKBP12 distribution and levels increase considerably in several areas of the brains of patients suffering from Parkinson's disease, Alzheimer's disease, and dementia with Lewy bodies [41–43]. HFKBP12 concentrations in the central nervous system are up to 50 times those in the immune system. Recent studies have shown that hFKBP12 and hFKBP52 are also abundant in the spinal cord [44].

It is difficult to discriminate the direct effects of PPIases on neuronal proteins from their chaperone activity in maintaining the function of receptor heterocomplexes since steroid receptors, the Ryr calcium channel, and IP3 receptors are also present in neurons. In particular, FKBP-controlled disruption of the steroid receptor heterocomplex is known to mediate neurite elongation via a "gain of function" [7,45]. The neuroregenerative activity of FK506 also involves the MAP kinase pathway, which mediates the neurotrophic activity of FK506. Moreover, hFKBP12 acts as an inhibitor of TGF-β, which causes cell cycle arrest [46].

FK506 is neuroprotective against global and focal ischemia in toxic chemical models and mechanical injury to a peripheral nerve. Its neuroprotective effects against ischemia seem to be dependent upon calcineurin inhibition but are independent in nonischemic diseases since FK506-related compounds devoid of immunosuppressive activity are equally active [6–8].

The presence in many neurodegenerative diseases of neurofibrillary tangles containing α-synuclein colocalized with hFKBP12 and other molecular chaperones, in particular hsp70 and hsp40, suggests that the presence of intraneuronal inclusions might be related to the CTI activity of PPIases. Moreover, the presence of hFKBP12 in filamentous τ lesions might be interpreted as an attempt to counteract the hyperphosphorylation of paired helical fragments usually found in neurofibrillary tangles [42,47].

Alzheimer's disease as well as other neurodegenerative disorders such as Pick's disease, corticobasal degeneration and supramolecular palsy are characterized by an accumulation of hyperphosphorylated τ protein and subsequent formation of paired-helical fragments which form neurofibrillary tangles [8,48]. Phosphorylation of Thr231-Pro232 and Ser235-Pro236 decreases the rate of uncatalyzed *trans* to *cis* isomerization and thus reduces binding of τ protein to tubulin. Therefore, τ proteins become more available for aggregation as PHF. Although Pin1 accelerates CTI of phosphorylated Ser/Thr-Pro sequences, its binding to the *trans* conformer of τ protein (either via its catalytic site or a WW domain) results in sequestering of Pin1 and a dramatic fall in soluble PPIase [47,49,50].

Prion diseases such as bovine spongiform encephalopathy, scrapie in sheep, kuru, and Creutzfeldt–Jakob disease, which are all characterized by irreversible aggregation of a proteinase conformer of protein PrP^c (called PrP^{Sc}), might also involve CTI of one or several amino acyl prolyl sequences [8]. Residue Pro101 of the prion protein has been clearly implicated in the formation of β-sheet fibrils [51] as well as in the model Ure2 protein isolated from *Saccharomyces cerevisiae* which contains a prion domain that folds very much like that of the prion protein [52]. However, the exact role of PPIases in prion diseases remains unclear, though recent reports show that when cells are cultured with the immunosuppressant CsA there is accumulation of PrP-like aggregosomes resembling those of scrapie [53].

12.2.6
PPIases and Cell Multiplication

Due to its specificity for phosphorylated Ser-Pro and Thr-Pro sequences, Pin1 plays an important part in the phosphorylation pathways which control cell division. Pin1 is required for DNA replication in *Xenopus laevis* and the involvement of the PPIase active site (in particular Cys109) rules out the participation of the Pin1 WW domain. Although the Pin1 WW domain exhibits affinity for the Thr^P-Pro sequences, this is not sufficient to support cell growth in inactive Pin1 variants. This suggests that this domain, which interacts with a multitude of phosphoproteins, is not the prime cause of the biological activity of Pin1 [47,54].

Studies of isomer-specific dephosphorylation by protein phosphatase 2a (PP2a) demonstrated that this enzyme is unable to dephosphorylate the *cis*-Ser^P/Thr^P-Pro moieties in model peptides [55,56]. A similar specificity was reported for the protein kinase Erk2, which only phosphorylates *trans*-Ser/Thr-Pro moieties [57]. Therefore, Pin1 is likely to play a regulatory function in the phosphorylation/ dephosphorylation cascade which controls cell replication [58].

The role of hFKBP12 in the receptor–protein kinase heterocomplex suggests that it also plays an important part in the regulation of intracellular mechanisms. This is so for the TGF-β receptor and epidermal growth factor (EGF) receptor. In the latter case, hFKBP12 inhibition of a protein tyrosine kinase [59] in many ways resembles the inhibition of Ser/Thr-Pro phosphorylation. In particular, FK506 and rapamycin enhance the phosphorylation of the EGF receptor, causing a decrease in cell growth.

Cyclophilins also participate in the regulation of transcription, by interacting with transcriptional inducers. As an example, the prolactin:CypB(wtCyp23) complex directly interacts with Stat5, thus enhancing Stat5 binding to DNA [60]. However, other modes of gene regulation are also possible: Pin1, Ess1, and hCyp18 might regulate transcription by interacting with chromatin [61]. Nuclear cyclophilin binds DNA in a zinc-dependent manner in macrophages and is suspected to recognize specific DNA sequences directly [62]. Therefore, some PPIases seem able to interfere directly with the genome in addition to mechanisms involving phosphorylation/dephosphorylation processes or receptor internalization.

The implication of PPIases in cancer, both as biochemical markers and as inducers and enhancers, is an important question and is the subject of a growing number of studies [63–65]. The constantly lengthening list of cancers involving enhanced PPIase levels precludes exhaustivity, but some relevant and recent examples are given in Table 12.1. It is noteworthy that PPIases are not overexpressed in all tumors and may also display anti-invasive properties [66].

In summary, PPIases seem to play more than a secondary part in cancer induction and progression. Although their exact function is not fully understood, some are possible biological markers for cancer detection and novel targets for anticancer therapeutics. Pin1 is particularly attractive for these purposes since it is expressed at very low levels in normal tissues while its expression is normally associated with cell division. A statistical study in a variety of human cancers has shown that Pin1 overexpression is a prevalent and specific event in human cancers [64] as it increases the transcription of several target genes, including cyclin D1 and c-myc, and cooperates with Ras signaling [83].

Table 12.1 Some examples of upregulated PPIases detected in human cancer cell lines and their putative specific functions in cancer progression.

PPIase	Cancer	Putative function	Ref.
hCyp18	Colorectal cancer	Unknown	67
	Prostate metastasis to bone	Unknown	68
	Lung cancer	Unknown	69
COAS2	Several aggressive metastatic chemotherapy-resistant tumors	Overexpressed almost exclusively in these tumors	70
HCypB (hCyp22.7)	Pancreatic cancer	Unknown	71
hCyp40 and FKBP52	Breast cancer	Chaperone for estrogen receptor	72
HCypD (hCyp22)	Breast, ovarian, uterine cancers	Apoptosis repressor	73
HFKBP12	Prostate cancer	Unknown	74
hFKBP51	Prostate cancer	Increases androgen receptor transcription	75
HFKBP51/52	Breast cancer	Control of progestin and glucocorticosteroid receptor-mediated transcription	76
	Idiopathic myelofibrosis	Dysregulation of apoptosis	77
hFKBP65	Colorectal cancer	Folding of other tumor-associated proteins	78
Pin1	Murine breast cancer	Enhances cyclin D1 transcription and stabilization	79
	Hepatocellular carcinoma	Upregulation of β-catenin	80
	Prostate cancer	Unknown – marker of elevated risk of recurrence	81
	Colorectal cancer	Concomitant overexpression of Pin1 and β-catenin	82
	Squamous cell carcinoma	Overexpression correlates with cyclin D1 overexpression	83

12.2.7
Implication of PPIases in Apoptosis

In 1999, Lee and coworkers [84] reported that rotamase activity is involved in Cu/Zn mutant superoxide dismutase-1-induced apoptosis of neuronal cells. Since then, several studies have highlighted the role played by certain PPIases in cell death. In particular, hCypD (a mitochondrial variant of hCyp18) is a component of the mitochondrial permeability transition pore (MPTP), a key player in apoptosis and necrosis [85] and, in more generally, in the response to intracellular stress [86]. The exact role of PPIases is still a matter of discussion [85,87], but their CTI activity is clearly implicated in the formation of MPTP [84] and in several other apoptotic processes, including control of the activity of the tumor suppressor and apoptosis inducer p53 [47,88]. CsA and SFA, a potent inhibitor of hCypD, block the opening of MPTP [89], protecting the cell from apoptosis, a result which seems to be in apparent contradiction with previous findings suggesting that CsA induces apoptotic cell death [90]. The involvement of PPIases in apoptotic pathways is complex and may overlay other cellular dysfunctions such as necrosis [85] and neurodegeneration [47,91]. Basak and coworkers recently showed that HP0175, a PPIase secreted by *Helicobacter pylori*, induces apoptosis of gastric epithelial cells via activation of the caspase pathway, a determinant of the apoptotic process [92].

12.2.8
PPIases and Infectious Diseases

PPIases are also found in many unicellular organisms such as bacteria, fungi and parasites and, in most cases, are essential for maintaining membrane integrity, multiplication, virulence, and infectivity of these organisms [2,93,94]. In particular, deletion of PPIase genes attenuates the virulence of pathogenic strains such as *Salmonella typhimurium* (FKBPs), *Salmonella enterica* (SurA), *Legionella pneumophila* (FKBP25), and *Cyanobacterium synechococcus* (Cyp15) [2]. For example, the macrophage infectivity potentiator (MIP) has been shown to be essential for infection by *Legionella pneumophila*, the causative agent of severe and often fatal Legionnaire's disease [95]. It has a FKBP-like PPIase activity which seems to be essential for virulence and is inhibited by FK506 [96,97].

Bacterial PPIases can also display toxin-like activity and are able to act as pathogenic agents. In plants, a eukaryotic cyclophilin is essential for the activation of AvrRpt2, a cysteine protease from *Pseudomonas syringae*, which is delivered into plant cells [98].

PPIases play an important role in the infectivity of several protozoan parasites such as *Leishmania* and *Trypanosoma*. *Leishmania major* possesses a cyclophilin (Cyp19) whose inhibition by CsA leads to a complete loss of infectivity, suggesting a central role of the enzyme in the development of the parasite [99]. *Trypanosoma cruzi* macrophage infectivity potentiator (TcMIP) possesses a catalytic core which displays a PPIase activity very similar to that of FKBPs and is inhibited by FK506-

related compounds. Resolution of the crystal structure of TcMIP raises the possibility of designing new inhibitory drugs for the treatment of trypanosomiasis [100]. Immunosuppressants display interesting antifungal properties as well [101].

PPIases also play a key role in cell entry and in the replication of several pathogenic viruses, as well as in the formation of mature virions. The vesicular stomatitis virus New Jersey serotype, which causes major diseases in animals, particularly cattle, contains several host proteins, including cyclophilin A, which are essential for viral replication. CsA and overexpression of catalytically inactive mutants of CypA drastically inhibit gene expression of the virus New Jersey serotype but have little effect on the virus Indiana serotype [102]. Although it is presently the subject of controversial publications and is still misunderstood, the involvement of PPIases – notably hCyp18 – in the HIV-1 life cycle is one of the most famous examples of the hijacking of host proteins by a lentivirus [8].

Cyclophilin is involved in several steps of the viral infection process [103]. It is incorporated inside new virus particles [104] together with several other host proteins [105–107] and a small part of it pokes out of the viral membrane. It is therefore able to attach to specific molecules on the target cell like glycosaminoglycan [108] and CD147 [109,110], a cell-surface protein that is essential for the attachment of HIV-1 to T cells. It might also control the gp120:CCR5 interaction, which mediates the first step of HIV-1 entry into cells [111–113], but the involvement of PPIase activity is unclear since residues 12–17 of the N-terminus of CCR5 are essential for hCyp18 binding [114].

HCyp18 also binds specifically to the capsid protein (CA) which is one of the proteolytic products of the precursor polyprotein Gag. Here again, the exact function of cyclophilin remains ambiguous [115,116] since it is able to catalyze CTI of the [Gly89-Pro90] sequence of the immature polyprotein Gag and forms a stable complex [117–120] the structure of which has been resolved by X-ray crystallography [11,121,122]. It is noteworthy that hCyp18 displays a higher affinity for other Gly-Pro sequences on Gag which could be secondary sites of interaction [123–125]. The influence of hCyp18:Gag interaction on the viral cycle is not fully understood. Cyclophilin might modulate the posttranslational cleavage of Gag by the viral protease [124] and is also suspected to control the conformation of the CA protein and, by extension, the transition from a spherical to a conical mature viral core [126]. About 250 molecules of hCyp18 are integrated into the viral core where they are suspected to destabilize the structure in order to facilitate core opening [127,128] and to allow the release of the viral RNA inside infected T cells.

Recently, hCyp18 was shown to interact with Vpr, an HIV-1 accessory protein [129] which carries "nuclear localization signals" and facilitates the nuclear entry and action of the "preintegration complex" (made of viral RNA, reverse transcriptase and integrase) and hence regulates the integration of the HIV genome into the host cell's genome [130]. Preliminary data support the involvement of a peptidyl prolyl CTI in the folding and function of the viral protein Vpr [131] and explain the previously observed interaction between Vpr and hCyp18 [130].

In summary, the involvement of cyclophilin in the HIV-1 replication cycle opens up an exciting field of research into novel anti-AIDS drugs that might be administered in synergy with classical multidrug regimens. However, several points remain to be clarified, in particular: (1) What is the exact role of the host cell cyclophilin relative to the viral embedded cyclophilin [132]? (2) Are there several isoforms of cyclophilin (and by extension other PPIases) that could play essential but distinct roles in the HIV-1 replication cycle [103,133,134]? (3) Are all the various putative functions of cyclophilin relevant? The recent finding that hCyp18 putatively protects HIV-1 from Ref-1 restriction factor might herald advances in new therapeutic strategies for the eradication of AIDS [135].

Very recently, Jiang and coworkers showed that the nucleocapsid protein of SRAS (severe respiratory acute syndrome) coronavirus tightly binds to hCyp18 (K_d ranging from 6 to 160 nmol L^{-1}). A combination of bioinformatic methods and mutagenesis experiments suggested that the interaction takes place at an unusual sequence Ala306-Glu307 through the formation of several H-bonds in a way very similar to that of the Gag CA:hCyp18 complex [136].

12.3
Structure and SAR studies of PPIases: Structural Evidence and Putative Catalytic Mechanism

12.3.1
Generalities

Among the plethora of NMR and crystallographic structures available in the Brookhaven Protein Data Bank, most concern either cyclophilin alone or associated with various substrates and effectors (81 structures). FKBP structures have essentially been resolved alone or bound to their effectors, including inhibitors (45 structures including MIP). Pin1, Par 14, and ESS1 (respectively 8, 4, and 1 structures) and the trigger factor (7 structures) are also represented [12].

Presently, the large secondary amide peptide bond *cis-trans* isomerase (APIase) Hsp70 (DnaK) has only been partially investigated and does not give reliable structures that could account for a catalytic mechanism and a particular substrate specificity. The recent solution structure of the DnaK[393–507] fragment complexed with a heptapeptide shows that the peptide substrate-binding domain does not contain helical structures, a particular superfold of the PPIase family, since helices are found in cyclophilins and FKBP substrate-binding domains [137]. Moreover, in contrast to PPIases, association of unfolded polypeptides with DnaK usually gives tight Michaelis complexes that necessitate the hydrolysis of ATP and the presence of two accessory proteins (DnaJ and GrpE) to release the product. Although DnaK is a heat shock protein that has not been yet implicated in particular diseases, this characteristic feature (in addition to the fact that it does not bind Xaa-Pro sequences) might be an important basis for the design of specific inhibitors. Moreover, DnaK is able to prevent protein aggregation in contrast to archetypical PPIases. This suggests that the function of APIases and PPIases is dependent on additional domains which modulate their selectivity, in particular TRP motifs in multidomain cyclophilins and FKBPs. The existence of extended protein sequences which control a particular location of PPIases is also a determinant for substrate/inhibitor selectivity [1,2,9].

12.3.2
Cyclophilins and FKBPs: Similar Molecular Basis for Distinct Catalytic Mechanisms

The PPIase domains of cyclophilin and FKBPs do not display any similarities whereas members within each subfamily show a high degree of homology in sequence and in three-dimensional structure, suggesting that conservation of the overall shape of the active site and of certain residues is essential for PPIase activity. Moreover, both enzymes feature structural differences from parvulins, in particular Pin1 (Fig. 12.2) [1,9,12].

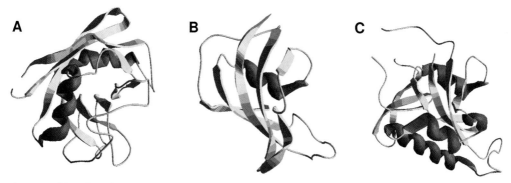

Fig. 12.2 Ribbon three-dimensional structures of hCyp18 and hFKBP12 and partial structure of Pin1 (fragment [40–44] is missing).

The archetypal cyclophilin A hCyp18 has an eight-stranded antiparallel β-sandwich terminated at both ends by two helices. The binding site is divided into two functionally independent subsites [138]: S1–S1' is covered by H-bond donors and acceptors except in the narrow hydrophobic pocket that accommodates the Pro pyrrolidine ring and a proximate S2'–S3' subsite which is mainly hydrophobic and is thought to be accessible to solvent and relatively bulky ligands. This structure is well conserved throughout the Cyp family. The narrow "proline pocket" means that cyclophilins are poorly permissive to ligand modifications of the pyrrolidine moiety. In particular, hCyp18 does not tolerate ring contraction (azetidine or aziridine) nor introduction of bulky substituents on the 5-membered ring at C4 and C5 (in particular alkylprolines), with the exception of fluoroprolines [8,139]. In this respect, the high-affinity binding of SFA to hCyp18 suggests that the 6-membered ring fits inside the active site in a way that is radically different from that of proline-containing compounds [31] The PPIase activity is also strongly affected by modification of the proline puckering which perturbs important interactions, suggesting that the spatial positioning of the nitrogen is essential for efficient binding. Trp121 also plays an important role in the binding of substrates, ligands and peptide inhibitors. In particular, the Trp/Phe substitution in *E. coli* cyclophilin accounts for its low sensitivity to the immunosuppressant CsA. Conversely, Phe/Trp mutation completely restores the binding of CsA. Moreover, bovine cyclophilin is inhibited by CsA though Trp is replaced by a His residue, which underlines the importance of H-bonding on this position for CsA binding. Many X-ray and NMR structures of cyclophilin complexed with CsA and analogs are available and the cyclophilin- and calcineurin-binding motifs of the undecapeptide have been delineated. Most residues interacting with the inhibitor in the CypA:CsA complex have also been implicated in the binding of peptide effectors and substrates, confirming the importance of a limited number of critical amino acids. The relatively solvent-exposed subsite S_1–S_1' explains the slight specificity of hCyp18 for this position, provided the side-chain is not too bulky [8,123]. Residues Arg55, Gln63 and Asn102 seem to be essential for both binding and catalysis [140,141] but also involve a larger network of hydrogen bonds which is highly conserved in cyclophilins from various species [142] (Fig. 12.3).

A particular feature of Cyp:ligand complexes is that the Xaa-Pro amide bond of peptide substrates may be found in either the *cis* or *trans* conformation. In fact, all complexes with di- and tetrapeptides crystallized in an exclusive *cis* conformation, though both conformers have been found to bind to cyclophilin to give stable complexes due to the existence of a highly mobile equilibrium. Conversely, all Gag-derived peptides possess a *trans* conformation whereas the protein undergoes a cyclophilin-catalyzed CTI process. Here again, the macrolide SFA does not bind to Trp121, suggesting an atypical pharmacophore model for SFA binding (Fig. 12.3) [31].

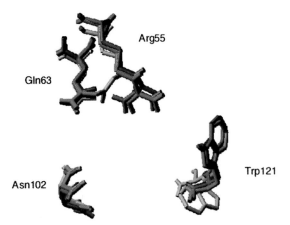

Fig. 12.3 Overlay of X-ray structures of hCyp18 complexed with Suc-Ala-Ala-Pro-Phe-pNA (red), CsA (blue), Gag CA[81–105] fragment (green), and SFA (cyan) using the Swiss PDB Viewer Software 3.7b2, Glaxo-Wellcome). Only essential residues which contract H-bonds with the ligand are shown.

FKBPs also adopt a highly conserved fold made of a five-stranded antiparallel β-sheet wrapped around a short α-helix linked together with flexible loops. The FKBP active site is made of mainly hydrophobic residues which are highly conserved throughout the FKBP family. They might favor the formation of intramolecular H-bonds inside the peptide substrates in addition to enzyme–substrate interactions with Asp37 and Val55 (in hFKBP12). As a consequence, FKBPs prefer Xaa-Pro sequences when Xaa is not too bulky a hydrophobic amino acid (e.g. Leu, Phe). In turn, the structural analysis of FKBP:macrolide complexes has shown that FKBP isoforms hFKBP12 and hFKBP12.6 bind rapamycin via identical interactions and display similar structures [143]. Subtle differences may account for particular specificities, but all FKBPs, including multidomain enzymes such as hFKBP51 and hFKBP52, adopt similar folds in their catalytic core. The multiplication of hydrophobic interactions versus intermolecular H-bonds might account for the broad permissivity of FKBPs toward ligands since they bind a wide variety of molecules, from highly constrained macrolides such as FK506 to very short molecules. This has opened the way to the design of a rich variety of ligands that bind FKBP with nanomolar affinities [6,144]. Surprisingly, there is no resolved structure of an FKBP:substrate complex, but substrate binding has been modeled by analogy with inhibitors. In particular, the crystal structure of the cytoplasmic domain of the TGF-β receptor in complex with hFKBP12 showed that the PPIase binds directly to the atypical Leu195-Leu196 moiety, the side-chains of which are accommodated inside the hydrophobic pocket in a way that is very similar to that of FK506 [145].

The exact catalytic mechanism of these PPIases is not fully understood. Many hypothetical processes have been proposed and most of them are supported by relevant data in small molecule models (see Chapter 8). All data rule out a nucleophile-assisted CTI that involves a tetrahedral covalent intermediate, and experimental arguments as well as calculations suggest that CTI is initiated by the H-bond-assisted tetrahedralization of the proline nitrogen by the Arg55 guanidinium moiety, while the Gln63 γ-amide and Asn102 backbone amide might assist the subsequent rotation of the deconjugated (keto-amine state) peptide bond [8,12,140] (Fig. 12.4). Therefore, CTI might occur via a concerted mechanism which involves limited side-chain motion [146] as well as more extended protein vibrations [141,142].

In contrast to cyclophilin, the hydrophobic FKBP active site is anticipated to bind the substrate in a type VIa β-turn (whereas hCyp18 induces a typeVIb β-turn conformation) which favors the formation of intramolecular H-bonds and thus might facilitate the isomerization process via the partial tetrahedralization of the proline nitrogen [8,12]. Although this hypothesis is supported by numerous examples of self-assisted isomerization in constrained peptide models [147,148] and folding tetrahydrofolate reductase (see Chapter 8), the recent observation by Fischer and coworkers that the active sites of hCyp18 and hFKBP12 have marked structural similarities (superimposition of several residues which coat the active sites of both enzymes) shows that identical or equivalent side-chains adopt similar conformations [12]. However, some significant differences in term of geometries

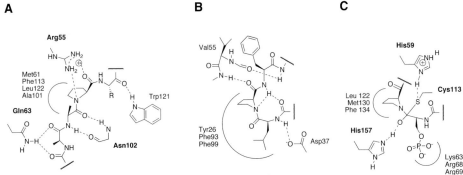

Fig. 12.4 Putative transition states of hCyp18- (A) hFKBP12- (B) and Pin1-catalyzed CTI (C); residues putatively implicated in the catalytic process are in bold.

(in particular for hCyp18 Arg55 and hFKBP12 Arg42) might explain why the two families of enzymes have distinct mechanisms. Three-dimensional structural relationships might therefore reflect similarities in terms of function and substrates rather than an evolutionary process toward a common mechanism (Fig. 12.4).

12.3.3
Parvulins

Among the different structures of Pin1 available, only one shows this PPIase bond to the Ala-Pro dipeptide, which fits inside the putative catalytic site [149]. In fact, Pin1 is highly specific for Ser^P-Pro and Thr^P-Pro sequences and, to a less extent, for the Glu-Pro moiety. Therefore, the relevance of this structure for understanding the Pin1-catalyzed CTI mechanism is controversial, though it gives important structural information (Fig. 12.4). A similar situation has been observed with hCyp18:dipeptide complexes whose structures are not identical to those of larger substrates in terms of functional interactions and positioning inside the active site [150,151]. Starting from the structure of the Pin1:Ala-Pro complex and mutagenesis experiments, Ranganathan and coworkers proposed a basically different picture. In contrast with other PPIases, the Pin1 transition state might involve a covalent tetrahedral intermediate through nucleophilic attack from Cys113. In this scheme, His59 and His157 act as H-bond acceptor and H-bond donor, respectively, in a general acid–base catalysis [149]. The importance of the cysteine residue is supported by the inactivation of Pin1 by juglone, which irreversibly reacts with Cys113, although the real effect of cysteine alkylation is still unclear [152,153].

Here again, Fischer's team demonstrated that unexpected similarities exist between hCyp18 and Pin1 active site residues, though Pin1 Arg68 strongly deviates from the overall orientation of hCyp18 Arg55 [12]. However, the existence of a cluster of positively charged residues in the region that is anticipated to bind the phosphoric ester moiety of the substrate strongly suggests that the evolutionary convergence of PPIase active sites only reflects substrate similarities and does not preclude the coexistence of distinct mechanisms.

Other NMR and X-ray structures were obtained from Pin1 either in the free state [154] or in complex with phosphorylated Ser/Thr-Pro sequences that bind the Trp-Trp domain on the enzyme [155,156]. However, binding of phosphorylated sequences to this domain seems to be disconnected from PPIase activity, though it participates in the specific binding of hyperphosphorylated proteins [47].

Other parvulins also give important structural and mechanistic information. In particular, Pin1At, the Pin1 homolog from *Arabidopsis thaliana*, displays a high degree of similarity with Pin1 despite the absence of the Trp-Trp site. In particular, all residues defined as essential in Pin1 are represented in Pin1At. In this enzyme, however, binding of the substrate seems to cause the Cys70 side-chain to protrude outside the catalytic site, making it unlikely to interact with the amino acyl proline peptide bond [157].

The human parvulin Par14, which exhibits a particular specificity for positively charged Xaa-Pro sequences, differs considerably from Pin1. In particular, the cluster of positively charged side-chains is replaced by two negatively charged residues, a situation that is consistent with the binding of a positively charged sequence. In addition, Cys113 is replaced by Asp74 while a serine residue is exchanged with an important phenylalanine [158]. However, although Pin1 and other parvulins are structurally related, they may act via basically different catalytic mechanisms.

12.4
PPIase Inhibitors: From In Vitro Inhibitors to Novel Therapeutics

12.4.1
Natural PPIase Inhibitors and Their Analogs

CsA, the most potent peptide inhibitor of many members of the cyclophilin family ($K_{i\ hCyp18}$ = 6 nmol L^{-1}), is a cyclic 11-mer that contains exclusively hydrophobic amino acids, most of them N-methylated. Moreover, it displays an atypical MeBmt residue which is essential for CsA activity though it does not interact with cyclophilins. The interaction network that stabilizes Cyp:CsA complexes was resolved by NMR and X-ray crystallography more than 10 years ago and, as anticipated, involves five H-bonds and a set of hydrophobic contacts in a way very similar to that observed in Cyp:substrate complexes (Fig. 12.5) [9,159]. In particular, Arg55, Gln63, and Asn102, which are likely to play a direct role in the catalytic process, are involved in the interaction. Trp121 also binds CsA via a H-bond with the MeLeu9 carbonyl, an interaction also observed with many peptide substrates where Trp121 binds the S2' carbonyl [11,12]. The large Abu pocket which accommodates Abu2 seems to be able to accept larger side-chains (as seen in CsB, C, D, G, and H as well as synthetic analogs), although the site is filled with several water molecules which are responsible for a lower affinity and a dramatic fall in immu-

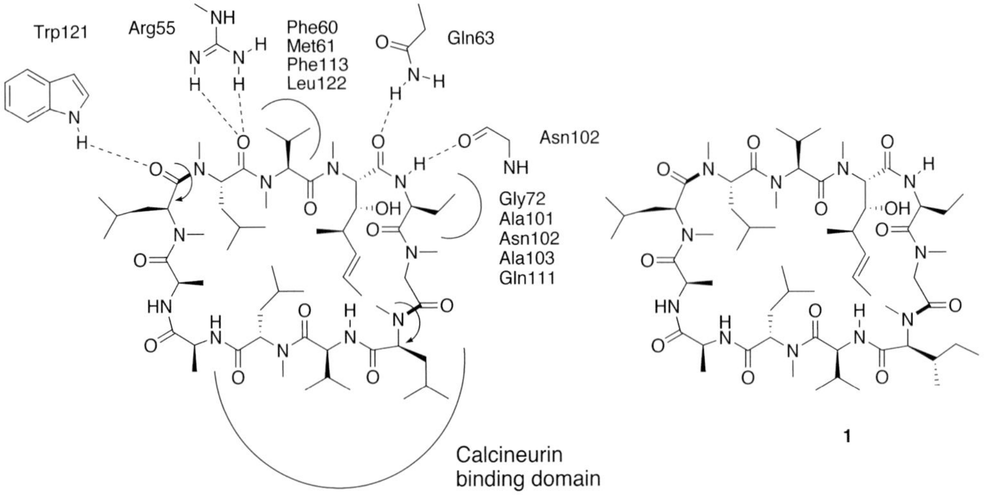

Fig. 12.5 Schematic representation of the interaction between CsA and cyclophilin hCyp18; structure of CsA analog 1 (SDZ NIM 811). CsA bonds that isomerize during hCyp binding are indicated with rotating arrows.

nosuppressive potency [160]. The opposite side of the molecule (in particular residues 4, 5, and 6) protrudes from the protein surface and is available to interact with calcineurin in order to induce immunosuppression [6,8,9] (Fig. 12.5).

Consequently, a large number of nonimmunosuppressive analogs have been prepared by modification of either the MeBnt1 or MeLeu4 side-chain (see Fig. 12.1) or simple alkylation of the secondary amide nitrogen of Val5. As an example, MeIle4 cyclosporine 2 inhibits hCyp18 as well as CsA but has an immunosuppressive effect 0.1% of that of CsA [8]. Other compounds modified at position 3, 7 or 8 (these positions are between the cyclophilin-binding domain and the calcineurin-binding domain) retained a significant affinity for CypA and had decreased immunosuppressive effects. Curiously, 3 simultaneous modifications gave a more potent inhibitor of hCyp18 (K_i = 3 nmol L^{-1}) which exhibited an immunosuppressive activity comparable to that of CsA [8,9,159,161]. The nonadditive effects of modifications might be the consequence of constraints that lock the cyclic peptide. However, NMR experiments have shown that CsA exists as multiple conformers in DMSO, suggesting the existence of an amide *cis-trans* equilibrium. In fact, CsA must isomerize from *cis* to *trans* at MeLeu9-MeLeu10 prior to binding to Cyp to form a collisional complex (K_d = 4 μmol L^{-1}) which undergoes a dramatic time-dependent ($t_{1/2}$ = 30 min) conformational change (all-*trans*) to give a tighter complex (K_i = 6 nmol L^{-1}) [8,162,163] (Fig. 12.5).

CsA is mainly used to prevent allograft rejection, but also shows interesting anti-HIV-1 activity which has prompted an intensive search for nonimmunosuppressive analogs [164]. Conversely, Fischer and coworkers have recently described C_α-branched CsA analogs that inhibit calcineurin independently of hCyp18 binding [165]. However, to the best of our knowledge, use of CsA and its analogs is still limited to posttransplantation chemotherapy, owing to their liver and kidney toxicity [166,167]. Moreover, CsA has been shown to initiate apoptosis [90] and to facilitate certain cancers [32,168].

Other cyclic peptides such as cyclolinopeptide c(Pro-Pro-Phe-Phe-Leu-Ile-Ile-Leu-Val (CLA) and its analogs [169], antamanides, and cycloamanides [170] have been shown to bind tightly to hCyp18 and display strong immunosuppressive activities [10]. In particular, CLA was reported to be a potent inhibitor of cyclophilin ($K_i = 7$ nmol L^{-1}) [30]. The biological activity of these peptides is not clearly related to their hydrophobic character since sulfonated CLA also binds cyclophilin [171]. Although their mode of action at the molecular level is far less documented than that of CsA, these compounds might be useful in designing novel cyclophilin inhibitors.

Linear peptides derived from the capsid domain of the Gag polyprotein [111,123] and the V3 loop of the Gp120 protein [111] from HIV-1, which all have a Gly-Pro moiety, have been shown to bind to hCyp18 with submicromolar to submillimolar affinities despite additional interactions relative to CsA (in particular H-bonds with Asn71, Gly72, His54, Ala102, and His126) [150,172,173], and some of them inhibit the PPIase activity of this enzyme [111]. In particular, fluorimetric titration experiments showed that the modified peptide *i*BuCO-His-Ala-Gly-Pro-Ile-NHBn binds to hCyp18 with an apparent affinity constant of 3 μmol L^{-1} and the PPIase kinetic assay indicated that it reversibly inhibits the enzyme with an IC$_{50}$ of 6 μmol L^{-1} [174]. Docking experiments suggest that this peptide interacts inside the catalytic site in a way that is common to substrate peptides and CsA [175]. Therefore, all attempts to introduce nonisosteric Gly-Pro mimetics led to a complete loss of affinity [174]. Other peptides containing the Gly-Pro sequence are also cyclophilin inhibitors: a 9-mer selected from a phage display library binds to hCyp18 ($K_d = 50$ μmol L^{-1}) and inhibits the PPIase activity [176], while Ac-Pro-Gly-Pro-Phe-NH$_2$ exerts powerful neurotrophic activity on chick sensory neurons at 1 mmol L^{-1}, an effect which is related to PPIase inhibition [177].

In 1999, a new class of potent immunosuppressive molecules called sanglifehrins (SFs) was isolated from *Streptomyces* A92-308110. Of the 20 SFs isolated, SFA is the most abundant and has an affinity for hCyp18 in a cell-free assay that is 60 times that of CsA and exhibits potent immunosuppressive activity on T cells and dendritic cell lines [27,28]. SFA has a unique chemical structure that consists of a 22-membered macrocycle including a short peptide sequence composed of valine, *m*-tyrosine and piperazic acid (Fig. 12.1). The degradation studies of Sedrani and coworkers showed that the pendant spirolactam system is not essential for activity. Preparation of a library of analogs demonstrated that the free piperazic motif and the *m*-hydroxyphenyl group both play a critical role in binding [31]. This was rationalized with the structural elucidation of hCyp18:SFA complex which showed that SFA binds cyclophilin through the formation of a network of multiple H-bonds and hydrophobic contacts that resemble those observed with CsA (Figs. 12.6 and 12.7) [178], except for Trp121 which seems to interact in an unusual way [179]. Overlay of the two structures suggests that the peptide moieties of both molecules have similar interaction patterns. However, SFA does not bind calcineurin or FKBP12-rapamycin-associated protein (FRAP) when bound to cyclophilin and it seems that a third PPIase-dependent immunosuppressive pathway does exist [27].

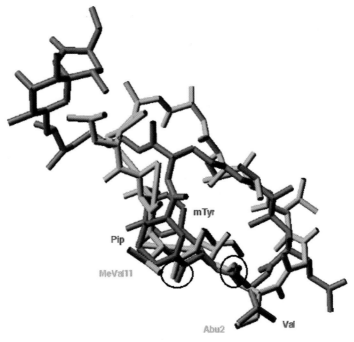

Fig. 12.6 Overlay of X-ray structures of SFA (red) and CsA (blue) shows similarities in the conformations of both peptide moieties that interact tightly with the hCyp18 binding site.

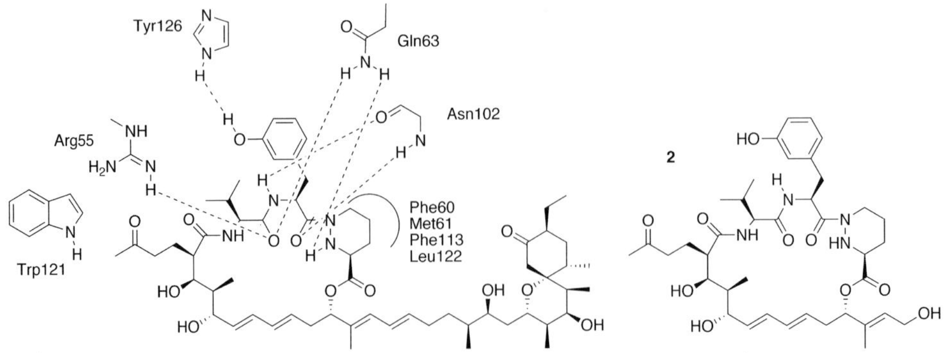

Fig. 12.7 Schematic representation of the interaction between SFA and cyclophilin hCyp18; structure of SFA analog **2**.

Although CsA has a complex chemical structure, several potentially interesting analogs have emerged and display affinities that are equivalent to that of SFA. For example, compound **2** inhibited the PPIase activity of hCyp18 with $IC_{50} = 5.7$ nmol L^{-1} despite a simplified structure relative to the parent molecule [31].

Macrolides such as cymbimicins A and B have been reported to bind to cyclophilin and to produce strong immunosuppressive effects, although their mode of action has not been elucidated [180]. Other nonpeptide macrocyclic compounds such as FK506, RAPA, and ascomycin selectively inhibit FKBPs and there is no cross-reaction with CsA though they share some structural similarities with it [9,181]. In particular, the time-dependent inhibition of hFKBP12 by FK506 was suggested to be the result of the CTI of the ketoamido pipecolinyl moiety [182]. The structure of the hFKBP12:FK506 complex was resolved by X-ray crystallography and NMR, and the structural requirements for PPIase inhibition on the one hand, and for immunosuppression on the other, have been delineated [183]. The hemiketal-ketoamide pipecolinate fragment seems to mimic a Leu-Pro moiety. The six-member ring penetrates deeply into the hydrophobic active site while the orthogonal dicarbonyl motif is suspected to be an analog of the twisted transition state [8]. The hemiketal structure interacts with a cluster of hydrophobic sidechains. In addition, the carbonyl amide and the hemiketal hydroxide contract a series of H-bonds with Tyr82 and Asp37. There are many similarities between FK506 and CsA when interacting at the active site of hFKBP12 and hCyp18, respectively. In particular, the effector-binding chain protrudes outside the binding site and is available for calcineurin binding. As anticipated, reduction in size of this chain causes a complete loss of immunosuppressive effect, as observed with synthetic analog **3** (Fig. 12.8) [9].

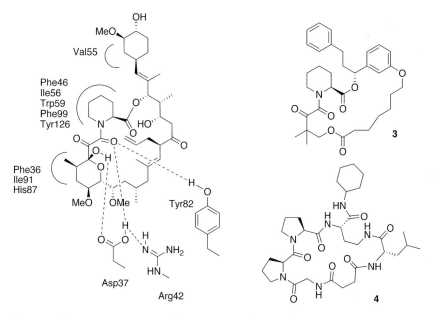

Fig. 12.8 Schematic representation of the interaction between FK506 and hFKBP12; structure of nonimmunosuppressive analog **3**.

The synthetic peptide cESA **4** (Fig. 12.8) characterized by the coexistence of a *cis-cis* and *trans-cis*-Gly-Pro-Pro sequence was expected to be an analog of FK506. However, little deviation of the structure relative to the macrolide chain interacting with FKBP as well as conformational heterogeneity might account for the apparent 1000-fold lower affinity [184].

Quinones such as juglone and plumbagine (Fig. 12.9) were the first molecules found to inhibit parvulins, in particular Pin1 [153,185]. They react with Cys113 at the catalytic site, which is irreversibly inactivated. The consequence of the nucleophilic attack of the thiol group is still unclear and several hypotheses have been proposed: (1) Cys113 might be essential for CTI and therefore cannot catalyze isomerization anymore after alkylation; (2) juglone-Cys113 adduct might hinder the binding site; (3) covalent binding of juglone is suspected to destabilize the active site and to cause a partial unfolding of the protein [152]. Interestingly, this was also observed with tetraoxobenzophenanthrolines, which are able to inhibit Pin1 and Par14 in a competitive and noncovalent way [186].

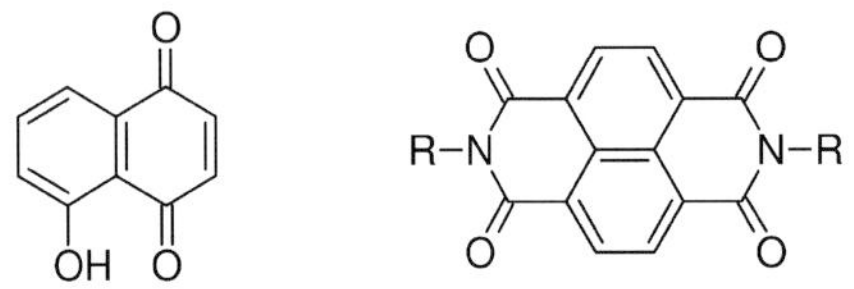

Fig. 12.9 Pin1 inhibitors juglone **5** and tetraoxobenzophenanthrolines **6**.

12.4.2
Mechanism-Based Inhibitors

Resolution of the structures of several PPIase:effector complexes by X-ray crystallography and NMR as well as SAR studies have opened the way to the development of pseudopeptide and peptidomimetic inhibitors which bind PPIases in the 1–50 µmol L^{-1} range. The ketoamide motif redundant to all natural FK506-like macrolides was proposed to be a mimic of the twisted transition state since both carbonyl groups lie in orthogonal planes. Therefore, two independent research groups synthesized Alaψ(COCO-NH)Pro-containing peptides, which were used to produce catalytic antibodies that displayed PPIase activity and were able to inhibit both FKBP and hCyp18 with an IC_{50} of about 10–15 µmol L^{-1} [187]. Tetrapeptide **7** containing the Glyψ(PO(OEt))Pro motif was able to inhibit hCyp18 selectively (Fig. 12.10), suggesting that the phosphonamide group is a transition state analog of the enzyme [188]. Conversely, the corresponding phosphinates, characterized by a CH_2/N substitution in the five-member ring, do not have any effect on the enzymatic activity [189].

The slowly isomerizable thioxoamide bond also provides an interesting basis for the design of potent inhibitors of PPIases. Fischer's group reported that a thioxo analog of the substrate moderately inhibited the PPIase activity of both hFKBP12 and Pin1. Introduction of a phosphoserine residue at P1 gives a good Pin1 inhibitor ($IC_{50} = 4$ µmol L^{-1}) provided at least three amide bonds are conserved (Fig. 12.10). It is noteworthy that replacement of the L-Ser(PO_3H_2) by its D-enantiomer also causes a potent PPIase inhibition of Pin1 ($IC_{50} = 1$ µmol L^{-1}) and a complete resistance to phosphatase activity in cell lysate [190].

Nonisomerizable substrate analogs containing an ethylene group instead of the amide bond also display interesting activities towards hFKBP12. Recently, Etzkorn and coworkers described a Ser$^P\psi$((Z)CH=CH)Pro- peptide **9** which inhibits Pin1 in the low micromolar range (Fig. 12.10). The *cis* isomer is 23 times more potent ($K_i = 1.74$ µmol L^{-1}) than the corresponding *trans* isomer in competitive inhibition of Pin1. Moreover, this compound exhibits an antiproliferative activity toward a human ovarian cancer cell line [191]. In fact, Pin1 and hFKBP12 are permissive to large modifications of the pyrrolidine ring of proline (see compound **9**) whereas hCyp18 does not tolerate substitutions at C4 and C5 except with fluorine, nor nitrogen suppression or other modification that may affect both the H-bond network and the pyrrolidine puckering [8].

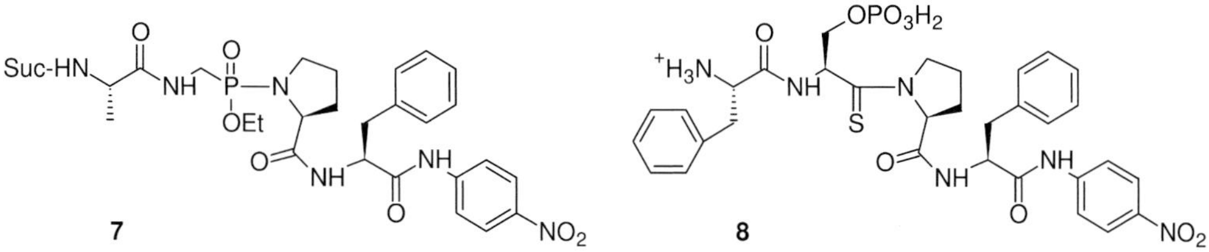

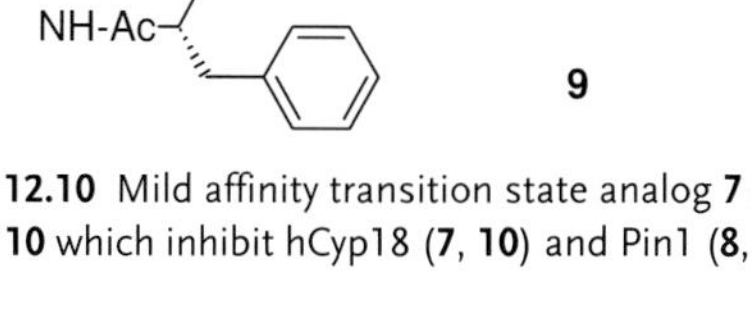

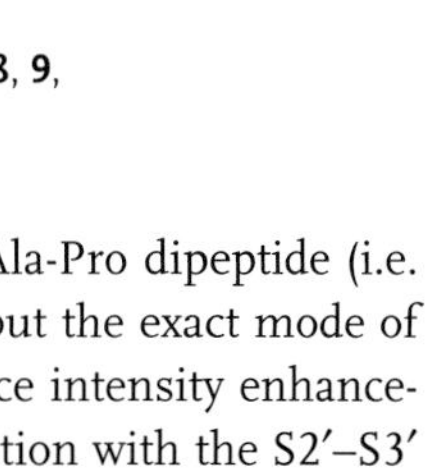

Fig. 12.10 Mild affinity transition state analog **7** and substrate analogs **8, 9**, and **10** which inhibit hCyp18 (**7, 10**) and Pin1 (**8, 9**).

A series of constrained ground-state analogs of a *cis*-Gly/Ala-Pro dipeptide (i.e. compound **10**) displayed micromolar affinities for hCyp18, but the exact mode of interaction with the PPIase is not clear, although fluorescence intensity enhancement at 350 nm strongly suggests the existence of an interaction with the S2′–S3′ subsite. However, no data regarding an inhibition of the enzymatic activity were reported [192].

12.4.3
Library Screening Versus in Silico Design: Current Status and Future Prospects

Considering the formal demonstration that PPIases are directly implicated in major diseases and are interesting targets for novel therapeutics, a growing number of synthetic molecules have been reported to inhibit PPIase activity. Based on the partition of FKBP's binding site, which recognizes only limited motifs on FK506, the combinatorial synthesis of acyclic FKBP inhibitors produced a huge number of active molecules devoid of immunosuppressive effects and exhibiting neuroprotective and neuroregenerative activities. Most of them were developed by the Hamilton's group at Guilford Pharmaceutical and show chemical motifs that clearly differ from the canonical ketoamide motif, such as amine **12**, sulfonamide **13**, and urea **14** (Fig. 12.11). Compounds V-10367 **11** and GPI-1046 **15** displayed nerve-regenerating properties comparable to that of FK606 and were active orally [7].

In 1996, Fesik and coworkers reported an alternative and versatile strategy (SAR by NMR) for the direct identification of hFKBP acyclic inhibitors exhibiting nanomolar affinities by simple connection of two independent modules that display moderate affinities [193]. Stabilization of the effector:protein complex resulted from the additive ΔG of each module alone plus an additional entropic contribution [194].

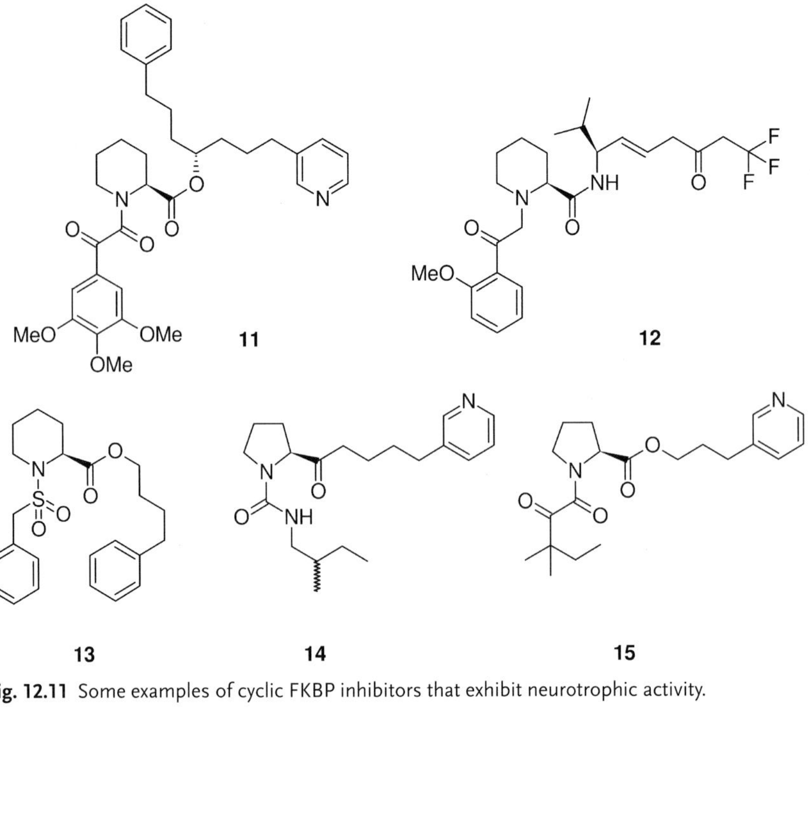

Fig. 12.11 Some examples of cyclic FKBP inhibitors that exhibit neurotrophic activity.

Systematic screening of large collections of small molecules has provided novel PPIase inhibitors with unexpected structures. Once again, most of the leads that have emerged are directed towards hFKBPs and Pin1, which are much more permissive to modifications of their ligands than cyclophilins. In particular, many FKBP ligands (Fig. 12.12) greatly differ from the FK506-derived pharmacophore. This is the case of sulfonylindazole **16** [195], which inhibits hFKBP52 in the micromolar range. Curiously, heterocyclic compounds, such as phenylsulfonylindoline **17** [196], were found to inhibit simultaneously CypB and IL-2 secretion, while simple heterocyclic compounds were patented as Pin1 inhibitors usable for the treatment of cancer [197]. The antibiotic cycloheximide also displayed FKBP-binding properties and alkylation of the imide nitrogen strongly reduced the cytotoxicity of molecule **18** without affecting its potency. Lactone **19**, which does not possess the characteristic phosphate motif recognized by Pin1, was recently patented as a novel inhibitor of this enzyme (Fig. 12.12) [198].

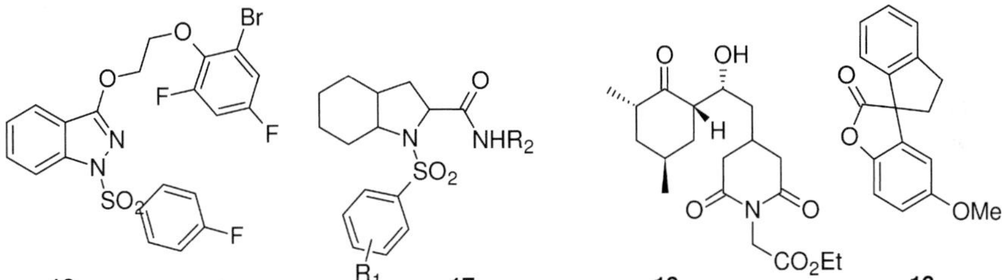

Fig. 12.12 Chemical structures of some PPIase inhibitors selected by screening of libraries.

Screening of virtual libraries through molecular docking has also given new leads whose structure diverges considerably from the peptide/pseudopeptide motifs usually found in immunophilin ligands (Fig. 12.13). The steroid 5-β-pregnan-3,20-dione **20** inhibits hFKBP12 ($K_d = 7$ μmol L^{-1}), but the biological significance of the binding remains to be elucidated [199]. Other compounds characterized by a hydrophobic core, such as thioether **21** [200] and diazabicyclo[3.3.1]nonane **22** [201], were designed to bind hFKBP12 in the micromolar to submicromolar range. In particular, the bicyclic motif of **21** is expected to make van der Waals contacts that are very similar to those observed in the FKBP:FK506 complex. The bicyclic motif was also found in the recently patented hFKBP12 inhibitor diamide **23** ($K_i = 18$ nmol L^{-1}) which stimulates growth and proliferation of neurons (Fig. 12.13) [202].

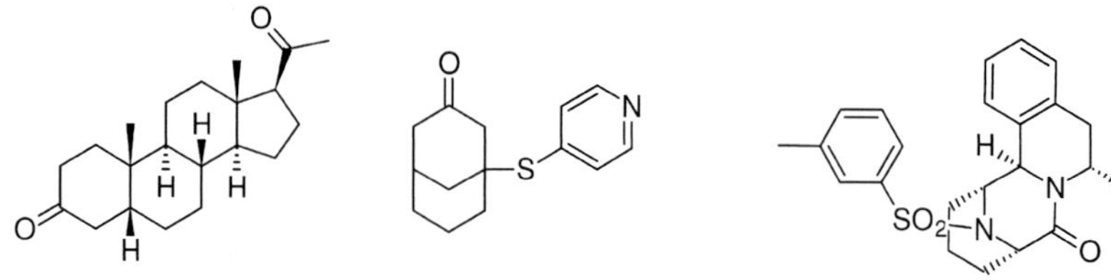

20   21   22

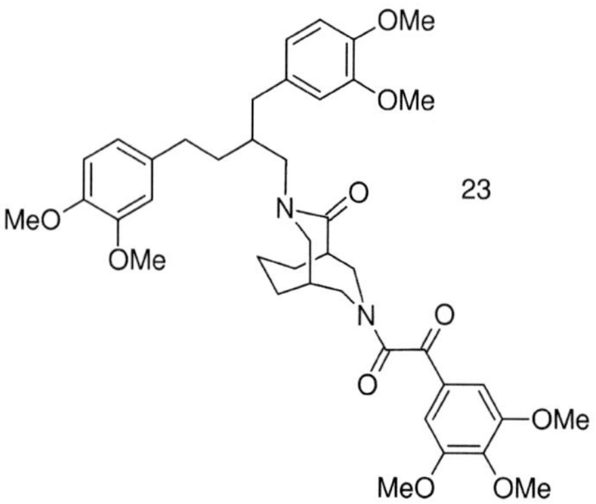

23

Fig. 12.13 Chemical structures of some PPIase inhibitors selected by virtual screening.

The finding that aromatic diamines (Fig. 12.14) fit inside the narrow hydrophobic pocket of cyclophilin [203] has opened the way to the development of achiral
polycyclic compounds which inhibit PPIase activity with submicromolar affinities.
The symmetric bis-urea **24** (IC$_{50}$ = 0.49 μmol L^{-1}) was the most potent inhibitor in

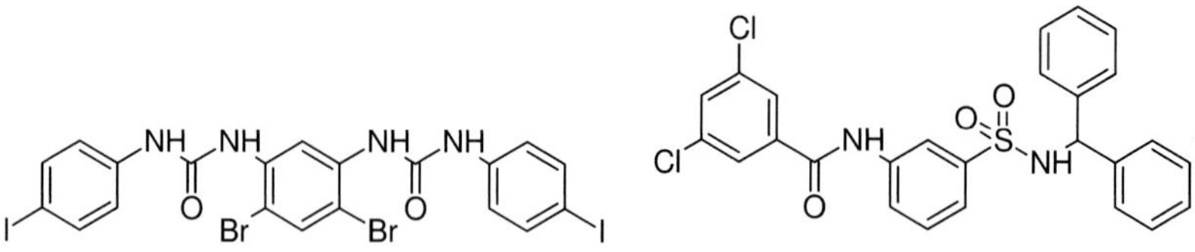

24   25

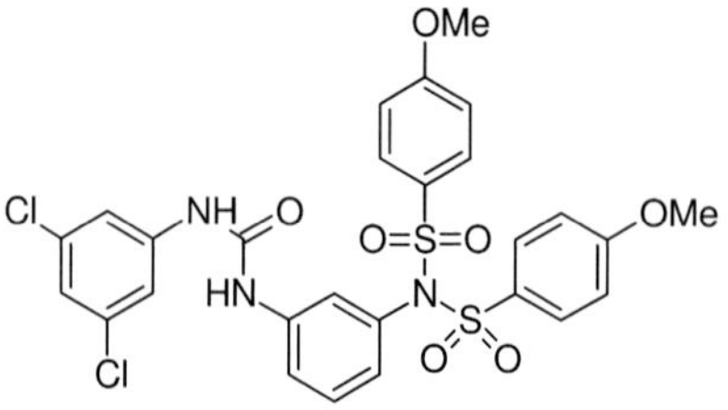

26

Fig. 12.14 Polyphenyl aniline derivatives that inhibit hCyp18 in the micromolar range.

this series and promoted significant neurite outgrowth at 10 nmol L^{-1} concentration (EC$_{50}$ about 100–1000 nmol L^{-1}) in spinal motor neurons previously treated with the glutamate reuptake inhibitor threo-hydroxyaspartate. Other nonsymmetrical polyphenyl compounds such as **25** [204] and **26** (Fig. 12.14) [205] also showed interesting biochemical properties, suggesting that a large number of variations on the aniline moiety might be used to develop a novel generation of nonpeptide cyclophilin inhibitors. The most intriguing cyclophilin inhibitor, however, probably remains the gold (Au^I) complex AuP(Et)$_3$Cl which blocks Cyp3 (*C. elegans*) PPIase activity through the binding of the metal to the active site histidine nitrogen, as clearly shown in the X-ray structure of the protein soaked in a solution of gold complex [206].

In silico docking studies have been used to design several novel molecules that effectively inhibit PPIases (Fig. 12.15). Surprisingly, none of them contain a proline residue, and although information about the mode of interaction is lacking, all possess a hydrophobic motif, which might be buried inside the active site. This is the case of the dimedone amino acid derivative **27** which binds hCyp18 with quite good affinity (K_d = 2 μmol L^{-1}) [207]. Very recently, a series of peptides derived from pepticinnamin E exhibited potent inhibition of Pin1 (**28**: IC$_{50}$ = 0.6 μmol L^{-1}) and induced apoptosis in transformed cell lines. Interestingly, these inhibitors seem to destabilize hPin1 in a juglone-like manner but without covalently modifying the protein [208] and might constitute a basis for the design of potent Pin1 inhibitors.

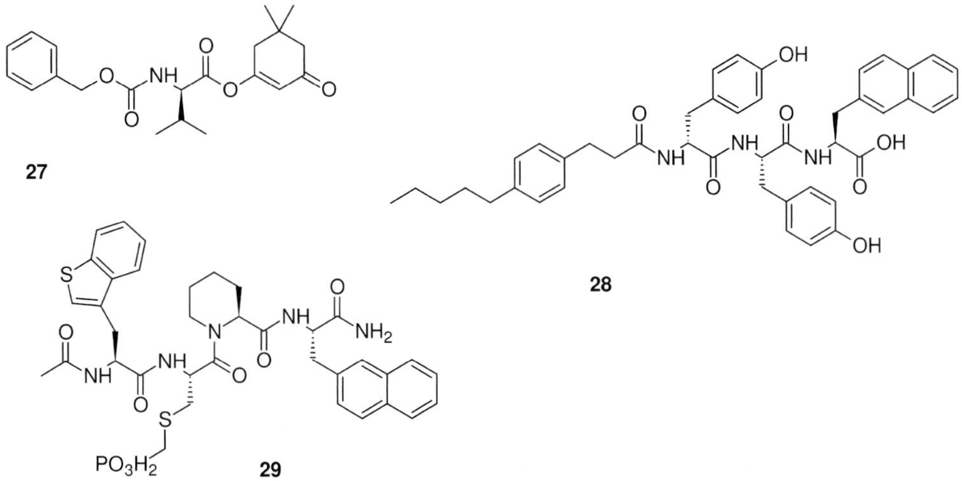

Fig. 12.15 Amino acid derivatives and peptides binding to hCyp18 (compounds **27** and **28**) and Pin1 (compound **29**).

12.5
Conclusion and Perspectives

In the past decade, numerous studies have highlighted the essential role of PPIases in a growing number of diseases. Beyond immunosuppression, PPIases have been shown to be involved in major biological processes including cell multiplication, gene expression, viral and bacterial infectivity as well as regulation of the function of receptors and ion channels. They are consequently implicated in a large number of illnesses such as cancer, genetic diseases, neurological disorders as well as infections by pathogenic agents such as HIV-1.

The design of selective and potent inhibitors of PPIases is of interest and numerous molecules have been designed or selected from chemical libraries with a view to curing these major diseases. The study of Pin1, which is clearly distinct from other members of the PPIase family on the basis of structure, binding site, catalytic mechanism, and biological implications, has opened up new perspectives in the biological chemistry of PPIases. The recent discoveries of the secondary amide peptide bond *cis-trans* isomerase (APIase) DnaK [209] and of a novel class of FK506 and cyclosporine-sensitive PPIase [210] are also major advances in this field.

Although most patents and publications still concern the neurotrophic effects of FK506 analogs, there are an increasing number of molecules with unprecedented structures. Most of them have relatively simple motifs, however, and should be seen as leads for the development of active compounds rather than real therapeutics, since none of them exhibit potencies comparable to that of complex peptides and macrolides. In this respect, the design of simple and acyclic analogs of SFA might provide potent inhibitors of hCyp18 with simplified structures. In the future, one major challenge for the medicinal chemistry of PPIases will be the selective inhibition of prokaryotic enzymes without affecting their eukaryotic equivalents with a view to designing the antibiotics of tomorrow.

References

1 A. Galat, *Curr. Top. Med. Chem.* **2003**, *3*, 1315–1347.
2 G. Fischer, T. Aumüller, *Rev. Physiol. Biochem. Pharmacol.* **2003**, *148*, 105–150.
3 C. Schiene-Fischer, C. Yu, *FEBS Lett.* **2001**, *495*, 1–6.
4 K. Dolinski, S. Muir, M. Cardenas, J. Heitman, *Proc. Natl Acad. Sci. USA* **1997**, *94*, 13093–13098.
5 J. Colgan, M. Asmal, J. Luban, *Genomics* **2000**, *68*, 167–178.
6 G. S. Hamilton, J. P. Steiner, *J. Med. Chem.* **1998**, *41*, 5119–5143.

7 B. J. Gold, J. E. Villafranca, *Curr. Top. Med. Chem.* **2003**, *3*, 1368–1375.
8 C. Dugave, L. Demange, *Chem. Rev.* **2003**, *103*, 2475–2532.
9 J. Dornan, P. Taylor, M. D. Walkinshaw, *Curr. Top. Med. Chem.* **2003**, *3*, 1392–1409.
10 C. Dugave, *Curr. Org. Chem.* **2002**, *6*, 1397–1431.
11 H. Ke, Q. Huai, *Front. Biosci.* **2004**, *9*, 2285–2296.
12 J. Fanghänel, G. Fischer, *Front. Biosci.* **2004**, *9*, 3453–3478.

13 G. Fischer, T. Tradler, T. Zarnt, *FEBS Lett.* **1998**, *426*, 17–20.

14 M. T. G. Ivery *Med. Res. Rev.* **2000**, *20*, 452–484,.

15 T. Ratajczak, B. K. Ward, R. F. Minchin, *Curr. Top. Med. Chem.* **2003**, *3*, 1348–1357.

16 C. Schiene, G. Fischer, *Curr. Opin. Struct. Biol.* **2000**, *10*, 40–45.

17 Y. Takaki, T. Muta, S. Inwanaga, *J. Biol. Chem.* **1997**, *272*, 28615–28621.

18 S. A. Helekar, J. Patrick, *Proc. Natl Acad. Sci. USA* **1997**, *94*, 5432–5437.

19 S. E. Lehnart, F. Huang, S. O. Marx, A. R. Marks *Curr. Top. Med. Chem.* **2003**, *3*, 1383–1391.

20 X. Meng, X. Lu, C. A. Morris, M. T. Keating, *Genomics* **1998**, *52*, 130–137.

21 A. M. Siverstein, M. D. Galigniana, K. C. Kanelakis, C. R. Radanyi, J.-M. Renoir, W. B. Pratt, *J. Biol. Chem.* **1999**, *274*, 36980–36986.

22 O. Tchaicheeyan, *FASEB J.* **2004**, *18*, 783–789.

23 J. Clardy, *Proc. Natl Acad. Sci. USA* **1995**, *92*, 56–61.

24 S. Matsuda, S. Koyasu, *Curr. Top. Med. Chem.* **2003**, *3*, 1358–1367.

25 K. N. Brazin, R. J. Mallis, D. B. Fulton, A. H. Andreotti, *Proc. Natl Acad. Sci. USA* **2002**, *99*, 1899–1904.

26 L. Min, D. B. Fulton, A. H. Andreotti, *Front. Biosci.* **2005**, *10*, 385–397.

27 G. Zenke, U. Strittmatter, S. Fuchs, V. J. Quesniaux, V. Brinkmann, W. Schuler, M. Zurini, A. Enz, A. Billich, J.-J. Sanglier, T. Fehr, *J. Immunol.* **2001**, *166*, 7165–7171.

28 L.-H. Zhang, J. O. Liu *J. Immunol.* **2001**, *166*, 5611–5618.

29 P. Gallo, M. Saviano, F. Rossi, V. Pavone, C. Pedone, R. Ragone, P. Stiuso, G. Colonna, *Biopolymers* **1995**, *136*, 273–281.

30 P. Gallo, F. Rossi, C. Pedone, G. Colonna, R. Ragone, *J. Biochem.* **1998**, *124*, 880–885.

31 R. Sedrani, J. Kallen, L. M. Martin Cabrejas, C. D. Papgeorgiu, F. Senia, S. Rohrbach, D. Wagner, B. Thai, A.-M. Jutzi Eme, J. France, L. Oberer, G. Rihs, G. Zenke, J. Wagner, *J. Am. Chem. Soc.* **2003**, *125*, 3849–3859.

32 G. J. Nabel, *Nature* **1999**, *397*, 471–472.

33 M. G. Chelu, C. I. Danila, C. P. Gilman, S. L. Hamilton, *Trends Cardiovasc. Med.* **2004**, *14*, 227–234.

34 T. H. Davies, E. R. Sanchez, *Int. J. Biochem. Cell Biol.* **2005**, *37*, 42–47.

35 P. Kumar, P. J. Mark, B. K. Ward, R. F. Minchin, T. Ratajczak, *Biochem. Biophys. Res. Commun.* **2001**, *284*, 219–225.

36 T. Ratajczak, P. J. Mark, R. L. Martin, R. F. Minchin, *Biochem. Biophys. Res. Commun.* **1996**, *220*, 208–212.

37 P. C. Waldemeier, K. Zimmermann, T. Qian, M. Tintlenot-Blomley, J. J. Lemasters, *Curr. Med. Chem.* **2003**, *10*, 1485–1406.

38 X. Guo, J. F. Dillman III, V. L. Dawson, T. M. Dawson, *Ann. Neurol.* **2001**, *50*, 6–16.

39 K. Pong, M. M. Zaleska, *Curr. Drug Targets CNS Neurol. Disord.* **2003**, *2*, 349–356.

40 H. Kato, T. Oikawa, K. Otsuka, A. Takahashi, Y. Itoyama, *Brain Res. Mol. Brain Res.* **2000**, *84*, 58–66.

41 B. G. Gold, J. G. Nutt, *Curr. Opin. Pharmacol.* **2002**, *2*, 82–86.

42 M. Avramut, C. L. Achim, *Curr. Top. Med. Chem.* **2003**, *3*, 1376–1382.

43 N. Tanaka, N. Ogawa, *Curr. Pharm. Des.* **2004**, *10*, 669–677.

44 Y. Manabe, H. Warita, T. Murakami, M. Shiote, T. Hayashi, N. Omori, I. Nagano, M. Shoji, K. Abe, *Brain Res.* **2002**, *935*, 124–128.

45 B. G. Gold, E. Udina, D. Bourdette, X. Navarro, *Neurol. Res.* **2004**, *26*, 371–380.

46 B. Aghdasi, K. Ye, A. Resnick, A. Huang, H. C. Ha, X. Guo, T. M. Dawson, V. L. Dawson, S. H. Snyder, *Proc. Natl Acad. Sci. USA* **2001**, *98*, 2425–2430.

47 G. Wulf, G. Finn, F. Suizu, K. P. Lu, *Nat. Cell Biol.* **2005**, *7*, 435–441.

48 K. P. Lu, Y. C. Liou, I. Vincent, *Bioessays* **2003**, *25*, 174–181.

49 J. Lim, K. P. Lu, *Biochim. Biophys. Acta* **2005**, *3*, 311–322.

50 N. L. Daly, R. Hoffmann, L. Otvos Jr., D. J. Craik, *Biochemistry* **2000**, *39*, 9039–9046.

51 H. Inouye, J. Bond, M. A. Baldwin,
H. L. Bal, S. B. Prusiner, D. A. Kirschner,
J. Mol. Biol. **2000**, *300*, 1283–1296.

52 D. Galani, A. R. Fersht, S. J. Perrett,
Mol. Biol. **2002**, *315*, 213–227.

53 E. Cohen, A. Taraboulos, *EMBO J.* **2003**,
22, 404–417.

54 K. P. Lu, *Trends Biochem. Sci.* **2004**, *29*,
200–209.

55 X. Z. Zhou, O. Kops, A. Werner, P. J. Lu,
M. H. Shen, G. Stoller, G. Küllertz,
M. Stark, G. Fischer, K. P. Lu, *Mol. Cell.*
2000, *6*, 873–883.

56 K. P. Lu, Y. C. Liou, X. Z. Zhou, *Trends
Cell Biol.* **2002**, *12*, 164–172.

57 M. Weiwad, G. Küllertz,
M. Schutkowski, G. Fischer, *FEBS Lett.*
2000, *478*, 39–42.

58 K. P. Lu, *Cancer Cell.* **2003**, *25*, 174–181.

59 M. Lopez-Ilasaca, C. Schiene,
G. Küllertz, T. Tradler, G. Fischer,
R. Wetzker, *J. Biol. Chem.* **1998**, *273*,
9430–9434.

60 M. A. Rycyzyn, C. V. Clevenger, *Proc.
Natl Acad. Sci. USA* **2002**, *99*,
6790–6795.

61 X. Y. Wu, C. B. Willcox,
G. Devashahayam, R. L. Hackett,
M. Arevalo-Rodriguez, M. E. Cardenas,
J. Heitman, S. D. Hanes, *EMBO J.* **2000**,
19, 3727–3738.

62 U. Krummrei, R. Bang, R. Schmidtchen,
K. Brune, H. Bang, *FEBS Lett.* **1995**,
371, 47–51.

63 Q. Yao, M. Li, H. Yang, H. Chai,
W. Fisher, C. Chen, *World J. Surg.* **2005**,
29, 276–280.

64 L. Bao, A. Kimzey, G. Sauter,
J. M. Sowadski, K. P. Lu, D.-G. Wang,
Am. J. Pathol. **2004**, *164*, 1727–1737.

65 I. Zucchi, E. Mento, V. A. Kuznetsov,
M. Scotti, V. Valsecchi, B. Simionati,
E. Vicinanza, G. Valle, S. Pilotti,
R. Reinbold, P. Vezzoni, A. Albertini,
R. Dulbecco, *Proc. Natl Acad. Sci. USA*
2004, *101*, 18147–18152.

66 S. Fong, L. Mounkes, Y. Liu,
M. Maibaum, E. Alonzo, P.-Y. Desprez,
A. D. Thor, M. Kashani-Sabet, R. J. Debs,
Proc. Natl Acad. Sci. USA **2003**, *100*,
14253–14258.

67 C. Melle, D. Osterloh, G. Ernst,
B. Schimmel, A. Bleul, F. Von Egeling,
Int. J. Mol. Med. **2005**, *16*, 11–17.

68 H. Andersen, O. N. Jensen, E. F. Eriksen,
Eur. J. Cancer **2003**, *39*, 989–995.

69 B. A. Howard, R. Furumay, M. J. Campa,
Z. N. Rabbani, Z. Vujaskovic,
X.-F. Wang, E. F. Patz Jr., *Cancer Res.*
2005, *65*, 8853–8860.

70 L. A. Meza-Zepeda, A. Forus, B. Lygren,
A. B. Dalgren, L. H. Godager,
A. P. South, I. Marenholz, M. Lioumi,
V.-A. Florenes, G. M. Maelandsmo,
M. Serra, D. Mischke, D. Nizetic,
J. Ragoussis, M. Tarkkanen,
J. M. Nesland, S. Knuutila,
O. Myklebost, *Oncongene* **2002**, *21*,
2261–2269.

71 P. Mauri, A. Scarpa, A.-C. Nascimbeni,
L. Benazzi, E. Parmagnani, A. Mafficini,
M. Della Peruta, C. Bassi, K. Miyazaki,
C. Sorio, *FASEB J.* **2005**, *19*, 1125–1127.

72 A. Gougelet, C. Bouclier, V. Marsaud,
S. Maillard, S. O. Mueller, K. S. Korach,
J.-M. Renoir, *Steroid Biochem. Mol. Biol.*
2005, *94*, 71–81.

73 A. Schubert, S. Grimm, *Cancer Res.*
2004, *64*, 85–93.

74 H. G. Van der Poel, C. Hhanrahan,
H. Zhong, J. W. Simons, *Urol. Res.*
2003, *30*, 380–386.

75 P. G. Febbo, M. Lowenberg,
A. R. Thorner, M. Brown, M. Loda,
T. R. Golub, *J. Urology* **2005**, *173*,
1772–1777.

76 S. Le Bihan, V. Marsaud, C. Mercier-
Bodard, E.-E. Baulieu, S. Mader,
J. H. White, J.-M. Renoir, *Mol. Endocri-
nol.* **1998**, *12*, 986–1001.

77 O. Bock, M. Neusch, G. Buesche,
M. Mengel, H. Kreipe, Eur. J. *Haematol-
ogy* **2004**, *72*, 239–244.

78 S. H. Olesen, L. L. Christensen,
F. B. Sorensen, T. Cabezon, S. Laurberg,
T. F. Orntoft, K. Birkenkamp-Demtro-
eder, *Mol. Cell Proteomics* **2005**, *4*,
534–544.

79 G. Wulf, P. Garg, Y.-C. Liou, D. Iglehart,
K. P. Lu, *EMBO J.* **2004**, *23*, 3397–3407.

80 R. Pang, J. Yuen, M. F. Yuen, C. L. Lai,
T. K. W. Lee, K. Man, R. T. P. Poon,
S. T. Fan, C. M. Wong, I. O. L. Ng,
Y. L. Kwong, E. Tse, *Oncogene* **2004**, *23*,
4182–4186.

81 G. Ayala, D. Wang, G. Wulf, A. Frolov,
R. Li, J. Sodawski, T. M. Wheeler,

K. P. Lu, L. Bao, *Cancer Res.* **2003**, *63*, 6244–6251.

82 C.-J. Kim, Y.-G. Cho, Y.-G. Park, S. W. Nam, S.-Y. Kim, S.-H. Lee, N.-J. Yoo, J.-Y. Lee, W-S. Park, *World J. Gastroenterol.* **2005**, *11*, 5006–5009.

83 H. M. Ashita, S. Mori, K. Motegi, M. Fukumoto, T. Uchida, *Oncol. Rep.* **2004**, *10*, 455–461.

84 J.-P. Lee, H. C. Palfrey, V. P. Bindokas, G. D. Ghadge, L. Ma, R. J. Miller, R. P. Roos, *Proc. Natl Acad. Sci. USA* **1999**, *96*, 3251–3256.

85 Y. Li, N. Johnson, M. Capano, M. Edwards, M. Crompton, *Biochem. J.* **2004**, *383*, 101–109.

86 L. Andreeva, R. Heads, C. J. Green, *Int. J. Exp. Pathol.* **1999**, *80*, 305–315.

87 D-T. Lin, J. D. Lechleiter, *J. Biol. Chem* **2002**, *277*, 31134–31141.

88 P. Zacchi, M. Gostissa, T. Uchida, C. Salvagno, F. Avolio, S. Volinia, Z. Ronai, G. Blandino, C. Schneider, G. Del Sal, *Nature* **2002**, *419*, 853–357.

89 S. Clarcke, G. P. McStay, A. P. Halestrap, *J. Biol. Chem.* **2002**, *277*, 34793–34799.

90 K. Kitagaki, H. Nagai, S. Hayashi, T. Totsuka, *Eur. J. Pharmacol.* **1997**, *337*, 283–289.

91 H. Terada, M. Matsushita, Y.-F. Lu, T. Shirai, S.-T. Li, K. Tomizawa, A. Moriwaki, S. Nishio, I. Date, T. Ohmoto, H. Matsui, *J. Neurochem.* **2003**, *87*, 1145–1151.

92 C. Basak, S. K. Pathak, A. Bhattacharyya, S. Pathak, J. Basu, M. Kundu, *J. Immunol.* **2005**, *174*, 5672–5680.

93 H. Fliri, *Expert Opin. Invest. Drugs* **1996**, *5*, 1003–1011.

94 S. F. Goethel, M. A. Marahiel, *Cell. Mol. Life Sci.* **1999**, *55*, 423–436.

95 A. Riboldi-Tunnnicliffe, B. Konig, S. Jessen, M. S. Weiss, J. Rahfeld, J. Hacker, G. Fischer, R. Hilgenfeld, *Nat. Struct. Biol.* **2001**, *8*, 779–783.

96 J. H. Helbig, B. Konig, H. Knospe, B. Bubert, C. Yao, C. P. Christian, A. Riboldi-Tunnicliffe, R. Hilgenfeld, E. Jacobs, J. Hacker, G. Fischer, *Biol. Chem.* **2003**, *384*, 125–137.

97 R. Koehler, J. Fanghaenel, B. Koenig, E. Lueneberg, M. Frosch, J.-U. Rahfeld, R. Hilgenfeld, G. Fischer, J. Hacker,

M. Steinert, *Infect. Immun.* **2003**, *71*, 4389–4397.

98 G. Coaker, A. Falick, B. Staskawicz, *Science* **2005**, *308*, 548–550.

99 C. Rascher, A. Pahl, A. Pecht, K. Brune, W. Solbach, H. Bang, *Biochem. J.* **1998**, *334*, 659–667.

100 P. J. B. Pereira, C. M. Vega, E. Gonzales-Rey, R. Fernandez-Carazo, S. Macedo-Ribeiro, F.-X. Gomis-Ruth, A. Gonzales, M. Coll, *EMBO Rep.* **2002**, *3*, 88–94.

101 J. R. Blankenship, W. J. Steinbach, J. R. Prfect, J. Heitman, *Curr. Opin. Invest. Drugs* **2003**, *4*, 192–199.

102 S. Bose, M. Mathur, P. Bates, N. Joshi, A. K. Banerjee, *J. Gen. Virol.* **2003**, *84*, 1687–1699.

103 D. Minder, J. Boni, J. Schupbach, H. Gehring, *Arch. Virol.* **2002**, *147*, 1531–1542.

104 A. C. Saphire, M. D. Bobardt, P. A. Gallay, *Immunol. Res.* **2000**, *21*, 211–217.

105 C. Gurer, A. Cimarelli, J. Luban, *J. Virol.* **2002**, *76*, 4666–4670.

106 C. Cartier, B. Hemonnot, C. Devaux, L. Briant, *Curr. Top. Virol.* **2003**, *3*, 1–15.

107 M. Hammarstedt, H. Garroff, *J. Virol.* **2004**, *78*, 5686–5697.

108 Y.-Y. Zhang, T. Hatziioannou, T. Zang, D. Braaten, J. Luban, S. P. Goff, P. D. Bieniasz, *J. Virol.* **2002**, *76*, 6332–6343.

109 T. Pushkarsky, V. Yurchenko, C. Vanpouille, B. Brichacek, L. Vaisman, S. Hatakeyama, K. II. Nakayama, B. Sherry, M. I. Bukrinsky, *J. Biol. Chem.* **2005**, *280*, 27866–27871.

110 T. Pushkarsky, G. Zybarth, L. Dubrovsky, V. Vyacheslav, H. Tang, H. Guo, B. Toole, B. Sherry, M. Bukrinsky, *Proc. Natl Acad. Sci. USA* **2001**, *98*, 6360–6365.

111 M. Endrich, H. Gehring, *Eur. J. Biochem.* **1998**, *252*, 441–444.

112 C. Callebaut, J. Blanco, N. Benkirane, B. Krust, E. Jacoto, G. Guichard, N. Seddiki, J. Svab, E. Dam, S. Muller, J.-P. Briand, A. G. Hovanessian, *J. Biol. Chem.* **1998**, *273*, 21988–21997.

113 F. Yarovinsky, J. F. Andersen, L. R. King, P. Caspar, J. Aliberti, H. Golding, A. Sher, *J. Biol Chem.* **2004**, *279*, 53635–53642.

114 H. Golding, S. Khurana, F. Yarovinsky, L. RL. King, G. Abdoulaeva, L. Antonsson, C. Owman, E. J. Platt, J. F. Andersen, A. Sher, *J. Biol. Chem.* **2005**, *280*, 29570–29577.

115 T. Hatziioannou, D. Perez-Caballero, S. Cowan, P. D. Bieniasz, *J. Virol* **2005**, *79*, 176–183.

116 E. Sokolskaja, D. M. Sayah, J. Luban *J. Virol.* **2004**, *78*, 12800–12808.

117 D. A. Bosco, E. Z. Eisenmasser, S. Pochapsky, W. I. Sundquist, D. Kern, *Proc. Natl Acad. Sci. USA* **2002**, *99*, 5247–5252.

118 A. C. S. Saphire, M. D. Bobard, P. A. Gallay, *J. Virol.* **2002**, *76*, 4671–4677.

119 A. C. S. Saphire, M. D. Bobard, P. A. Gallay, *J. Virol.* **2002**, *76*, 2255–2262.

120 L. Dietrich, L. S. Ehrlich, T. J. LaGrassa, D. Ebbets-Reed C. Carter, *J. Virol.* **2001**, *75*, 4721–4733.

121 H. Ke, Q. Huai, *Biochem. Biophys. Res. Commun.* **2003**, *311*, 1095–1102.

122 C. Tang, Y. Ndassa, M. F. Summers, *Nat. Struct. Biol.* **2002**, *9*, 537–543.

123 D. A. Bosco, D. Kern, *Biochemistry* **2004**, *43*, 6110–6119.

124 M. M. Endrich, P. Gehrig, H. Gehring, *J. Biol. Chem.* **1999**, *274*, 5326–5332.

125 M. Schutkowski, M. Drewello, S. Wöllner, M. Jakob, U. Reimer, G. Scherer, A. Schierhorn, G. Fischer, *FEBS Lett.* **1996**, *394*, 289–294.

126 M. BonHomme, S. Wong, C. Carter, S. Scarlata, *Biophys. Chem.* **2003**, *105*, 67–77.

127 M. BonHomme, C. Carter, S. Scralata, *Biophys. J.* **2005**, *88*, 2078–2088.

128 I. Scholz, B. Arvidson, D. Huseby, E. Barklis, *J. Virol.* **2005**, *79*, 1470–1479.

129 T. Kino, G. N. Pavlakis, *DNA Cell Biol.* **2004**, *23*, 193–205.

130 K. Zander, M. P.Sherman, U. Tessmer, K. Bruns, V. Wray, A. T. Prechtel, E. Schubert, P. Henklein, J. Luban, J. Neidleman, W. C. Greene, U. Schubert, *J. Biol. Chem.* **2003**, *278*, 43202–43213.

131 K. Bruns, T. Fossen, V. Wray, P. Henklein, U. Tessmer, U. Schubert, *J. Biol. Chem.* **2003**, *278*, 43188–43201.

132 D. M. Sayah, E. Sokolskaja, L. Berthoux, J. Luban, *Nature* **2004**, *430*, 569–573.

133 S. Misumi, N. Takamune, S. Shoji, *Proteomics: Biomedical and Pharmaceutical Applications*, H. Hondermarck, Ed. Kluwer Academic, Dordrecht, **2004**, 339–365.

134 S. Misumi, T. Fuchigami, N. Tkamune, I. Takahashi, M. Takama, S. Shoji, *J. Virol.* **2002**, *76*, 10000–10008.

135 G. J. Towers, T. Hatziioannou, S. Cowan, S. P. Stephen, J. Luban, P. D. Bieniasz, *Nat. Med.* **2003**, *9*, 1138–1143.

136 C. Luo, H. Luo, S. Zheng, C. Gui, L. Yue, C. Yu, T. Sun, P. He, J. Chen, J. Shen, X. Luo, Y. Li, H. Liu, D. Bai, J. Shen, Y. Yang, F. Li, J. Zuo, R. Hilgenfeld, G. Pei, K. Chen, X. Shen, H. Jiang, *Biochem. Biophys. Res. Commun.* **2004**, *321*, 557–565.

137 S. Y. Stevens, S. Cai, M. Pellecchia, E. R. P. Zuiderweg, *Protein Sci.* **2003**, *12*, 2588–2586.

138 L. Demange, M. Moutiez, K. Vaudry, C. Dugave, *FEBS Lett.* **2001**, *505*, 191–195.

139 L. Demange, PhD dissertation **2001**, University of Paris XI Orsay, France.

140 S. Hur, T. C. Bruice, *J. Am. Chem. Soc.* **2002**, *124*, 7303–7313.

141 E. Z. Eisenmesser, D. A. Bosco, M. Akke, D. Kern, *Science* **2002**, *295*, 1520–1523.

142 P. K. Agarwal, A. Geist, A. Gorin, *Biochemistry* **2004**, *43*, 10605–10618.

143 C. C. S. Deivanayagam, M. Carson, A. Thotakura, S. V. L. Narayana, R. S. Codavarapu, *Acta Crystallogr. D.* **2000**, *56*, 266–271.

144 S. B. Shuker, P. J. Hajduk, R. P. Meadows, S. W. Fesik, *Science* **1996**, *274*, 1531–1534.

145 M. Huse, Y-G. Chen, J. Massagué, J. Kuriyan, *Cell* **1999**, *96*, 425–436.

146 G. Li, Q. Cui, *J. Am. Chem. Soc.* **2003**, *125*, 15028–15038.

147 P. K. Pallaghy, W. He, E. C. Jimenez, B. M. Olivera, R. S. Norton, *Biochemistry* **2000**, *39*, 12845–12852.

148 D. L. Rabenstein, T. Shi, S. Spain, *J. Am. Chem. Soc.* **2000**, *122*, 2401–2402.

149 R. Ranganathan, K. P. Lu, T. Hunter, J. P. Noel, *Cell* **1997**, *89*, 875–886.

150 Y. Zhao, H. Ke, *Biochemistry* **1996**, *35*, 7356–7361.

151 Y. Zhao, H. Ke, *Biochemistry* **1996**, *35*, 7362–7368.

152 L. Hennig, C. Christner, M. Kipping, B. Schelbert, K. P. Rücknagel, S. Grabley, G. Küllertz, G. Fischer, *Biochemistry* **1998**, *37*, 5953–5960.

153 S.-H. Chao, A. L. Greenleaf, D. H. Price, *Nucleic Acids Res.* **2001**, *29*, 767–773.

154 E. Bayer, S. Goettsch, J. W. Mueller, B. Griewel, E. Guiberman, L. M. Mayr, P. Bayer, *J. Biol. Chem.* **2003**, *278*, 26183–26193.

155 M. A. Verdecia, M. E. Bowman, K. P. Lu, T. Hunter, J. P. Noel, *Nat. Struct. Biol.* **2000**, *7*, 639–643.

156 R. Wintjens, J.-M. Wieruszeski, H. Drobecq, P. Rousselot-Pailley, L. Buée, G. Lippens, I. Landrieu, *J. Biol. Chem.* **2001**, *276*, 25150–25156.

157 I. Landrieu, J. M. Wieruszeski, R. Wintjens, D. Inze, G. Lippens, *J. Mol. Biol.* **2002**, *320*, 321–332.

158 E. Sekerina, J. U. Rahfeld, J. Muller, J. Fanghänel, C. Rascher, G. Fischer, P. Bayer, *J. Mol. Biol.* **2000**, *301*, 1003–1017.

159 J. Kallen, V. Mikol, M. Taylor, M. D. Walkinshaw, *J. Mol. Biol.* **1998**, *283*, 435449.

160 C. Papageorgiou, H.-P. Weber, R. French, X. Borer, *Bioorg. Med. Chem. Lett.* **1995**, *5*, 213–218.

161 V. Mikol, J. Kallen, M. D. Walkinshaw, *J. Mol. Biol.* **1998**, *283*, 451–461.

162 J. L. Kofron, P. Kuzmic, V. Kishore, G. Gemmecker, S. W. Fesik, D. H. Rich, *J. Am. Chem. Soc.* **1992**, *114*, 2670–2675.

163 M. Köck, H. Kessler, D. Seebach, A. Thaler, *J. Am. Chem. Soc.* **1992**, *114*, 2676–2686.

164 R. Huss, *Immunol. Today* **1996**, *17*, 259–260.

165 Y. Zhang, R. Raumgrass, M. Schutkowski, G. Fischer, *ChemBioChem.* **2004**, *5*, 1006–1009.

166 S. Watanabe, S. Tsuruoka, S. Vijayakumar, G. Fischer, Y. Zhang, A. Fujimura, Q. Al-Awqati, G. J. Schwartz, *Am. J. Physiol.* **2005**, *288*, F40-F47.

167 J. M. Grinyo, J. M. Cruzado, *Transplant. Proc.* **2004**, *36*, 240S–242S.

168 M. Hojo, T. Morimoto, M. Maluccio, T. Asano, K. Morimoto, M. Lagman, T. Shimbo, M. Suthanthiran, *Nature* **1999**, *397*, 530–534.

169 E. Benedetti, C. Pedone *J. Peptide Sci.* **2005**, *11*, 268–272.

170 I. Z. Siemion, M. Cebrat, Z. Wieczorek, *Arch. Immunol. Ther. Exp.* **1999**, *47*, 143–153.

172 F. F. Vajdos, S. Yoo, M. Houseweart, W. I. Sundquist, C. P. Hill, *Protein Sci.* **1997**, *6*, 2297–2307.

173 S. Yoo, D. G. Myszka, C.-Y. Yeh, M. McMurray, C. P. Hill, W. I. Sundquist, *J. Mol. Biol.* **1997**, *269*, 780–795.

174 Q. Li, M. Moutiez, J-B. Charbonnier, K. Vaudry, A. Ménez, E. Quéméneur, C. Dugave, *J. Med. Chem.* **2000**, *43*, 1770–1779.

175 M. Cui, X. Huang, X. Luo, J. M. Briggs, R. Ji, K. Chen, J. Shen, H. Jiang, *J. Med. Chem.* **2002**, *45*, 5249–5259.

176 K. Piotukh, W. Gu, M. Kofler, D. Labudde, V. Helms, C. Freund, *J. Biol. Chem.* **2005**, *280*, 23668–23674.

177 G. S. Hamilton, L. Wei, J. P. Steiner, *World Patent 2004204340*, **2004**.

178 J. Kallen, R. Sedrani, G. Zenke, J. Wagner, *J. Biol. Chem.* **2005**, *280*, 21965–21971.

179 T. J. Pemberton, J. E. Kay, *FEBS Lett.* **2003**, *555*, 335–340.

180 T. Fehr, V. Quesniaux, J.-J. Sanglier, L. Oberer, L. Gschwind, M. Ponelle, W. Schilling, S. Wehrli, A. Enz, G. Zenke, W. Schuler, *J. Antibiot.* **1997**, *50*, 893–899.

181 Y. Odagaki, J. Clardy, *J. Am. Chem. Soc* **1997**, *119*, 10253–10254.

182 T. Zarnt, K. Lang, H. Burtsscher, G. Fischer *Biochem. J.* **1995**, *305*, 159–164.

183 M. T. Ivery, L. Weiler, *Bioorg. Med. Chem.* **1997**, *5*, 217–232.

184 L. D. D'Andrea, M. Mazzeo, C. Isernia, F. Rossi, M. Saviano, P. Gallo, L. Paolillo, C. Pedone, *Biopolymers* **1997**, *42*, 349–361.

185 S. Sugie, K. Okamoto, K. M. Rahman, T. Tanaka, K. Kawai, J. Yamahara, H. Mori, *Cancer Lett.* **1998**, *127*, 177–183.

186 T. M. Uchida, M. Takamiya, M. Takahashi, H. Miyashita, H. Ikeda,

T. Terada, Y. Matsuo, M. Shirouzu,
S. Yokoyama, F. Fujimori, T. Hunter,
Chem. Biol. **2003**, *10*, 15–24.

187 L. Ma, L. C. Hsieh-Wison, P. G. Schultz,
Proc. Natl Acad. Sci. USA **1998**, *95*,
7251–7256.

188 L. Demange, M. Moutiez, C. Dugave,
J. Med. Chem. **2002**, *45*, 3928–3933.

189 L. Demange, C. Dugave *Tetrahedron
Lett.* **2001**, *42*, 6295–6297.

190 Y. Zhang, S. Füssel, U. Reimer,
M. Schutkowski, G. Fischer, *Biochemistry* **2002**, *41*, 11868–11871.

191 X. J. Wang, B. Xu, A. B. Mullins,
F. K. Neiler, F. A. Etzkorn, *J. Am. Chem.
Soc.* **2004**, *126*, 15533–15542.

192 H. C. Wang, K. Kim, R. Bakthiar,
J. P. Germanas, *J. Med. Chem.* **2001**, *44*,
2593–2600.

193 S. B. Shuker, P. J. Hajduk,
R. P. Meadows, S. W. Fesik, *Science*
1996, *274*, 1531–1534.

194 H. Kessler, *Angew. Chem. Int. Ed. Eng.*
1997, *36*, 829–831.

195 N. Hoefgen, K. Brune, R. Schindler,
H. Poppe, *Canadian Patent 2,271,831*,
1999.

196 D. Reichelt, B. Kutscher, I. Szelenyi,
H. Pope, G. Quinkert, K. Brune,
H. Bang, H. Deppe *US Patent
6,251,932*, **2001**.

197 T. D. McKee, R. K. Suto, *World Patent
2003074550*, **2003**.

198 B. Hernandez Alvarez, G. Fischer,
M. Braun, A. Hessamian-Alinejad,
H. Fliri, *World Patent 2003093258*, **2003**.

199 P. Burkhard, U. Hommel, M. Sanner,
M. D. Walkinshaw, *J. Mol. Biol.* **1999**,
287, 853–858.

200 R. E. Babine, T. M. Bleckman,
C. R. Kissinger, R. Showalter,

L. A. Pelletier, C. Lewis, K. Tucker,
E. Moomaw, H. E. Parge,
J. E. Villafranca, *Bioorg. Med. Chem. Lett.*
1995, *5*, 1719–1724.

201 C. Guo, C. E. Angelli-Szafran,
N. S. Barta, S. L. Bender, C. F. Bigge,
B. W. Caprathe, A. Chatterjee, J. Deal,
L. Dong, L. K. Fay, X. Hou, R. A. Hudjack
Jr., *World Patent 2002089806*, **2002**.

202 G. M. Dubovchnik, P. D. Provencal,
US Patent 6,818,643, **2004**.

203 Y.-Q. Wu, S. Belyakov, C. Choi,
D. Limburg, B. E. Thomas IV, M. Vaal,
L. Wei, D. E. Wilkinson, A. Holmes,
M. Fuller, J. McCormick, M. Connolly,
T. Moeller, J. Steiner, G. Hamilton,
J. Med. Chem. **2003**, *46*, 1112–1115.

204 Y.-Q. Wu, S. Belyakov, G. S Hamilton,
D. Limburg, J. P. Steiner, M. Vaal,
L. Wei, D. Wilkinson, *World Patent WO
2002059080*, **2002**.

205 J. P. Steiner, S. G. Hamilton, *US Patent
6,593,362*, **2005**.

206 J. Zou, P. Taylor, J. Dornan,
S. P. Robinson, M. D. Walkinshaw,
P. J. Sadler, *Angew. Chem. Int. Ed. Engl.*
2000, *39*, 2931–2934.

207 M. D. Walkinshaw, P. Taylor,
N. J. Turner, S. L. Flitsch, *World Patent
WO 2002048178*, **2002**.

208 E. Bayer, M. Thutewohl, C. Christner,
T. Tradler, F. Osterkamp, H. Waldmann,
P. Bayer, *Chem. Commun.* **2005**,
516–518.

209 C. Schiene-Fischer, J. Habazettl,
F.X. Schmid, G. Fischer, *Nat. Struct.
Biol.* **2002**, *9*, 419–424.

210 B. Adams, A. Miyenko, R. Kumar,
B. Sailen, *J. Biol. Chem.* **2005**, *280*,
24308–24314.

13

Other *Cis-Trans* Isomerizations in Organic Molecules and Biomolecules

Muriel Gondry and Christophe Dugave

13.1
Introduction

The concept of *cis-trans* (*Z-E*) isomerism, originally used for the description of the relative geometry of olefins, has been extended to many other functions which feature a double bond character (pseudo double bonds), such as amides, as well as single bonds with a partial or complete limited rotation due to steric or stereoelectronic effects. *Cis-trans* isomerization (CTI) therefore exists in non-π-bonded or overcrowded molecules that switch from a given stable conformational state to another. This is the case of biaryl compounds which have been utilized in organic chemistry as the basis of molecular switches and rotors [1,2]. Nature has also exploited CTI of single bonds to increasing molecular diversity, in particular with the bulky thyroxin, a thyroid hormone, and the well-known disulfide bond which plays a critical role in the structure of peptides and in the conformation of proteins.

Other particular groups that contain a permanent or transient π-bond have been found in biomolecules, such as Schiff bases, dehydroamino acids, and dehydropeptides, which all display particular chemical and biological properties. Organic and medicinal chemists have also exploited the large panel of *cis*/*trans*-isomerizable groups which can be used for synthetic purposes and for biologic and therapeutic applications.

13.2
Cis-Trans Isomerization around Single Bonds

13.2.1
Cis-Trans Isomerism of Aryl Compounds

Cis-trans isomerism around single bonds has been widely used in supramolecular chemistry, in particular with bulky biaryl compounds such as the metal-chelating agents bipyridines and sterically hindered binaphthyl motifs. In these compounds, a moderate barrier to rotation exists due to the resonance of the two con-

cis-trans Isomerization in Biochemistry. Edited by Christophe Dugave
Copyright © 2006 WILEY-VCH Verlag GmbH & Co. KGaA, Weinheim
ISBN: 3-527-31304-4

jugated aryl moieties and hence two *cis* (or cisoid) and *trans* rotamers (or transoid) may coexist in solution.

In simple compounds, the *trans* form of bipyridines is favored due to nitrogen electronic doublet repulsion, and addition of a metal cation initiates the CTI process to give the *cis* conformer, which is stabilized by the coordination of the metal (Fig. 13.1A). This concept was mainly utilized for the design of cation-tunable molecular switches that isomerize through a fully reversible pathway [3]. In fact, non-complexed *cis*-bipyridines deviate greatly from planarity (Fig. 13.1B) and protonation strongly influences the geometry of the molecule [4,5].

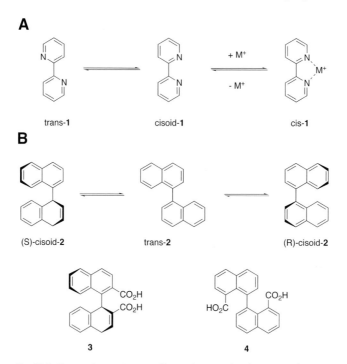

Fig. 13.1 *Cis-trans* isomerization of bipyridine **1** and aryl compounds **2–4**.

The transoid–cisoid equilibrium in crowded binaphthyl compounds generates two enantiomeric cisoid forms which may interconvert. For these compounds, CTI requires heating to proceed (i.e. $\Delta G^{\ddagger} = 23.5$ kcal mol^{-1} for compound **2**, $t_{1/2} = 14.5$ min at 50 °C). Introduction of substituents increases the barrier to rotation and hence stabilizes the chiral configuration [6]. For example, (S)-1,1'-binaphthyl-2,2'-dicarboxylic acid **3** could not be racemized at 175 °C in DMF (Fig. 13.1B). Depending on substituents, racemization may be favored by steric/electrostatic repulsions and CTI of dicarboxylate **4** occurs at a lower temperature ($\Delta G^{\ddagger} = 24.4$ kcal mol^{-1} for compound **2**, $t_{1/2} = 51.5$ min at 50 °C).

Thyroid hormones display a canonic diaryl ether motif characterized by two or more iodine atoms which restricts the relative positioning of the aromatic cycles.

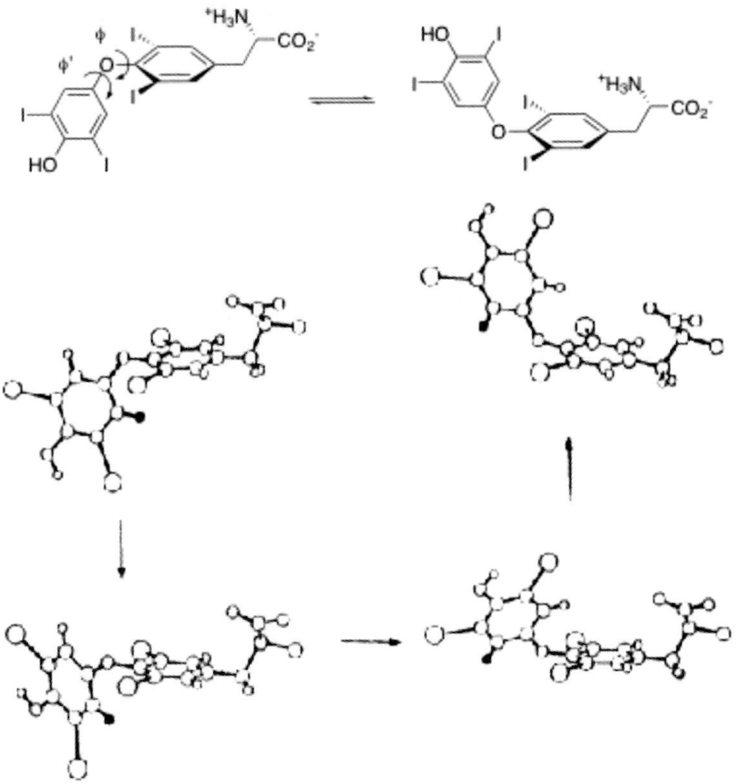

Fig. 13.2 Sketch of interconversion of transoid to cisoid rotamer of thyroxin **5** around the ϕ and ϕ' dihedral angles [7].

In solution, thyroxin exists as two distinct cisoid and transoid isomers in a roughly 50:50 ratio (Fig. 13.2) which results from oxygen doublet–iodine interactions as well as iodine–iodine repulsion.

Crystallographic resolution of the thyroid receptor:thyroxin complex has demonstrated that the active conformation of thyroxin **5** is the transoid form and that the cisoid isomer is not bound and cannot isomerize inside the binding site. Variable temperature ^{1}H NMR experiments have shown that the CTI process seems to involve a relative motion of the aromatic cycles via a concerted rotation about ϕ and ϕ' (Fig. 13.2) rather than a rotation of the alanyl moiety ($\Delta G^{\ddagger} = 8.7$– 8.9 kcal mol^{-1}) [7].

13.2.2
Disulfide Bonds

Disulfide bonds are true single bonds, the rotation of which is mainly limited by interactions of the sulfur lone pair. Consequently, limitation of conformational freedom around the C–S–S–C bond enables the existence of cisoid, rectangular

[8,9], transoid and *trans* conformations [10], though several intermediary conformations may coexist [11]. A similar tendency was also observed in trisulfide (C–S–S–S–C)-containing compounds, which have been reported to adopt a preferential helical (transoid) conformation [12]. In disulfides, the transoid and *trans* conformations ($\pm 100° \leq \omega \leq 180°$) are by far the most abundant in acyclic systems as well as vicinal disulfide turns, whereas the *cis* rotamer does not usually exist due to strong electronic interactions (Fig. 13.3).

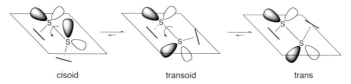

cisoid   transoid   trans

Fig. 13.3 Cisoid–transoid–*trans* isomerism about the disulfide bond.

The cisoid conformation is only found in compounds that feature particular strains [13] or contain unusually long S–S bonds such as (TMS)$_3$–S–S–(TMS)$_3$ [14] and when the barrier to rotation is relatively low, in particular with 4,4′-dithiopyridine [15,16], a motif currently used for the assembly of bimetallic supramolecular structures [17]. In contrast, bis(diphenylthiophosphoryl)disulfide showed a cisoid geometry whereas bis(diphenoxythiophosphoryl)disulfide was in a transoid conformation, all of which suggests a pretty high barrier to interconversion. As anticipated, the S2–S2′ bond in **6** (2.027 Å) was shorter than the S2–S3 bond in **7** (2.070 Å) [18].

Disulfide bonds play an important part in the structure and stabilization of peptides and proteins. They are usually formed by spontaneous oxidation of the corresponding sulfide, but the reaction is catalyzed by a family of ubiquitous proteins called protein disulfide isomerases (PDIs) through a complex sequence of reactions that involves many partially folded intermediates [19]. This explains how the unfavored cisoid conformation may be represented for a given disulfide bond [20] though the transoid and *trans* isomers are much more frequently observed, as are intermediary rotamers [21].

13.2.3
Amide Surrogates with Restricted Rotation of a σ-Bond

Amide surrogates such as aminimides, phosphonamides, and sulfonamides (Fig. 13.4) display high conformational flexibilities and have been used to mimic the tetrahedral transition state of proteases. The intriguing aminimides display interesting biological properties [22] and have been introduced in peptidomimetic inhibitors of elastase [23] and HIV-1 protease [24] and are found in the diuretic agent besulpamide. In these compounds, the high resonance O=C–N⁻ $\leftrightarrow$ ⁻O–C=N fixes the amide bond in *trans* ($\omega = 180°$) while energy minima have been found for $\phi = 60, 180$ and $300°$ with an energy barrier of about 8.5 kcal mol⁻¹.

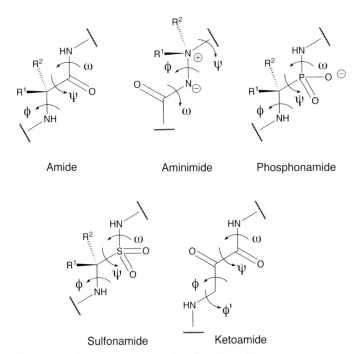

Fig. 13.4 Amide surrogates aminimides, phosphonamides, and sulfonamides and associated dihedral angles.

The P–N bond of phosphonamides and the S–N bond of sulfonamides have a lower π-bond character than amides. Consequently, free rotation of the ω dihedral angle is restricted due to the deconjugation with a π-system (2p(N) $\rightarrow$ 3d(P or S) and free electron doublet interactions, and they may exist as mixtures of multiple diastereomeric rotamers though they do not display a *cis-trans* isomerism [25–27]. The nonisostere ketoamide surrogate displays a particular structure with the ketone carbonyl orthogonal relative to the planar amide bond ($\psi = -90°$). To our knowledge, nothing is known about the possible rotation about the O=C–C=O bond, which may exist as two enantiomeric forms. Phosphonamides and keto-amide surrogates have been utilized for the design of PPIase transition state analogs and, in such a way, indirectly confirm the deconjugated tetrahedral transition state proposed for amide CTI [14].

13.3
C=N-containing Compounds

13.3.1
Oximes and Nitroso Compounds

Oximes and oxime ethers exist as a mixture of *E* and *Z* isomers with a relatively low difference of $\Delta G°$ and a moderate energy barrier to isomerization (<10 kcal mol^{-1}) [28]. They show some similarities with imines and may interconvert at room temperature, spontaneously, by an acid- or base-catalyzed isomerization involving a nitronium ion, and photochemically [29,30]. Oxime ethers have been employed as amide surrogates in peptides where they display a marked *Z-E* isomerism which is mainly controlled by the formation of H-bonds, which stabilize a given isomer. As an example, the structure of pseudopeptide **6** was investigated by Fourier transform infrared spectroscopy (FTIR) and NMR spectroscopy which both showed that Z-**6** is folded in a *β*-like conformation by a strong bifurcate hydrogen bond whereas the *E* isomer adopts an extended conformation (Fig. 13.5) [31].

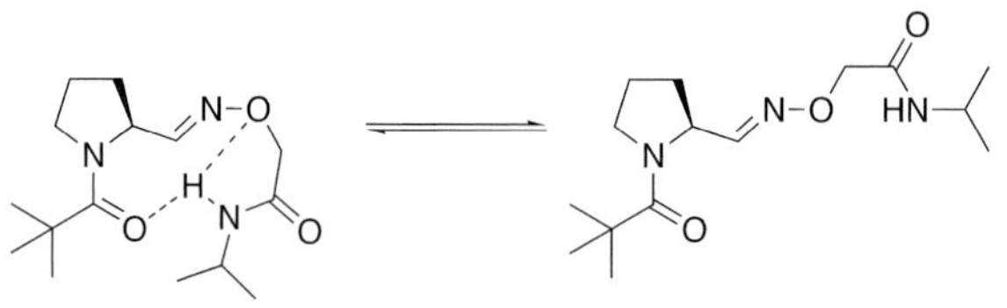

Z-**6**   *E*-**6**

Fig. 13.5 Structures of *Z*- and *E*-**6**.

C- and N- nitroso compounds may exist as two *Z* and *E* (major) isomers due to the tautomeric effect which restricts the rotation about the C–N and N–N bonds. Thus CTI can occur via free rotation of the tautomeric form, doublet inversion etc. [29,14]. In this respect, *C*-nitroso compounds are the tautomeric form of the corresponding oximes (CH–N=O → C=N–OH). The energy barrier is usually between 12 and 27 kcal mol^{-1} [32], however, many factors may influence CTI, such as substituents, solvents, and pH [33]. For example, alkyl-substituted C- and N-nitroso compounds usually exist as a mixture of *Z* and *E* isomers [33], whereas acylnitroso [34] and nitrosocarbamate [35] species have an exclusive *E* relative stereochemistry due to oxygen–oxygen electronic repulsion.

13.3.2
Imines and Schiff Bases

The imine group is widely employed in synthetic organic chemistry and combinatorial chemistry, and is found in many proteins bearing a prosthetic group as the standard linkage between a protein-derived amine moiety and a carbonyl from the prosthetic group. CTI of imines and their tautomeric isomers, enamines, has been studied for a long time and many CTI pathways have been observed, though the barrier to isomerization is pretty high ($\Delta G^{\ddagger} \approx 23$ kcal mol^{-1}) and the isomerization rate is slow on the NMR time-scale [36]. The imine/enamine tautomerism accounts for CTI in simple imines [37,38] since the X=Y–Z system is considered as a potential dipole [39]. The process is strongly accelerated in push–pull imines [40] and push–pull enamines [41] or in the presence of bases [42,43], acids and Lewis acids [44], polar solvents [45] and metals [46]. Photoisomerization of enal Schiff bases via a singlet state was also reported [47]. Locked enamines found in cyclidene dioxygen carriers isomerize spontaneously at room temperature [48] and the formation of the *trans* isomer has been proposed to cause a decreased oxygen affinity [49].

The main studies of Schiff base CTI in proteins concern the retinal-bound chromoproteins, in particular rhodopsin and bacteriorhodopsin, which are G protein-coupled photoreceptors (see Chapter 4). In rhodopsin, the main visual pigment in animals and algae, the 11-*cis*-retinal chromophore, is bound to a lysine residue of the protein via a Schiff base in a C^{15}=N *anti*-configuration. Upon absorption of a photon, the rhodopsin retinyl moiety isomerizes from 11-*cis*-15-*anti* to all-*trans*-15-*anti*, while bacteriorhodopsin isomerizes from an all-*trans-anti* to a 13-*cis-anti* configuration via an ultrafast CTI process which initiates a multi-intermediate isomerization pathway characterized by sequential modifications in the protein–chromophore interaction that trigger the signal [50]. The protein environment as well as the imine protonation state influence the position of CTI [51], controls the imine *cis-trans* interconversion [52] and, in bacteriorhodopsin, acts as a light-driven proton pump [53]. In particular, Touw et al. have reported that the C14–C15 bond has an enhanced double bond character, while C13–C14 and C15=N have a reduced double bond character [51]. Very recently, Sheves and coworkers have demonstrated that 9-*cis*-15-*syn*, 11-*cis*-15-*syn* intermediates (obtained from light induced isomerization of the meta III intermediate) undergo a thermal CTI to the corresponding 15-*anti* isomers while prolonged illumination of metarhodopsin III (*all-trans*-15-*syn*) isomerizes to the corresponding 15-*anti* isomers meta I (protonated) and meta II (neutral) (Fig. 13.6) [54,55].

It is noteworthy that despite the highly conjugated structure of the retinal Schiff base, back-isomerization necessitates the hydrolysis of the imine and the release of retinal from the opsin protein. Enzyme-catalyzed isomerization of retinal is an endothermic reaction ($\Delta\Delta G^{\circ} = +4$ kcal mol^{-1}) which involves several enzymes and is still under investigation. Tritium release experiments have unambiguously shown that back-isomerization must occur at the alcohol oxidation stage. This first requires the reduction of the aldehyde and further esterification of the allylic alcohol. CTI itself is catalyzed by an isomero-hydrolase via an addition/elimination/addition/isomerization process (Fig. 13.7) which might provide the energy to drive CTI ($\Delta\Delta G^{\circ} = -5$ kcal mol^{-1}). Conjugated addition of water to the substrate–enzyme covalent adduct enables CTI and release of the enzyme nucleophilic group [56].

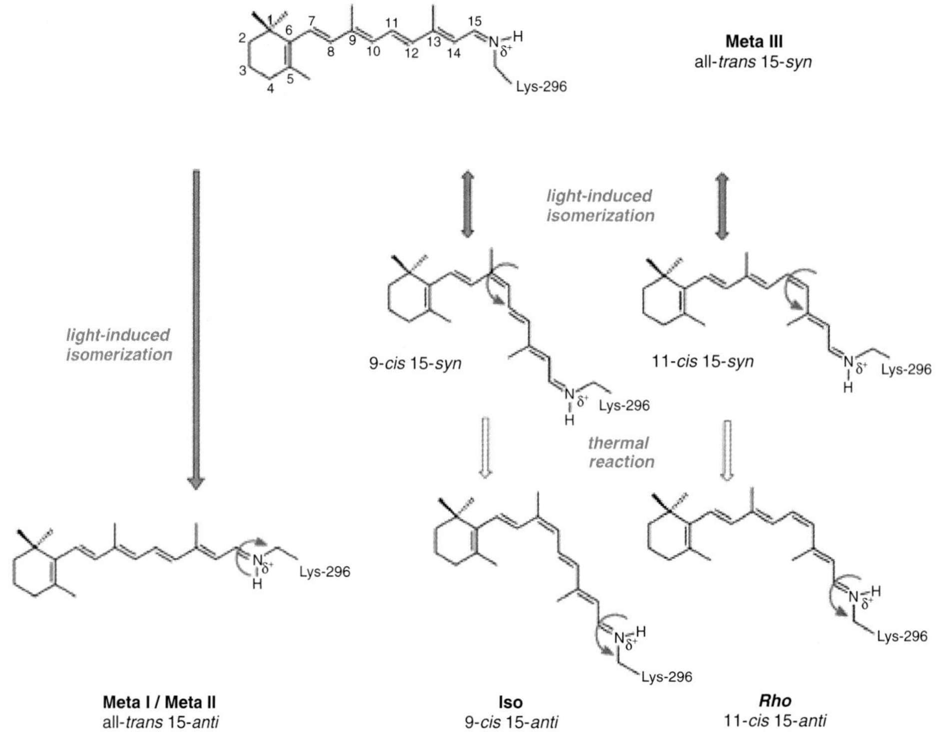

Fig. 13.6 Thermal and light-induced imine *cis-trans* isomerization in rhodopsin [55].

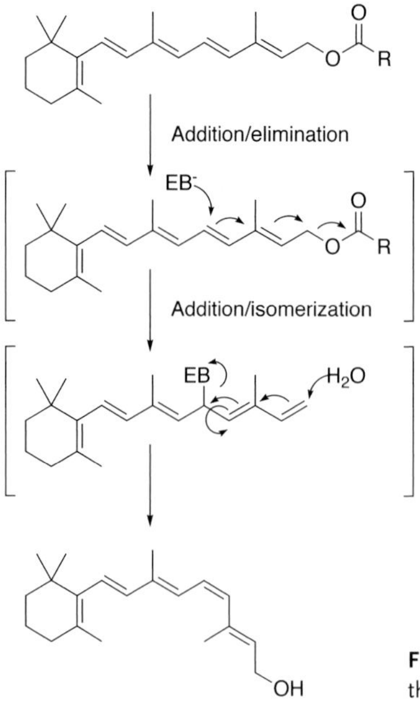

Fig. 13.7 Isomerization of 11-*trans*-retinol ester to the 11-*cis*-retinol catalyzed by the isomero-hydrolase.

Hydrazones are closely related to imines though the presence of an additional nitrogen atom which decreases the C=N double bond character and facilitates isomerization. Hydrazone CTI proceeds via photochemical and thermal pathways, the latter being accelerated by polar solvents, acid–base catalysis and electron-donating substituents. This suggests the conversion to the tautomeric form di-azene, which can rotate freely about the C–N bond. In most cases, the lowest ener-gy conformation is *E*, though H-bonds and molecular strains may affect the *Z:E* ratio [14].

In conjugated (aromatic) imines, the imine–enamine tautomerism (CH–C=N → C=C–NH) plays a critical role in the formation of H-bonds that may regulate CTI [33]. Enamines are also found in dehydroamino acids and dehydropeptides, which we tackle next.

13.4
Dehydroamino Acids and Dehydropeptides

13.4.1
Acryloyl Peptides, Acrylates and Related Molecules

Synthetic furylacryloyl peptides are commonly used as substrates for the determi-nation of the kinetic parameters of many proteolytic enzymes. However, the rapid photoisomerization of the acrylate moiety in plain daylight yields *cis* and *trans* iso-mers that are hydrolyzed with distinct catalytic efficiencies. This was clearly dem-onstrated with the hydrolysis of 3-(2-furylacryloyl)-Phe-Leu by carboxypeptidase Y [57] (Fig. 13.8).

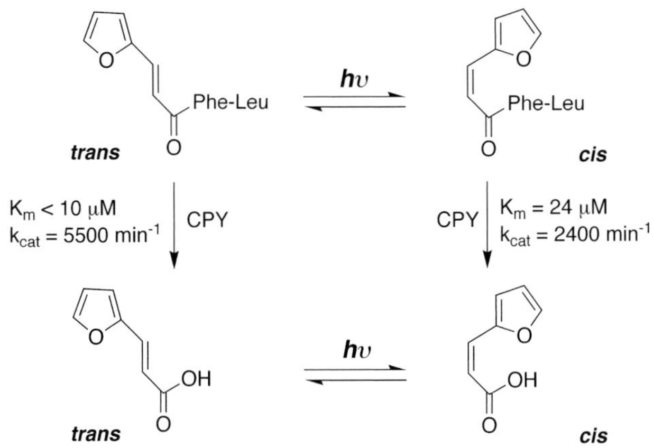

$K_m < 10~\mu M$
$k_{cat} = 5500~min^{-1}$

$K_m = 24~\mu M$
$k_{cat} = 2400~min^{-1}$

Fig. 13.8 *Trans* to *cis* photoisomerization of 3-(2-furylacryloyl)-Phe-Leu influences the catalytic efficiency of carboxypeptidase Y (CPY).

Natural acryloyl compounds can also undergo CTI catalyzed by enzymatic systems. In this way, the CTI of maleylacetoacetate, a degradation product of phenylalanine and tyrosine, to fumarylacetoacetate was achieved by the *cis-trans* isomerase glutathione system. The reaction seems to involve a nucleophilic attack at the C2 carbon rather than a conjugated addition and the subsequent rotation of the single bond [58].

Ionic interactions can catalyze CTI of acrylates and α,β-unsaturated carbonyl compounds such as the aroma component geranial (*E*-form) which isomerizes into neral (*Z*-form) in alkaline aqueous solution. This process is significantly accelerated by the zwitterionic forms of asparagine or glutamic acid which presumably stabilize the cation/enolate intermediate [59]. Norbadione A **7**, a pigment known to complex cesium cations, is a pulvinic acid derivative which undergoes a pH-dependent *Z-E* isomerization of the two pulvinic moieties, leading to four possible stereomeric forms (Fig. 13.9). Each diastereomer is associated with an individual protonation state: protonation of the enol favors the *E* configuration, whereas deprotonation causes an $E \rightarrow Z$ isomerization [60].

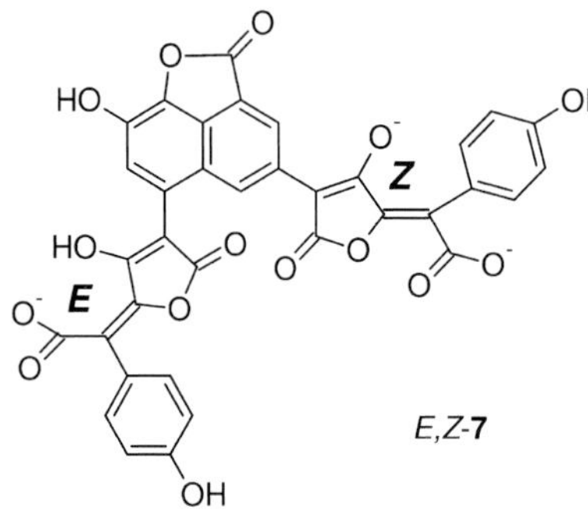

Fig. 13.9 One of the four possible isomers *E,Z*-norbadione A **7** in the protonated/deprotonated enol state.

Geldanamycin, an antibiotic which specifically binds to molecular chaperone heat shock protein 90 (HSP90), displays an enzyme-catalyzed CTI. The X-ray structure of bound geldanamycin indicates that the amide group bonded to the benzoquinone has a *cis* configuration, whereas that of unbound geldanamycin shows that the same amide group is in a *trans* configuration. A recent study using quantum chemical calculations and mutational analysis suggests that HSP90 itself catalyzes the CTI of geldanamycin according to the mechanism shown in Fig. 13.10 [61].

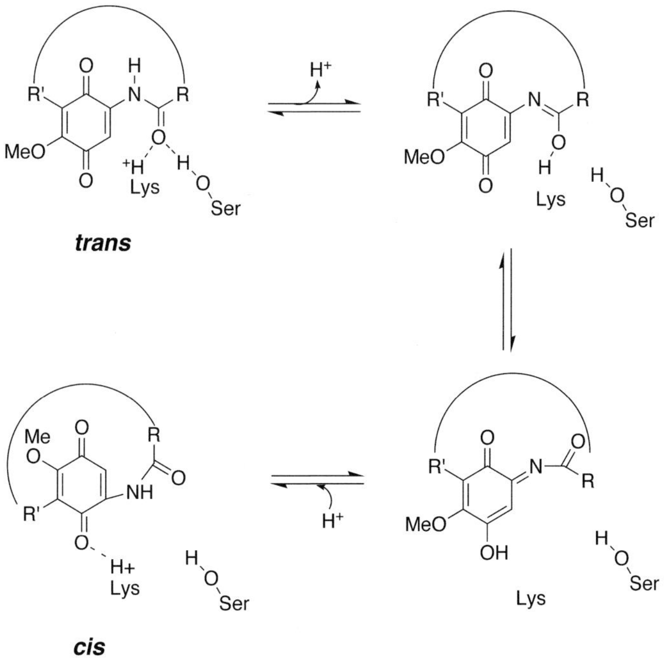

Fig. 13.10 Geldanamycin *trans* to *cis* isomerization catalyzed by Lys112 and Ser113 residues of HSP90. R and R′ indicate the portion of the ansa ring of geldanamycin.

13.4.2
Naturally Occurring Dehydroamino Acids and Dehydropeptides

αβ-Dehydroamino acids (Δ-residues) are widely distributed in naturally occurring peptides of prokaryotic and lower eukaryotic origin [62–65]. In some cases, they are key intermediates in the biosynthesis of microbial peptide metabolites in which they are masked by further conjugated additions to the double bond. For example, lantibiotics like epidermin from *Staphylococcus epidermidis* or nisin from *Lactococcus lactis* feature thioether bis-amino acid motifs called lanthionines and 3-methyl-lanthionine, resulting respectively from the addition of the thiol group of a cysteinyl residue to the double bond of ΔAla and Δbutyrine [66,67]. Δ-Amino acids are also found in proteins but few examples have been reported to date: ΔAla is present in the 4-methylidene-imidazole-5-one moiety found in the active site of several enzymes such as histidine ammonia lyase (HAL), phenylalanine ammonia lyase and tyrosine aminomutase [68,69]; ΔTyr belongs to the *p*-hydroxy-benzylidene-imidazolidone chromophore of green fluorescent protein (GFP) [70] (Fig. 13.11). The mechanism of photoisomerization associated with the GFP chromophore is detailed in Chapter 5.

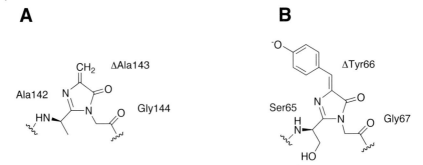

Fig. 13.11 Structures of the electrophilic motif of the HAL active site (A) and the GFP chromophore (B).

Natural dehydro-metabolites display a large diversity of structures and various biological activities. Most of them have a cyclic structure and they are predominantly formed of Z-dehydroamino acids (Δ^Z-residues). For example, the diketopiperazine albonoursin **8** produced by *Streptomyces noursei* contains both Δ^ZPhe- and Δ^ZLeu residues [71] (Fig. 13.12) and exhibits an antibiotic activity towards some *Bacillus* species [72]. Phenylahistin **9** isolated from *Aspergillus ustus* is an isoprenylated Δ^ZHis-containing diketopiperazine which possesses antitumor activity [73] (Fig. 13.12). Tentoxin **10** produced by *Alternaria alternata* is a Δ^ZPhe-containing cyclotetrapeptide [74] which exhibits phytotoxic activity by causing chlorosis in the seedlings of many plants [75] (Fig. 13.12).

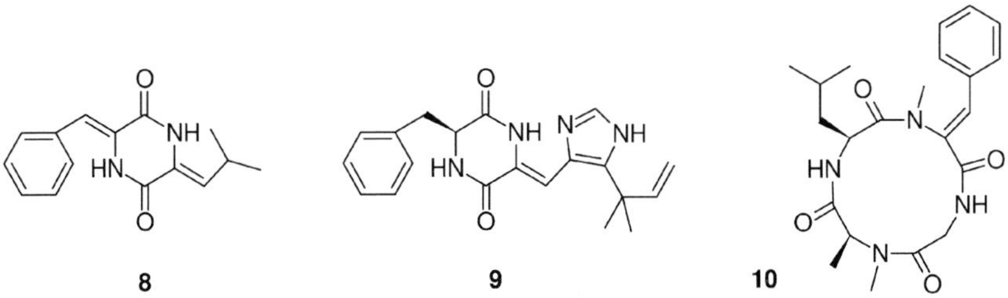

Fig. 13.12 Δ^Z-residues in albonoursin (**8**), phenylahistin (**9**), and tentoxin (**10**).

Many other examples of Δ^Z-peptides are listed in the literature and suggest that Δ^Z-amino acids are usually more stable than the E isomers [76]. However, some dehydropeptides like the neurotoxins roquefortine **11** and oxaline **12**, isolated from *Penicillium roqueforti* and *Penicillium oxalicum*, respectively, both contain a Δ^EHis residue (Fig. 13.13) [77]. Phomopsin A **13** (Fig. 13.13), an antimitotic agent produced by *Phomopsis leptostromiformis*, is also a Δ^E-peptide as it contains Δ^EAsp and Δ^EIle residues [78,79].

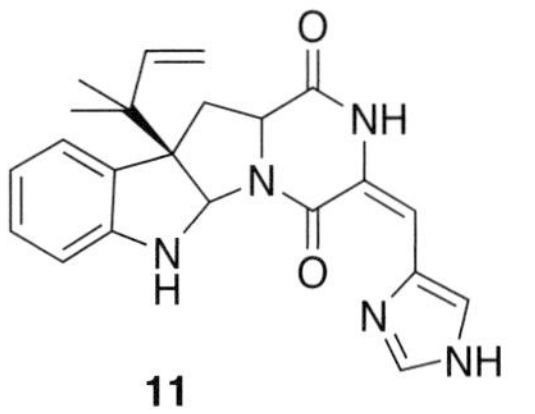

Fig. 13.13 Δ^E-residues in roquefortine (11), oxaline (12), and phomopsin A (13).

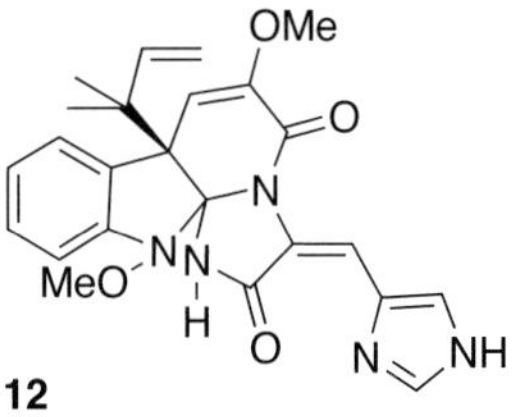

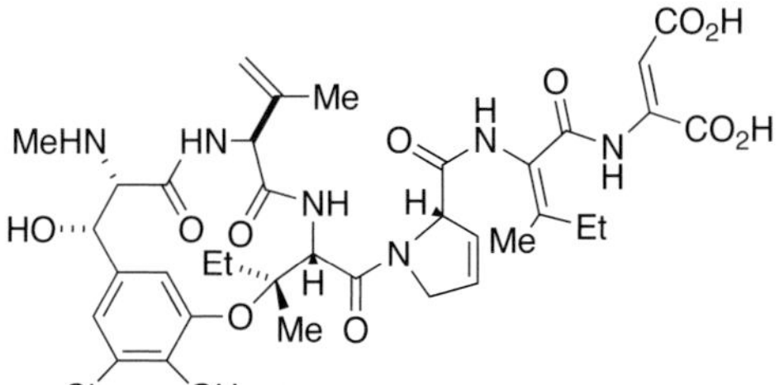

The widespread occurrence and diversity of natural dehydropeptides suggests that many specific enzymes are involved in the biosynthesis of Δ-amino acyl residues. However, amino acyl $\alpha\beta$-dehydrogenase activities remain poorly documented and only five enzymes have been characterized [80]. These are structurally unrelated and use different mechanistic pathways (Fig. 13.14). First, L-tryptophan 2′,3′-oxidase from *Chromobacterium violaceum* was postulated to catalyze the direct $\alpha\beta$-dehydrogenation of a tryptophan residue via a *cis*-elimination leading to a Δ^ZTrp (Fig. 13.14, pathway a) [81,82]. A similar mechanism was described for cyclopeptine dehydrogenase from *Penicillium cyclopium* which catalyzes the formation of a Δ^ZPhe residue at an intermediary stage of the biosynthesis of various alkaloids [83,84]. Similarly, Epi D from *Staphylococcus epidermidis* catalyzes the $\alpha\beta$-dehydrogenation of a cysteinyl residue to produce the C-terminal Δ^ZCys of the lantibiotic epidermin [85,86]. On the other hand, pristinamycin synthase from *Streptomyces pristinaespiralis* catalyzes the transient β-hydroxylation of a D-prolyl residue, immediately followed by a dehydration reaction leading to the Δ^ZPro motif of pristinamycin II$_A$ (Fig. 13.14, pathways b–b′) [87]. Finally, cyclic dipeptide oxidase from *Streptomyces noursei* was postulated to use a two-step mechanism involving the transient formation of an intermediate imine followed by its subsequent rearrangement to yield the ΔPhe or ΔLeu found in albonoursin 8 (Figs 13.12 and 13.14, pathways c–c′). The Δ-products were predominantly obtained in Z configurations but the other isomers (*Z–E, E–Z, E–E*) were also produced; this might arise from a free rotation around the C$_\beta$ carbon prior to the formation of the α–β double bond [80]. This mechanism was originally hypothesized by Bycroft to explain the frequent occurrence of Δ- and D-amino acids in microbial peptides [88].

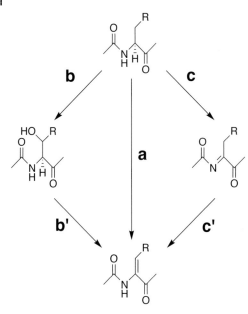

Fig. 13.14 Possible pathways for enzymatic synthesis of Δ-residues in peptides [80].

Natural dehydropeptides have interesting biological activities and, therefore, the production of synthetic dehydropeptides appears to be very attractive for the generation of new potential pharmacological tools and drugs. The enzyme-based synthesis of dehydroamino acyl moieties is currently poorly developed due to our limited knowledge of amino acyl αβ-dehydrogenases. However, the enzymatic synthesis of Δ^ZTrp-containing peptides was successfully achieved using L-tryptophan 2′,3′-oxidase [82,89]. In the same way, various $\Delta^{Z,Z}$-, $\Delta^{Z,E}$-, $\Delta^{E,Z}$-, and $\Delta^{Z,E}$-diketopiperazines were synthesized using cyclic dipeptide oxidase [80]. The chemical preparation of dehydropeptides is far more developed than enzymatic synthesis, and thus has enabled the extensive study of chemical and biological properties of dehydropeptides.

13.4.3
Synthetic Dehydroamino Acids and Dehydropeptides

Numerous studies dealing with dehydropeptides have used model Δ^ZPhe-peptides owing to their easy chemical synthesis [76]. Preparative procedures are now available to produce various dehydroamino acids such as ΔLeu [90] and ΔVal [91] and efficient stereoselective methods are described for the synthesis of Δ^E-amino acids [92–94]. In particular, photoisomerization of Δ^ZPhe is a straightforward route to produce Δ^EPhe-containing peptides [95,96]. For example, UV irradiation of the tripeptide Boc-Ala-Δ^ZPhe-Val-OMe gave Boc-Ala-Δ^EPhe-Val-OMe (Fig. 13.15) [97].

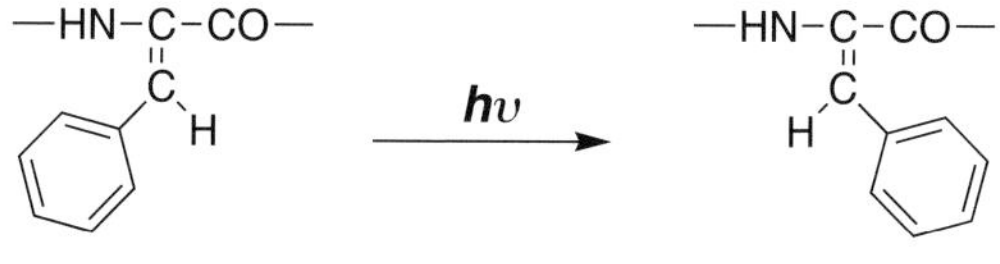

Fig. 13.15 Photoisomerization of Boc-Ala-Δ^ZPhe-Val-OMe.

The incorporation of a dehydroamino acyl residue inside a peptide generally causes dramatic changes in the chemical, conformational and biological properties of the peptide. Generation of the α–β unsaturation generally increases the hydrophobicity of the peptide and introduces structural modifications and constraints which account for the improved resistance of these peptides to proteolysis [98]. Moreover, the dehydroamino acid moiety acts as a fairly reactive Michael acceptor which readily reacts with soft nucleophiles such as thiol or amine groups [3,4] (see Section 13.4.2). Therefore, many biologically active peptide analogs containing Δ-amino acids have been described [99,100]. In many cases, receptors frequently discriminate between Z or E isomers. For example, [ΔLeu^2]-enkephalin [64] exhibits a higher affinity for the μ-receptors in brain [101] while [Δ^ZPhe^4]-enkephalin displays an enhanced interaction with the δ-receptors [102,103]. The Δ^ZPhe residue of phytotoxin tentoxin **10** (Fig. 13.12) is crucial for chlorosis since its natural saturated precursor dihydrotentoxin has almost no chlorotic effect [104]. Moreover, the Δ^EPhe-tentoxin isotentoxin, produced by UV irradiation, seems not to result in chlorosis [95]. In the same way, in phomalide E-**14**, a cyclic depsipeptide phytotoxin produced by *Phoma lingam* which contains an E-configured 2-amino-2-butenoic (Aba) motif (Fig. 13.16), the double bond configuration has been shown to be crucial for toxicity [105]. Interestingly, isomerization of the toxic phomalide into the nontoxic Z isomer isophomalide Z-**14** could explain the host-selective phytotoxicity since the CTI process would not occur in susceptible plants [105].

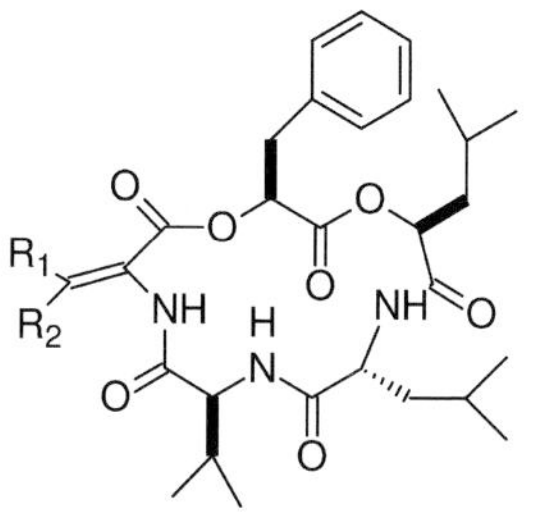

E-**14** Phomalide: R$_1$ = Me; R$_2$ = H

Z-**14** Isophomalide: R$_1$ = H; R$_2$ = Me

Fig. 13.16 Molecular structures of phomalide E-**14** and isophomalide Z-**14**.

The conformational effects introduced by Δ-amino acyl residues in synthetic dehydropeptides have also been extensively studied and were shown to generate and/or to stabilize particular secondary structures in peptides. Much more is known about the conformational preferences of peptides containing Δ^Z-residues

[106–109] than those containing their *E* isomers, although both conformers have been used as peptide conformation modifiers [110–112].

The effect of C_α–C_β unsaturations in dehydropeptides may also be critical in metal ion binding, in particular in stabilizing the metallopeptide complexes much more efficiently than the corresponding saturated peptides. In particular, the configuration of the Δ-amino acyl moiety influences the ability of the peptide to bind Cu(II) and Ni(II) ions [113]. For example, the *Z* isomer of Gly-ΔPhe-Gly has a higher affinity for Cu^{2+} than the *E* isomer [114].

Finally, synthetic dehydropeptides appear to be attractive tools for the design of new bioactive molecules. CTI of dehydropeptides modulates or drastically modifies the structural characteristics and biological activities of these molecules.

13.5
Phototunable Biomolecules Containing an Azobenzene Moiety

Arylazo and, to a less extent, triazene compounds, are well known to isomerize at higher wavelengths than the corresponding olefins (320–350 nm for diarylazo compounds compared with 290–300 nm for stilbene derivatives) due to the lower energy of their excited states and their ability to undergo CTI through several distinct processes including $S_0 \rightarrow S_2$ π–π^* rotation, $S_0 \rightarrow S_1$ n–π inversion as well as an ultrafast cleavage–recombination mechanism (cf. Chapter 2). Moreover, *cis* to *trans* interconversion usually occurs either by simple heating or irradiation above 400 nm and this double control enables a fine tuning of the configuration of the Ar–N=N–Ar motif. In general, however, isomerization is not complete and gives a mixture of isomers with a largely predominant *cis* or *trans* form. Moreover, the relatively slow isomerization rates might be a major drawback to the use of these compounds. Novel azobenzene derivatives requiring irradiation times under one second were described in the late 1990s [115]. Azo compounds are also usually easier to prepare than their C=C equivalents. As a consequence, a large collection of molecules bearing a diarylazo group including peptides, nucleic acids and polymers have been synthesized and open novel perspectives in biochemistry in terms of structural control and modulation of dynamics.

13.5.1
Phototunable Ligands

In the late 1960s, Erlanger and coworkers reported the photoregulation of bioactivity of an azo-containing analog of carbamylcholine, an agonist of the nicotinic acetylcholine receptor (AChR). In particular, the photochromic reagent was able to photoregulate the AChR-dependent membrane potential: the *trans* isomer proved to be a stronger inhibitor of innervated membrane depolarization than the *cis* isomer [116]. Photochromic activator of AChR **15** only acted in its *trans* form in the submicromolar affinity range, whereas the *cis* isomer was almost devoid of bioactivity [117] (Fig. 13.17). Use of phototunable depolarizing ligands provided important information on the acetylcholine binding site more than 25 years before the structure of AChR was resolved by Unwin et al. [118].

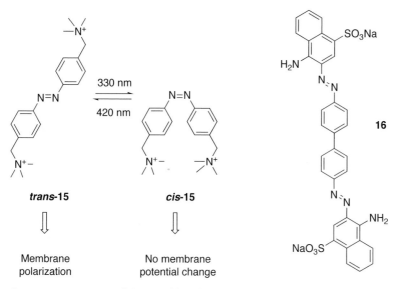

Fig. 13.17 Isomerization of phototunable AChR ligand **15** and structure of Congo Red **16**.

Azo compounds have been widely utilized as dyes and their use as additives in food is strongly challenged due to toxic side effects, such as those of tartrazine. On the other hand, diazenes and triazenes [119,120] have shown interesting biological activities as protein ligands. In particular, the old-fashioned dye and pH indicator Congo Red **16** binds strongly and specifically to amyloid aggregates [121,122], though to the best of our knowledge no study of the effect of CTI on such binding has been performed. Also, methyl yellow (4-dimethylaminoazo-benzene) and analogs are known to activate the yeast aryl hydrocarbon receptor [123].

The coating of surfaces with photoswitchable RGD peptides was recently proposed as a new way to tune cell adhesion by changing the distance and relative orientations of ligands upon light irradiation [124]. Phototunable self-assembly of cylindrical peptides was investigated by Ghadiri and coworkers. A photoswitchable diarylazo moiety tethering two octapeptides was used to photoregulate the supra-molecular arrangement of the peptide. The peptide sequence was designed to form a planar ring which tends to stack via hydrogen-bonding to form very stable flat layers at the air–water interface. Therefore, the *trans* isomer was shown to form regular multimolecular systems which dissociate upon UV light irradiation to give a monomeric form stabilized by intramolecular H-bonds. The *trans* to *cis* isomerization was induced by UV light and the back process was triggered by visible light irradiation in a very reproducible way, leading to a remarkable variation of 70 Å of the molecular surface (Fig. 13.18) [125,126].

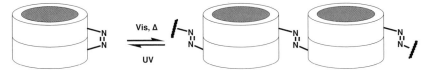

Fig. 13.18 Stylized sketch of *cis-trans* isomerization of photoswitchable hydrogen-bonding in self-organized cylindrical peptide systems.

13.5.2
Phototunable Conformation of Peptides

Large structural modifications caused by CTI of the diarylazo moiety have been used for the photoregulation of the conformation of peptides. In particular, cyclization of peptides such as PRP [127], RGD, GnRH, tuftsin and VIP through diazotization has been proposed for the development of novel photoswitchable peptide analogs [128]. This was done either by azoic coupling of Phe/Tyr analogs [127,129] or by introduction of an exogenous photoisomerizable cross-linker through thioether, disulfide or amide bonds [130]. In the latter case, judicious choice of motifs connecting the azobenzene core (alkyl, amide, carbamate, and urea) allowed fine adjustment of the conformation of a test peptide by a combination of heating and irradiation. Very recently, Hamm and coworkers reported the design and study of photoswitchable peptides which interconvert from a loop (*cis*-azo) conformation to an α-helix (*trans*-azo) conformation upon light irradiation (Fig. 13.19). Peptide conformational transitions were monitored by time-resolved IR spectroscopy [131].

The concept of photomodulation of peptide conformation was also applied to the photoinduction of a β-hairpin formed by 10 amino acids which mimic a 12-mer peptide inside which the original Pro-Gly sequence was replaced with a diarylazo moiety [132,133]. The rate of light-induced *cis* → *trans* isomerization of the peptide was only 30% slower than that of the free chromophore, suggesting

A

Ac-Glu-Ala-Cys-Ala-Arg-Glu-Ala-Ala-Ala-Arg-Glu-Ala-Ala-Cys-Arg-Gln-NH$_2$

AZO

B

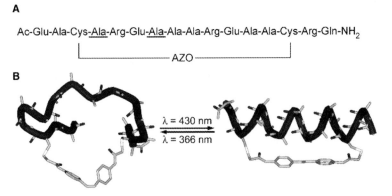

λ = 430 nm
λ = 366 nm

Fig. 13.19 Sequence (A) and schematic structure (B) of a photoswitchable diarylazo peptide adduct which interconverts from *cis* (loop) to *trans* (α-helix) [131].

that it is possible to investigate and control the folding and stability of hairpin structures. One of the most interesting applications of photoresponsive peptides was the development by Moroder and coworkers of cyclic octamers related to *E. coli* thioredoxin reductase containing a Cys-Ala-Thr-Cys sequence. Significant peptide backbone modifications were triggered upon light irradiation at 360 nm, resulting in a dramatic rise in redox potential (*trans*: E'_0 = –200 mV to *cis*: E'_0 = –146 mV). From this point of view, the *cis* isomer was very similar to the native enzyme PDI (E'_0 = –147/–159 mV) and has a disulfide isomerization activity reminiscent of the foldase properties of the natural protein disulfide isomerases. A structural study by NMR spectroscopy showed that the bicyclic peptide flip-flops between multiple conformational states with a helix-like structure involving the Cys-Ala-Thr-Cys motif and correlates very well with the variation of redox potential [21]. Large and reversible structural reorganizations were also observed with azobenzene-containing peptide dendrimers and photoinduced CTI seems to produce local geometrical modifications that are proportional to the size of the molecule [134] and which might be used for tuning of molecular recognition processes. Poly-lysine and poly-aspartate/glutamate oligomers grafted with diarylazo moieties also display some interesting properties which we shall just mention in passing as they fall outside the scope of this chapter [14]. In particular, photoisomerization of azobenzene-grafted poly-aspartate at wavelengths below 400 nm led to a dramatic amplification of the signal. Above 10% of azo groups in the *cis* configuration, the left-handed helix (100% *trans* form) switches to a 70% preference for a right-handed helix [1,135]. A similar transition was observed with an azobenzene-sulfonyl-grafted poly-lysine polymer which reversibly photoswitches from coil to *a*-helix [1]. This was also observed with oligomers bearing other photoisomerizable groups such as merocyanines [135]. Photoinduced stacking of azobenzene motifs was employed to control peptide aggregation/disaggregation, which modulates peptide solubility. Irradiation of the aggregated *trans* isomer around 350 nm caused a complete dissolution of the polymer, whereas irradiation of the solution completely reversed the tendency. In fact, photostimulated aggregation/dissolution processes are relevant to molecular mechanisms implicated in photoregulated biological processes such as phytochrome aggregation (cf. Chapter 4) [135].

13.5.3
Modifications of Proteins with Photoisomerizable Motifs

Several approaches to the photoregulation of proteins have been employed, including photoisomerizable ligands and inhibitors, casting the protein inside a photoisomerizable polymer, and chemical modification of the protein itself [136].

Several azobenzene protein effectors have been reported to act as photoregulators of enzymes and receptors. Photoregulation of cysteine and serine proteases was carried out using simple photoswitchable inhibitors which block the active site in their *trans* form only [137]. Conversely, the *cis* isomers of *a*-ketoester-based inhibitors were more potent inhibitors of *a*-chymotrypsin than the *trans* forms

[138,139]. Screening of a library of photoswitchable peptides selected a single peptide in which the *cis* and *trans* isomers bind the nuclear import receptor karyopherin *a* differentially [140]. In fact, photoisomerization of the azobenzene spacer, which acted as a phototelastic linker, regulated the ability of binding sequences to interact with the two distinct karyopherin *a* subsites simultaneously. A diarylazo derivative bearing an *N*-ethylmaleimide group on the one extremity and a quaternary ammonium motif on the other was employed for the specific modification of a cysteine residue properly positioned at the edge of the pore of the voltage-gated "Shaker" K+ channel. Photoirradiation of the azo moiety induced the displacement of the ammonium group from the center (*E* isomer) to the edge (*Z* isomer) of the pore and thus relieved pore blockade and allowed ion conduction. Reverse CTI upon long-wavelength light irradiation restored the blocked state [141].

Although the *trans* to *cis* photoisomerization of merocyanins leads to a reversible cyclization reaction (see Chapter 2), use of these photoswitchable motifs may be mentioned herein. A photoisomerizable FAD analog was employed to modify glucose oxidase in such a way that electron transfer resulting from glucose oxidation occurs in the spiropyran form and not in the merocyanin form, probably due to the variation of positioning of the FAD moiety [142].

Covalent incorporation of a photoactivatable molecular switch in proteins can be carried out either via the selective modification of the sequence or by specific derivatization of the active site. Phototunable type S ribonucleases that display enzymatic activity were obtained by noncovalent assembly of the S-protein (resulting from the action of subtilin on ribonuclease A) together with azobenzene derivatives of the complementary S-peptide synthesized chemically bearing a phenyl-azophenylalanine residue at various positions [143,144]. Several azobenzene photoisomerizable amino acids have been synthesized (see Ref. [140]) such as L-phenylazophenylalanine (Pap or azoAla) 17 which has been inserted inside numerous peptides and proteins, in particular streptavidin, using in vitro protein-synthesizing systems [145]. Pap and a Pap analog were incorporated into a specific position of the dimer interface of the restriction enzyme *Bam*HI, where they blocked the activity in the *trans* form. Photoirradiation at 366 nm and subsequent CTI induced the specific cleavage of DNA by the enzyme (Fig. 13.20) [146].

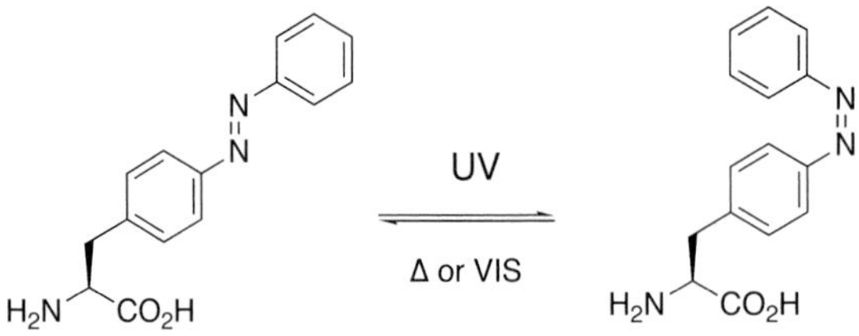

Fig. 13.20 Structures of *E*- and *Z*-Pap **17**.

The specific modification of the active site was also performed with an inverse substrate bearing a photoswitchable moiety. Acylation of trypsin allowed the photocontrol of deacylation rates: the *cis* form was shown to deacylate more quickly than the *trans* form [147]. The photoisomerizable motif may be grafted onto proteins via the selective modification of an accessible amino acid side-chain such as amidation of lysine [148] or thioalkylation of the endoglucanase 12A mutant N55C with a photoswitchable polymer which hides the active site when packed in the *cis* form and unfolds in the *trans* form in such a way that the substrate can access the active site upon specific light irradiation [149]. Ten years ago, Willner and coworkers reported the immobilization of *a*-chymotrypsin in polyacrylamide gels containing various photoswitchable copolymers and a dramatic rise in catalytic rates of substrate hydrolysis was observed upon isomerization from *trans* to *cis* [150].

13.5.4
Other Phototunable Biomolecules

Photoswitching of azobenzene motifs has been employed for the photoregulation of various self-assembly processes, in particular phospholipid analogs. Recently, a photoisomerizable phospholipid surrogate bearing two diarylazo moieties was shown to activate a bacterial mechanosensitive channel of large conductance upon UV irradiation [151]. It is noteworthy that in this case, CTI elicited a large reshaping of the phospholipid analog and hence causes a change in membrane lateral pressure which is suspected to trigger channel activation via a mechanism that might be very similar to that observed in bacteria.

Photoregulation of DNA duplex and triplex formation has been widely investigated by Komiyama and coworkers. The binding of the *trans* form of an 8/14-mer oligonucleotide bearing an azobenene motif to a single-strand DNA [152,153] or a double-strand DNA helix [154,155] to form respectively a duplex or a triplex was weakened by *trans* to *cis* isomerization by UV light irradiation (the melting temperature T_m fell by 10 °C) and resulted in a rapid dissociation of the phototunable oligonucleotide. Such photoswitchable oligonucleotides have also been used to photoregulate the transcription reaction of T7 RNA polymerase by impeding gene expression in the *trans* form and switching on transcription upon CTI [156]. DNA elongation from a primer DNA by DNA polymerase proceeded to completion only when the photomodulator oligonucleotides were in the *cis* form and dissociated from the larger DNA strand which is used for elongation [157]. Very recently, a series of photoswitchable oligonucleotides carrying azobenzene, spiropyran, or stilbene derivatives were shown to regulate electron transfer in DNA hairpins [158] and RNA digestion by DNA enzymes with improved RNA cleavage activity [159].

Other photoisomerizable molecules including azobenzene-capped *β*-cyclodextrins, dendrimers, aza-crown and calixarene compounds have been reported [14,160,161] and, though they are not used for studying biomolecules, they open up new perspectives in the field of supramolecular chemistry and help to understand basic concepts that have been validated in biochemistry.

In the first part of the twentieth century, azo-related compounds were proposed as therapeutic agents and in recent years they seem to have attracted attention as cytotoxic agents and photoactivatable bioconjugates usable for photodiagnosis and phototherapy. As an example, bis(2-phenylazopyridine)ruthenium^{II} complexes display cytotoxic antitumor activities that are comparable to that of cisplatin [162]. Azocompounds including azoxanthenes, azoacridines, and azocoumarins and their peptide bioconjugates (i.e. somatostatin, neurotensin, bombesin) have been patented recently as photosensitizers which act via the specific reaction of the corresponding radical species with tumor cells [163]. Azo dyes may also be used as biosensors, in particular for the detection of monosaccharides [164].

References

1 B. L. Feringa, R. A. van Delden, N. Koumura, E. M. Geerstema, *Chem Rev.* **2000**, *100*, 1789–1816.

2 B. L. Feringa, *Acc. Chem. Res.* **2001**, *34*, 504–513.

3 C. Kaes, A. Katz, M. W. Hosseini, *Chem. Rev.* **2000**, *100*, 3553–3590.

4 S. T. Howard, *J. Am. Chem. Soc.* **1996**, *118*, 10269–10274.

5 A. Hazell, *Polyhedron* **2004**, *23*, 2081–2083.

6 L. Pu, *Chem. Rev.* **1998**, *98*, 2405–2494.

7 B. M. Duggan, D. J. Craik, *J. Med. Chem.* **1997**, *40*, 2259–2265.

8 S. Capasso, L. Mazzarella, T. Tancredi, A. Zagari, *Biopolymers* **2004**, *23*, 1085–1097.

9 B. Donzel, B. Kamber, R. Schwyzer, *Helv. Chim. Acta* **2004**, *55*, 947–961.

10 O. Carugo, M. Cemazar, S. Zahariev, I. Hudáky, Z. Gáspári, A. Perczel, S. Pongor, *Protein Eng.* **2003**, *16*, 637–639.

11 H. E. Van Wart, H. A. Schegara, *Proc. Natl Acad. Sci. USA* **1977**, *74*, 13–17.

12 M. Ostrowski, J. Jeske, P. G. Jones, W. W. du Mont, *Chem. Ber.* **1993**, *126*, 1355–1359.

13 T. Sato, M. Wakabayashi, K. Hata, M. Kainosho, *Tetrahedron* **1971**, *27*, 2737–2755.

14 C. Dugave, L. Demange, *Chem. Rev.*, **2003**, *103*, 2475–2532.

15 H.-J. Chou, T. B. Rauchfuss, L. F. Szczepura, *J. Am. Chem. Soc.* **1998**, *120*, 1805–1811.

16 I. de Sousa Moreira, J. B. de Lima, D. W. Franco, *Coord. Chem. Rev.* **2000**, *196*, 197–217.

17 F. M. Tabellion, S. R. Seidel, A. M. Arif, P. J. Stang, *J. Am. Chem. Soc.* **2001**, *123*, 7740–7741.

18 A. C. Gallacher, A.A. Pinkerton, *Acta Cryst.* **1993**, *C49*, 1793–1796.

19 W. J. Wedemeyer, E. Welker, M. Nayaran, H. A. Scheraga, *Biochemistry* **2000**, *39*, 4207–4216.

20 F. Ivo, P. Malon, J. Karel, B. Karel, Proc. F. E. C. S. *International Conference on Circular Dichroism.* VCH Publishers, Weinheim, Germany, **1997**, 332.

21 A. Cattani-Scholz, C. Renner, C. Cabrele, R. Behrendt, D. Oesterhelt, L. Moroder, *Angew. Chem. Int. Ed.* **2002**, *41*, 289–292.

22 S. B. Fulton, A. P. Kaplan, J. C. Hogan, S. L. Gallion, *Methods Mol. Med.* **2000**, *23*, 527–537.

23 E. Peisach, D. Casebier, S. L. Gallion, P. Furth, G. A. Petsko, J. C. Hogan Jr, D. Ringe, *Science* **1995**, *269*, 66–69.

24 E. E. Rutenber, F. McPhee, A. P. Kaplan, S. L. Gallion, J. C. Hogan Jr., C. S. Craik, R. M. Stroud, *Biorg. Med. Chem.* **1996**, *4*, 1545–1558.

25 W. J. Moree, A. Schouten, J. Kroon, R. M. J. Liskamp, *Int. Peptide Protein Res.* **1995**, *45*, 501–507.

26 J. L. Radkiewicz, M. A. McAllister, E. Goldstein, K. N. Houk, *J. Org. Chem.* **1998**, *63*, 1419–1428.

27 L. Demange, M Moutiez, C. Dugave, *J. Med. Chem.* **2002**, *45*, 3928–3933.

28 P. I. Nagy, J. Koekoesi, A. Gergely, A. Racz, *J. Phys. Chem.* **2003**, *107*, 7861–7868.

29 H. Suginome, *CRC Handbook of Organic Photochemistry and Photobiology*, 2nd edn, W. Horspool, F. Lenci, Eds. CRC Press, Boca Raton, FL, USA, **2000**.

30 R. A. O'Ferall, D. O'Brien, *J. Phys. Org. Chem.* **2004**, *17*, 631–640.

31 R. Vanderesse, L. Thevenet, M. Marraud, N. Boggetto, M. Reboud, C. Corbier, *J. Peptide Sci.* **2003**, *9*, 282–299.

32 H. Roohi, F. Deyhimi, A. Ebrahimi, *Theochem* **2001**, *543*, 299–308.

33 M. H. Holschbach, D. Sanz, R. M. Claramunt, L. Infantes, S. Motherwell, P. R. Raithby, M. L. Jimeno, D. Herrero, I. Alkorta, N. Jagerovic, J. Elguero, *J. Org. Chem.* **2003**, *68*, 8831–8837.

34 B. T. Shireman, M. J. Miller, M. Jonas, O. Wiest, *J. Org. Chem.* **2001**, *66*, 6046–6056.

35 V. Benin, P. Kaszynski, G. Radziszewski, *J. Am. Chem. Soc.* **2002**, *124*, 14115–14126.

36 O. Tietze, B. Schiefner, B. Ziemer, A. Zschunke, *Fresenius's J. Anal. Chem.* **1997**, *357*, 477–481.

37 D. R. Boyd, W. B. Jennings, L. C. Waring, *J. Org. Chem.* **1986**, *51*, 992–995.

38 C. Cativiela, J. I. Garcia, J. A. Mayoral, L. Salvatella, *Theochem* **1996**, *368*, 57–66.

39 R. Grigg, G. Donegan, H. Q. N. Gunaratne, D. A. Kennedy, J. F. Malone, V. Sridharan, S. Thianpatanagul, *Tetrahedron* **1989**, *45*, 1723–1746.

40 D. K. Singh, S. N. Balasubrahmanyam, N. Prasad, *Indian J. Chem. Sect. B* **1990**, *29B*, 804–810.

41 R. R. Papallardo, E. Sanchez Marcos, M. F. Ruiz-Lopez, D. Rinaldi, J.-L. Rivail, *J. Am. Chem. Soc.* **1993**, *115*, 3722–3730.

42 R.; Richter, G. H. Temme, *J. Org. Chem.* **1978**, *43*, 1825–1827.

43 S. Liao, D. B. Collum, *Abstract of Papers, 224th ACS National Meeting, Boston, MA*. American Chemical Society, Washington DC, USA, **2002**.

44 J. E. Johnson, N. M. Morales, A. M. Gorczyca, D. D. Dolliver,

M. A. McAllister, *J. Org. Chem.* **2001**, *66*, 7979–7985.

45 B. Mompon, D. Loyaux, E. Kauffmann, A. M. Krstulovic, *J. Chromatogr.* **1986**, *363*, 372–381.

46 T. Yamagishi, E. Mizushima, H. Sato, M. Yamaguchi, *Chem. Lett.* **1998**, *27*, 1255–1256.

47 R. F. Childs, B. Dickie, *J. Chem. Soc. Chem. Commun.* **1981**, *24*, 1268–1268.

48 A. G. Kolchinski, N. W. D. H. Alcock, Busch, *Inorg. Chem.* **1997**, *36*, 2754.

49 A. G. Kolchinski, B. Korybut-Daskiewicz, E. V. Rybak-Akimova, D. H. Busch, N. W. Alcock, H. J. Clase, *J. Am. Chem. Soc.* **1997**, *119*, 4160–4171.

50 B. Bohran, M. L. Souto, H. Imai, Y. Shichida, K. Nakanishi, *Science* **2000**, *288*, 2209–2212.

51 S. I. E. Touw, H. J. M. de Groot, F. Buda, *J. Phys. Chem. B* **2004**, *108*, 13560–13572.

52 L. S. Brown, A. K. Dioumaev, R. Needleman, J. K. Lanyi, *Biochemistry* **1998**, *37*, 3982–3993.

53 U. Zadok, A. E. Asato, M. Sheves, *Biochemistry* **2005**, *44*, 8479–8485.

54 R. Vogel, F. Siebert, X.-Y. Zhang, G. Fan, M. Sheves, *Biochemistry* **2004**, *43*, 9457–9466.

55 R. Vogel, S. Lüdeke, I. Radu, F. Siebert, M. Sheves, *Biochemistry* **2004**, *43*, 10255–10264.

56 R. R. Rando, *Chem. Rev.* **2001**, *101*, 1881–1896.

57 A. Kanstrup, O. Buchardt, *Anal. Biochem.* **1991**, *194*, 41–44.

58 M. S. Finnin, S. Seltzer, *Can. J. Chem.* **1999**, *77*, 557–564.

59 W. A. M. Wolken, R. ten Have, M. J. van der Werf, *J. Agric. Food. Chem.* **2000**, *48*, 5401–5405.

60 P. Kuad, M. Borkovec, M. Desage-El Murr, T. Le Gall, C. Mioskowski, B. Spiess, *J Am Chem Soc* **2005**, *127*, 1323–1333.

61 Y. S. Lee, M. G. Marcu, L. Neckers, *Chem Biol* **2004**, *11*, 991–998.

62 E. Gross, *Handbook of Biochemistry and Molecular Biology*, CRC Press, Boca Raton, FL, USA, **1976**.

63 U. Schmidt, J. Häusler, E. Öhler, H. Poisel, *Progress in the Chemistry of*

Organic Natural Products, Springer Verlag, New York, **1979**.

64 K. Noda, Y. Shimohigashi, N. Izumiya, *The Peptides: Analysis, Synthesis, Biology*, Vol. 5, Academic Press, New York, **1983**.

65 C. H. Stammer, *Chemistry and Biochemistry of Amino Acids, Peptides and Proteins*, Marcel Dekker, New York, **1982**.

66 G. Jung, *Angew. Chem. Int. Eng. Ed.* **1991**, *30*, 1051–1068.

67 H. G. Sahl, R. W. Jack, G. Bierbaum, *Eur. J. Biochem.* **1995**, *230*, 827–853.

68 T. F. Schwede, J. Retey, G. E. Schulz, *Biochemistry* **1999**, *38*, 5355–5361.

69 S. D. Christenson, W. Liu, M. D. Toney, B. Shen, *J. Am. Chem. Soc.* **2003**, *125*, 6062–6063.

70 D. P. Barondeau, C. J. Kassmann, J. A. Tainer, E. D. Getzoff, *Biochemistry* **2005**, *44*, 1960–1970.

71 C.-G. Shin, M. Hayakawa, K. Mikami, J. Yoshimura, *Tetrahedron Lett.* **1977**, *10*, 863–866.

72 K. Fukushima, K. Yazawa, T. Arai, *J. Antibiot. (Tokyo)* **1973**, *26*, 175–176.

73 K. Kanoh, S. Kohno, J. Katada, Y. Hayashi, M. Muramatsu, I. Uno, *Biosci. Biotechnol. Biochem.* **1999**, *63*, 1130–1133.

74 W. L. Meyer, G. E. Templeton, C. I. Grable, R. Jones, L. F. Kuyper, R. B. Lewis, C. W. Sigel, S. H. Woodhead, *J. Am. Chem. Soc.* **1975**, *97*, 3202–3209.

75 R. D. Durbin, T. F. Uchytil, *Phytopathology* **1977**, *67*, 602–603.

76 T. J. Nitz, E. M. Holt, B. Rubin, C. H. Stammer, *J. Org. Chem.* **1981**, *46*, 2667–2671.

77 R. Vleggaar, P. L. Wessels, *J. Chem. Soc. Chem. Commun.* **1980**, 160–162.

78 C. C. J. Culvenor, J. A. Edgar, M. F. Mackay, C. P. Gorst-Allman, W. F. O. Marasas, P. S. Steyn, R. Vleggaar, P. L. Wessels, *Tetrahedron* **1989**, *45*, 2351–2372.

79 Y. Li, H. Kobayashi, Y. Tokiwa, Y. Hashimoto, S. Iwasaki, *Biochem. Pharmacol.* **1992**, *43*, 219–224.

80 M. Gondry, S. Lautru, G. Fusai, G. Meunier, A. Menez, R. Genet, *Eur. J. Biochem.* **2001**, *268*, 1712–1721.

81 R. Genet, C. Denoyelle, A. Menez, *J. Biol. Chem.* **1994**, *269*, 18177–18184.

82 R. Genet, P. H. Benetti, A. Hammadi, A. Menez, *J. Biol. Chem.* **1995**, *270*, 23540–23545.

83 E. S. A. Aboutabl, M. Luckner, *Phytochemistry* **1975**, *14*, 2573–2577.

84 E. S. A. Aboutabl, A. El Azzouny, K. Winter, M. Luckner, *Phytochemistry* **1976**, *15*, 1925–1928.

85 T. Kupke, S. Stevanovic, H. G. Sahl, F. Gotz, *J. Bacteriol.* **1992**, *174*, 5354–5361.

86 T. Kupke, M. Uebele, D. Schmid, G. Jung, M. Blaesse, S. Steinbacher, *J. Biol. Chem.* **2000**, *275*, 31838–31846.

87 D. Thibaut, N. Ratet, D. Bisch, D. Faucher, L. Debussche, F. Blanche, *J. Bacteriol.* **1995**, *177*, 5199–5205.

88 B. W. Bycroft, *Nature* **1969**, *224*, 595–597.

89 A. Hammadi, H. Lam, M. Gondry, A. Ménez, R. Genet, *Tetrahedron* **2000**, *56*, 4473–4477.

90 G. P. Zecchini, M. P. Paradisi, I. Torrini, G. Lucente, E. Gavuzzo, F. Mazza, G. Pochetti, M. Paci, M. Sette, A. D. Nola, G. Veglia, S. Traniello, S. Spisani, *Biopolymers* **1993**, *33*, 437–451.

91 G. Pietrzynski, B. Rzeszotarska, Z. Kubica, *Int. J. Pept. Protein Res.* **1992**, *40*, 524–531.

92 M. M. Stohlmeyer, H. Tanaka, T. J. Wandless, *J. Am. Chem. Soc.* **1999**, *121*, 6100–6101.

93 H. Sai, T. Ogiku, H. Ohmizu, *Synthesis* **2003**, 201–204.

94 K. Nakamura, T. Isaka, H. Toshima, M. Kodaka, *Tetrahedron Lett.* **2004**, *45*, 7221–7224.

95 B. Liebermann, R. Ellinger, E. Pinet, *Phytochemistry* **1996**, *42*, 1537–1540.

96 Z. Kubica, T. Kozlecki, B. Rzeszotarska, *Chem Pharm Bull (Tokyo)* **2000**, *48*, 296–297.

97 Y. Inai, S. Kurashima, T. Hirabayashi, K. Yokota, *Biopolymers* **2000**, *53*, 484–496.

98 Y. Shimohigashi, H. C. Chen, C. H. Stammer, *Peptides* **1982**, *3*, 985–987.

99 G. H. Fisher, P. Berryer, J. W. Ryan, V. Chauhan, C. H. Stammer, *Arch. Biochem. Biophys.* **1981**, *211*, 269–275.

100 Y. Shimohigashi, C. H. Stammer, *Int. J. Pept. Protein. Res.* **1982**, *20*, 199–206.

101 Y. Shimohigashi, C. H. Stammer, *Int. J. Pept. Protein Res.* **1982**, *19*, 54–62.

102 Y. Shimohigashi, M. L. English, C. H. Stammer, T. Costa, *Biochem. Biophys. Res. Commun.* **1982**, *104*, 583–590.

103 T. J. Nitz, Y. Shimohigashi, T. Costa, H. C. Chen, C. H. Stammer, *Int. J. Pept. Protein Res.* **1986**, *27*, 522–529.

104 J. Bland, J. V. Edwards, S. R. Eaton, A. R. Lax, *Pestic. Sci.* **1993**, *39*, 331–340.

105 D. E. Ward, A. Vazquez, M. S. Pedras, *J. Org. Chem.* **1996**, *61*, 8008–8009.

106 P. Mathur, S. Ramakumar, V. S. Chauhan, *Biopolymers* **2004**, *76*, 150–161.

107 G. Pietrzynski, B. Rzeszotarska, *Polish J. Chem.* **1995**, *69*, 1595–1614.

108 R. Jain, V. S. Chauhan, *Biopolymers* **1996**, *40*, 105–119.

109 T. P. Singh, P. Kaur, *Progress Biophys. Mol. Biol.* **1997**, *66*, 141–165.

110 M. A. Broda, D. Siodlak, B. Rzeszotarska, *J. Peptide Sci.* **2005**, *11*, 235–244.

111 M. Thormann, H.-J. Hofmann, *J. Mol. Struct (Theochem)* **1998**, *431*, 79–96.

112 G. Pietrzynski, B. Rzeszotarska, E. Ciszak, M. Lisowski, Z. Kubica, G. Boussard, *Int. J. Pept. Protein Res.* **1996**, *48*, 347–356.

113 J. Swiatek-Kozlowska, J. Brasun, L. Chruscinski, E. Chruscinska, M. Makowski, H. Kozlowski, *New J. Chem.* **2000**, *24*, 893–896.

114 J. Swiatek-Kozlowska, J. Brasun, M. Luczkowski, M. Makowski, *J. Inorg. Biochem.* **2002**, *90*, 106–112.

115 P. H. Rasmussen, P. S. Ramanujam, S. Hvilsted, R. H. Berg, *J. Am. Chem. Soc.* **1999**, *121*, 4738–4743.

116 W. J. Deal, B. F. Erlanger, D. Nachmansohn, *Biochemistry* **1969**, *64*, 1230–1234.

117 E. Bartels, N. H. Wassermann, B. F. Erlanger, *Proc. Natl Acad. Sci. USA* **1971**, *68*, 1820–1823.

118 N. H. Wassermann, E. Bartels, B. F. Erlanger, *Proc. Natl Acad. Sci. USA* **1979**, *76*, 256–259.

119 D. C. M. Chan, C. A. Laughton, S. F. Queener, M. F. G. Stevens, *J. Med. Chem.* **2001**, *44*, 2555–2564.

120 E. Carceller, J. Salas, M. Merlos, M. Giral, R. Ferrando, I. Escamilla, J. Ramis, J. Garcia-Rafanell, J. Forn, *J. Med. Chem.* **2001**, *44*, 3001–3013.

121 J. W. Kelly, *Curr. Opin. Struct. Biol.* **1996**, *6*, 11–17.

122 T. Miura, C. Yamamiya, M. Sasaki, Y. Suzuki, H. Takeuchi, *J. Raman Spectrosc.* **2002**, *33*, 530–535.

123 T. Kato, T. Matsuda, S. Matsui, T. Mizutani, K. Saeki, *Biol. Pharm. Bull.* **2002**, *25*, 466–471.

124 J. Auernheimer, C. Dahman, U. Hersel, A. Bausch, H. Kessler, A. *J. Am. Chem. Soc.* **2005**, *127*, 16107–16110.

125 M. S. Vollmer, T. D. Clark, C. Steinem, M. R. Ghadiri, *Angew. Chem. Int. Ed.* **1999**, *38*, 1598–1601.

126 C. Steinem, A. Janshoff, M. S. Vollmer, M. R. Ghadiri, *Langmuir* **1999**, *15*, 3956–3964.

127 Z. Szewczuk, A. Kubik, I. Z. Siemion, *Int. J. Pepide Protein Res.* **1988**, *32*, 98–103.

128 G. Fridkin, C. Gilon, *J. Peptide Res.* **2002**, *60*, 104–111.

129 I. Z. Siemion, Z. Szewczuk, Z. S. Herman, Z. Stachura, *Mol. Cell Biochem.* **1981**, *34*, 23–29.

130 N. Pozhidaeva, M.-E. Cormier, A. Chaudahari, G. A. Wooley, *Bioconjugate Chem.* **2004**, *15*, 1297–1303.

131 J. Bredenbeck, J. Helbing, J. R. Kumita, G. A. Wooley, P. Hamm, *Proc. Natl Acad. Sci. USA* **2005**, *102*, 2379–2384.

132 A. Aesmissegger, V. Kräutler, W. F. van Gunsteren, D. Hilvert, *J. Am. Chem. Soc.* **2005**, *127*, 2929–2936.

133 V. Kräutler, A. Aesmissegger, P. H. Hünenberger, D. Hilvert, T. Hansson, W. F. van Gunsteren, *J. Am. Chem. Soc.* **2005**, *127*, 4935–4942.

134 A. Cattani-Scholz, C. Renner, D. Oesterhelt, L. Moroder, *ChemBioChem.* **2001**, *2*, 542–549.

135 O. Pieroni, A. Fissi, N. Angelini, F. Lenci, *Acc. Chem. Res.* **2001**, *34*, 9–17.

136 R. H. Kramer, J. J. Chambers, D. Trauner, *Nat. Chem. Biol.* **2005**, *1*, 360–365.

137 P. R. Westmark, J. P. Kelly, B. D. Smith, *J. Am. Chem. Soc.* **1993**, *115*, 3416–3419.

138 A. J. Harvey, A. D. Abell, *Tetrahedron* **2000**, *56*, 9763–9771.

139 A. J. Harvey, A. D. Abell, *Bioorg. Med. Chem. Lett.* **2001**, *11*, 2441–2444.

140 S. B. Park, R. F. Standaert, *Bioorg. Med. Chem.* **2001**, *9*, 3215–3223.

141 M. Banghart, K. Borges, E. Isacoff, D. Trauner, R. H. Kramer, *Nat. Neurosci.* **2004**, *7*, 1381–1386.

142 R. Blonder, E. Katz, I. Willner, V. Wray, A. F. Bückmann, *J. Am. Chem. Soc.* **1997**, *119*, 11747–11757.

143 D. Liu, J. Karanicolas, C. Yu, Z. Zhang, G. A. Wooley, *Bioorg. Med. Chem. Lett.* **1997**, *7*, 2677–2680.

144 D. A. James, D. C. Burns, G. A. Wooley, *Protein Eng.* **2001**, *14*, 983–991.

145 T. Hohsaka, D. Kajihara, Y. Ashizuka, H. Murakanmi, M. Sisido, *J. Am. Chem. Soc.* **1999**, *121*, 34–40.

146 K. Nakayama, M. Endo, T. Majima, *Bioconjugate Chem.* **2005**, *16*, 1360–1366.

147 H. Sekizaki, A. Kamugai, K. Itoh, E. Toyota, K. Horita, Y. Noguchi, K. Tanizawa, *Biorg. Med. Chem. Lett.* **2003**, *13*, 3809–3812.

148 I. Willner, S. Rubin, A. Riklin, *J. Am. Chem. Soc.* **1991**, *113*, 3321–3325.

149 T. Shimoboji, E. Larenas, T. Fowler, S. Kulkarni, A. S. Hoffman, P. S. Stayton, *Proc. Natl Acad. Sci.* **2002**, *99*, 16592–16596.

150 I. Wiillner, S. Rubin, R. Shatzmiller, T. Zor, *J. Am. Chem. Soc.* **1993**, *115*, 8690–8694.

151 J. H. A. Folgering, J. M. Kuiper, A. H. De Vries, J. B. F. N. Engberts, B. Poolman, *Langmuir* **2004**, *20*, 6985–6987.

152 H. Asanuma, T. Ito, T. Yoshida, X. Liang, M. Komiyama, *Angew. Chem. Int. Ed.* **1999**, *38*, 2393–2395.

153 H. Asanuma, X. Liang, T. Yoshida, M. Komiyama, *ChemBiochem* **2001**, *2*, 35–38.

154 H. Asanuma, X. Liang, T. Yoshida, A. Yamazawa, M. Komiyama, *Angew. Chem. Int. Ed.* **2000**, *39*, 1316–1318.

155 X. Liang, H. Asanuma, M. Komiyama, *J. Am. Chem. Soc.* **2002**, *124*, 1877–1883.

156 H. Asanuma, D. Tamaru, A. Yamazawa, M. Liu, M. Komiyama, *ChemBiochem* **2002**, *3*, 786–789.

157 A. Yamazawa, X. G. Liang, H. Asanuma, M. Komiyama, *Angew. Chem. Int. Ed.* **2000**, *39*, 2356–2358.

158 F. D. Lewis, T. Wu, YY. Zhang, R. L. Letsinger, S. R. Greenfield, M. R. Wasielewski, *Science* **1997**, *277*, 673–676.

159 H. Asanuma, K. Makato, M. Daijiro, K. Takeshi, *World Patent WO 2005083073*, **2005**.

160 A. Harada, *Acc. Chem. Res.* **2001**, *34*, 456–464.

161 Y. Liu, Y.-L. Zhao, H.-Y. Zhang, Z. Fan, G.-D. Wen, F. Ding, *J. Phys. Chem. B.* **2004**, *108*, 8836–8843.

162 A. C. G. Hotze, M. Bacac, A. H. Velders, B. A. J. Jansen, H. Kooijman, A. L. Spek, J. P. Haasnoot, J. Reedjik, *J. Med. Chem.* **2003**, *46*, 1743–1750.

163 R. Rajogapalan, G. L. Cantrell, J. E. Bugaj, S. I. Achilefu, *US Patent 6,485,704*, **2003**.

164 N. DiCesare, J. R. Lakowicz, *Org. Lett.* **2001**, *3*, 3891–3893.

14

Cis-Trans Isomerism in Metal Complexes*

Alzir Azevedo Batista and Salete Linhares Queiroz

14.1
Introduction

The stereochemistry of coordination compounds is as old as coordination chemistry itself. The Swiss chemist Alfred Werner (1866–1919) derived the theory of coordination, published in 1893, to a large extent from stereochemical arguments.

Thus, the geometrical model for chemical compounds was necessary to give support to Werner's theory on coordination chemistry. A fundamental pillar in the development of coordination chemistry that gave support to Werner's theory was, without doubt, the discovery of isomers in metal complexes. This brilliant scientist used the octahedral coordination geometry of many metals as one of the very basic facts from which he deduced his theory. Werner's generalizing way of thinking was clearly demonstrated by the fact that besides the octahedron he also introduced square-planar geometry. Werner realized that an octahedral surrounding a central atom would explain the two isomers of complexes of type $[MA_4B_2]$ and square-planar geometry would explain those for complexes of type $[MA_2B_2]$. Indeed, at the time Werner proposed his theory, only one isomer of the $[CoCl_2(NH_3)_4]Cl$ complex was known, the green complex (Fig. 14.1A). Based on his theory Werner predicted that another isomer should exist and his discovery in 1907 of the violet complex (Fig. 14.1B) was crucial in convincing scientists who were still critical of his model for coordination chemistry.

Chemists were only totally convinced that Werner's point of view was to be taken seriously when, in 1911, he resolved six-coordinate compounds, such as $[CoCl(NH_3)(en)_2]Cl_2$, into enantiomers, which in accordance with his prediction, should be optically active if the groups are located at the vertices of an octahedron [1]. This was probably the major reason that Werner was awarded the Nobel Prize for chemistry two years later in 1913.

The cobalt atoms in Fig. 14.1 are stereogenic and the site symmetries are C_{2v} and D_{4h}, for (A) and (B), respectively. The term "stereogenic center" is at present used to designate a coordinated atom that was formerly called an asymmetric atom [2].

---

*)   Please find a list of abbreviations at the end of this chapter.

cis-trans Isomerization in Biochemistry. Edited by Christophe Dugave
Copyright © 2006 WILEY-VCH Verlag GmbH & Co. KGaA, Weinheim
ISBN: 3-527-31304-4

(A) – *cis*- [Co^{III}Cl$_2$(NH$_3$)$_4$]$^+$ **(B)** – *trans*- [Co^{III}Cl$_2$(NH$_3$)$_4$]$^+$

Fig. 14.1 The [Co^{III}Cl$_2$(NH$_3$)$_4$]$^+$ ion complex: (A) is green and (B) is violet.

To consider Werner's model for coordination compounds totally consistent, it is imperative to show not only the existence of the other isomer for complexes of the [CoCl$_2$(NH$_3$)$_4$]$^+$ type but also for others containing different ligands like [Co(NO$_2$)$_2$(en)$_2$]$^+$, and [Co(NH$_3$)$_2$(en)$_2$]$^{3+}$ or [Co(NH$_3$)$_4$(py)$_2$]$^{3+}$, where the species in the inner coordination sphere of the metal are neutral. Thus, for this purpose by about 1913 approximately two dozen cobalt isomers complexes had been synthesized and characterized [3]. Undoubtedly this material was enough to prove the veracity of Werner's belief and to give scientific basis to his coordination theory, as well as the existence of different isomers for hexacoordinated compounds of the type [MA$_4$B$_2$]. The structures of these complexes, which are called isomers, differ in the orientation of the two ligands around the metal center. According to Werner's model of isomerism, certain complexes such as [CoCl(OH)(NH$_3$)$_4$]Cl also present *cis-trans* isomers. Thus, isomers of a given compound are commonly defined as substances having the same stoichiometric composition, the same "molecular weight," and the same chemical formula but different structures, which behave generally differently with respect to most chemical and physical properties. This definition excludes polymorphism of solids, and molecules with different isotopic composition, which are called "isotopomers." Complex A in Fig. 14.1, which shows the two chlorines occupying adjacent positions is called the *cis* isomer while complex B, which shows the two chlorines occupying positions across from one another, or on opposite sides in the octahedron, is called the *trans* isomer. *Cis-trans* isomers are called geometrical isomers (or diastereomers). Specifically, diastereomers are a subset of stereoisomers. The other subset of stereoisomers are the enantiomers, sometimes called "optical isomers," owing to their ability to rotate the plane of linearly polarized light in opposite directions.

Four-coordinate complexes can present tetrahedron or square-planar geometries. While square-planar complexes allow *cis/trans* isomers, the same is not possible for tetrahedron complexes. Again, Werner was correct in his conclusions concerning the tetrahedron structure; he stated that only one isomer would be produced for this composition, since, if one of the four corners is occupied than the three remaining ones are equivalent. Werner studied a series of four-coordinate complexes of palladium and platinum and two isomers had been actually isolated for compounds of formula $[MA_2B_2]$. Since he was able to isolate two nonelectrolytes of formula $[PtCl_2(NH_3)_2]$, formed by reactions of $PtCl_2$ with NH_3 and HCl, and knowing that they cannot be tetrahedral, they should be, in fact, square-planar geometrical isomers. The dichlorodiamminoplatinum(II) complex is a classical example of this pair of isomers (Fig. 14.2).

(A) *cis*- $[PtCl_2(NH_3)_2]$, $\mu \neq 0$ **(B)** *trans*- $[PtCl_2(NH_3)_2]$, $\mu = 0$

Fig. 14.2 The $[PtCl_2(NH_3)_2]$ complex: (A) (*cis* isomer) is orange and (B) (*trans* isomer) is yellow-brown.

With the knowledge and understanding on isomerization of transition metals, the numbers of this kind of compound have grown enormously and many isomers of Pt(IV), Ir(III), Ir(IV), Cr(III), Ru(III), and Rh(III) have been obtained and characterized. In 1938 Breuel and Gleu [4,5] synthesized the $[RuCl_2(NH_3)_4]^+$ and $[RuBr_2(NH_3)_4]^+$ complexes, which are still used mainly as precursors for the synthesis of ruthenium compounds. Isomers of rhodium were first synthesized almost 25 years later, in 1960 by Anderson and Basolo [6], who obtained the two isomers of the $[RhCl_2(en)_2]^+$ complex. From the early 1960s until today the number of geometrical isomers synthesized and characterized by several techniques has increased tremendously. Some of these complexes have become very important in giving support for theories or even for practical applications.

As mentioned previously, in the 1950s many inorganic isomers had already been synthesized and well characterized. It was not by accident that at this time chemists started to look systematically at the reaction mechanisms of inorganic systems. It is true that inorganic chemistry is, by itself, multifaceted, since it involves more than 100 elements, most of which can assume several different oxidation states. Each of these oxidation states can be associated with a wide range of coordination numbers, and each coordination number may be associated with more than one geometry. In this case, the knowledge of the geometries that the transition elements can assume for each coordination number of a determinated metal was very helpful in understanding or developing theories in the field of inorganic chemistry. It was also useful for synthetic purposes; in spite of this widely studied area, it was not yet possible to achieve total understanding of the reaction mechanisms in inorganic chemistry. This can be attributed to the inherent difficulties involved in trying to systematize the reactions of more than 100 elements that can assume different oxidation states, coordination numbers, and geometries. In an ideal world, chemists would like to comprehend and to have control of chemical reactions involving a diverse variety of metals and ligands to produce only desired compounds. Despite these difficulties, some level of understanding has now been achieved for substitution in four-coordinate square-planar complexes that can be used for the synthesis of desired isomer.

Having been through an overview of chemistry since the beginning of the last century, starting with Werner's work until the present, we can see that isomerization is involved in all fields of inorganic chemistry.

The development of coordination theories, such as ligand valence theory (LVT), gives us a better understanding of the geometry of metal ions. Thus, it is well known that tetrahedral complexes (sp^3 hybridization) are preferred over higher coordination numbers if the central atom is small or the ligands large, such as halogens, for then steric effects take priority over the energy advantage of forming further metal–ligand bonds. For those ligands that can form π-bonds, accepting electrons from the metal, the d^8 configurations favor square-planar (dsp^2 hybridization) over the tetrahedral geometry. From this, it is easy to conclude that four-coordinated complexes containing d^8 configurations present *cis/trans* isomers mostly when in the presence of metals belonging to the second and third rows of the d block. These complexes are almost invariably square planar. The elements of the first row of the d block are capable of forming square-planar geometry when they are coordinated to strong bonding ligands. The crystal field (CF) theory and the molecular orbital (MO) theory explain that, in general, tetrahedral structures are not stabilized by a large crystal field stabilization energy (CFSE). Considering the above mentioned it is accurate to say that the study of the phenomenon of *cis/trans* isomerism of transition metals can find support in the coordination chemistry theories.

In the forthcoming discussion, it will be assumed that the reader is familiar with the basic concepts of ligand field theory in order to understand the preferences of a given central metal for certain coordination geometry. In general d^5 and d^6 complexes are octahedral and d^8 are square-planar.

Geometrical *cis/trans* isomers can be obtained in different ways as pure compounds or as a mixture and, when possible, they can be separated one from other. The synthesis of isomers can involve some difficulties and a great amount of work, since the stereochemistry of these isomers can be controlled by tiny changes in the reaction conditions which have to be carefully selected. Solvent, temperature, reaction time, and presence or not of light are generally the keys for the successful synthesis of the desired isomer. Thus, the synthesis of *trans*-[RuCl$_2$(bipy)(dppb)] can be easily carried out starting from [RuCl$_2$(dppb)(PPh$_3$)] in dichloromethane, at room temperature (dppb = 1,4-bis(diphenylphosphine)butane) [7]. However, if this reaction is promoted at reflux for 24 h, the *cis* isomer will be obtained. This means that although the *trans* isomer is kinetically preferred, the *cis* form is the thermodynamically more stable. In this special case the isomerization process also occurs in the presence of light. Moreover, the *cis* isomer reacts quickly with coordinating solvents, which is why they have to be avoided [8,9].

14.1.1
Trans Effect

Without doubt, no other factor has been studied more exhaustively than the effect of the *trans* ligand on the leaving group, which is known as the *trans* effect.

Depending on the nature of this ligand, it is possible to carry out the reaction to obtain the desired product and cause rate changes of many orders of magnitude. Furthermore, the *trans* effect can be wisely used in designing syntheses, mainly for four-coordinate complexes. For example, in the reaction shown in Fig. 14.3 only the *cis*-diamminodichloroplatinum(II) complex was found as product.

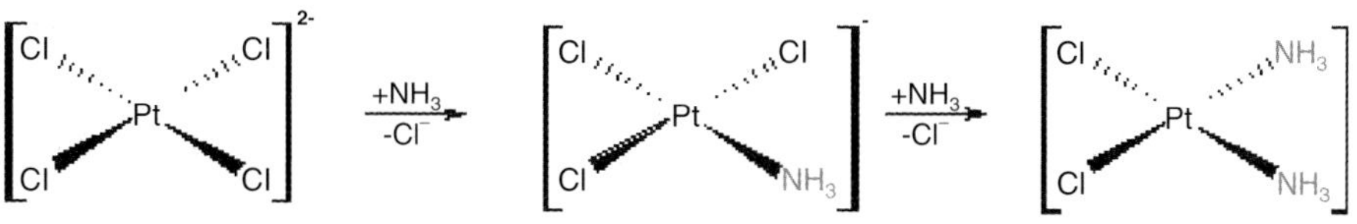

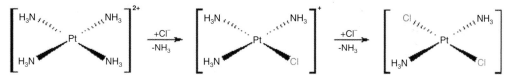

Fig. 14.3 *Trans* effect in the synthesis of the *cis*-diamminodichloroplatinum(II).

This reaction can be rationalized in terms of the following: in step 1 there is a simple displacement since all four groups present in the $[PtCl_4]^{2-}$ are identical. In this case only one compound, $[PtCl_3(NH_3)]^-$, is formed. In the second step, which gives direction to the reaction, two products could potentially be formed, but in practice only the *cis* product is found. Thus, it is easy to conclude that the product formed is by substitution of a *trans* ligand to a chlorine, independent of the nature of the nucleophile. This is a consequence of the *trans* effect, which can be defined as the labilization of *trans* ligands to certain other ligands. Indeed the ligand with the larger *trans* effect plays the dominant role in defining the product of the reaction; it is therefore called the *trans*-directing ligand [10].

To obtain *trans*-diamminodichloroplatinum(II), the other possible complex that theoretically could be formed in the above reaction, the rational approach, taking in account the *trans* effect, is to start with $[Pt(NH_3)_4]^{2+}$, as shown in Fig. 14.4.

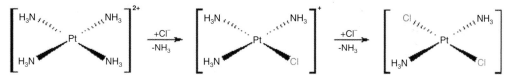

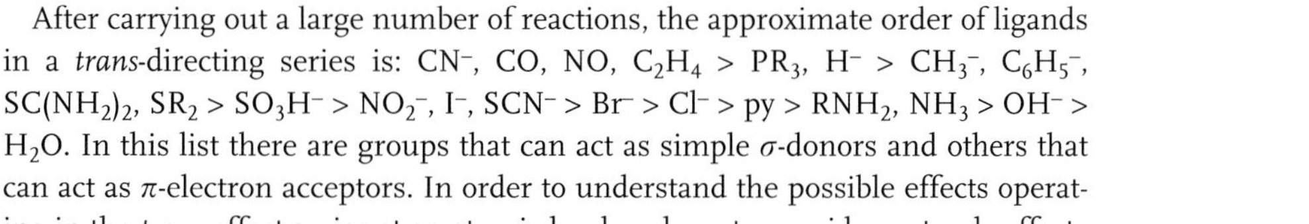

Fig. 14.4 *Trans* effect in the synthesis of the *trans*-diamminodichloroplatinum(II).

After carrying out a large number of reactions, the approximate order of ligands in a *trans*-directing series is: CN^-, CO, NO, C_2H_4 > PR_3, H^- > CH_3^-, $C_6H_5^-$, $SC(NH_2)_2$, SR_2 > SO_3H^- > NO_2^-, I^-, SCN^- > Br^- > Cl^- > py > RNH_2, NH_3 > OH^- > H_2O. In this list there are groups that can act as simple σ-donors and others that can act as π-electron acceptors. In order to understand the possible effects operating in the *trans* effect series at an atomic level we have to consider not only effects in the ground state of the metal center, but also effects that may operate in the transition state of the reaction. The use of the *trans* effect as a tool in the synthesis of coordination complexes is very useful and has to be kinetically controlled since an increase in rate of one reaction over another can arise from a destabilization of the ground state or a stabilization of the transition state. Because of this, the thermodynamically most stable isomer is not always produced in the reaction. Thus, pure π-bases and preferably small, such as hydride, ligands are strongly polarizing and are able to pull electron density towards themselves, depriving the *trans* group, thus increasing its lability. This destabilizes the ground state of the metal–*trans*-ligand bond.

The *trans* effect is also strongly present when the metal is coordinated to π-acids which are able to form back donation, accepting electron density from the metal center. In this case, the *trans* effect probably takes place because the π-acids can provide a pathway for the removal of electron density from the vicinity of the *trans* ligands and hence stabilize the transition state, facilitating the nucleophilic attack on the metal center.

The *trans*-[RuCl(dppe)$_2$(NO)]$^{2+}$ complex was synthesized and characterized by the authors and its bulk electrolysis, at –150 mV, in CH$_3$CN produces the *trans*-[Ru(CH$_3$CN)(dppe)$_2$(NO)]$^{3+}$ (dppe = 1,2-bis(diphenylphosphine)etane) [11]. The origin of the mono acetonitrile complex is perhaps the one-electron reduction of nitrosyl followed by a solvolysis reaction to give the detected product. The solvent complex formation observed here can be understood if it is assumed that the nitrosyl ligand has an electronic interaction through the metal to the *trans* chloride ligand, facilitating its dissociation. In other words, when a chloride ligand is coordinated *trans* to the NO^o group in the Ru^{II}-NO^o species the NO no longer behaves as a strong π-acid, and the chloride ligand is easily dissociated.

Presumably a *cis* effect also exists, but it is very small compared to the *trans* effect and does not significantly affect the direction of the reaction.

14.1.2
Protonation of the Leaving Group

The synthesis of four-coordinate compounds of d^8 metals is much simpler than for other ions, however, some effective strategies can be used for this proposal and one of them is the protonation of the leaving group. Thus, the strategy for synthesis of tetracoordinate complexes starting from hexacoordinate precursors involves the dissociation of the coordinate chelate ligand by an acid-catalyzed dissociation (Fig. 14.5).

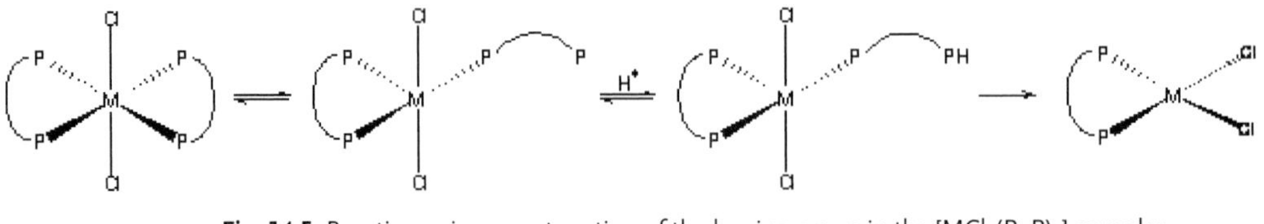

Fig. 14.5 Reaction using a protonation of the leaving group in the [MCl$_2$(P–P)$_2$] complex.

In this reaction, for the bidentate ligand to be dissociated in the presence of acid, the protonation of the pendant donor atom needs to compete with the ring-closure reaction by binding to the lone pair of electrons successfully blocking its re-coordination to the metal, leading to the loss of the chelate. This type of reaction is very general and is found in a wide range of metals, oxidation states, and coordination geometries.

There are many other routes of synthesis for *cis/trans* isomers but because they are not general it is impossible to list all of them here.

14.1.3
Separation or Purification of *Cis-Trans* Isomers

The separation of *cis* from *trans* isomers has always been a challenging problem for coordination chemists. Adsorption chromatography and thin layer chromatography (TLC) are valuable techniques that have been applied to the solution of this problem. The pioneers in this area were probably Linhard et al. [12], who separated *cis* from *trans* isomers of the $[Co(N_3)_2(NH_3)_4]^+$ ion on an alumina columm. *Cis* and *trans* isomers of cobalt(III) complexes can be successfully separated by using either acidic, basic, or buffer solutions as eluents and TLC on silica gel [13–16]. Overwhelming evidence shows that for octahedral complexes, developed with an acid or water-containing solvent, the *trans* isomer is more mobile than the corresponding *cis* isomer. This is a generalization, considering that the *cis* isomer can be retained more strongly on silica gel or on a resin bed than the *trans* isomer since the former can have two points of attachment to the silanol sites of the silica gel or to the acidic sites of an ion exchange resin, whereas the *trans* groups, being on opposite sites of the complex, can attach at only one point. If the mechanism is one of adsorption rather than of ion exchange, then this trend should also prevail since the *cis* isomer would be more polarized than the *trans* isomer.

This mechanism has been used to separate isomers, but contrary to the above argument, the reverse order of R_F values has also appeared in the literature [17–23]. The differences in R_F values for *cis* and *trans* isomers may originate in the differences in dipole moments, solubilities, ion-pair formation between the complex cation and the anion present in the species, stability, steric hindrance, adsorptivity and symmetry of the complexes. All these physical differences allow the use of the solubility of complexes to separate their isomers.

14.1.4
Identification of *Cis-Trans* Isomers

The identification of isomers is as important as their synthesis and purification. The technique to be used for this purpose depends on the species to be analyzed and if it is in solution or in solid state. Without doubt if there are crystals of the material to be analyzed the best way to do it is by X-ray diffraction, but there are other techniques that can be used, such as infrared and Raman spectroscopy, electronic paramagnetic spectroscopy, nuclear magnetic resonance, and electronic spectroscopy. The choice of which technique to use depends on the intrinsic property of the material to be analyzed, and whether it presents paramagnetic or magnetic properties.

Bearing in mind that the intensity and position of the spectral bands are related to the structure of the compound, the interpretation of the spectra of transition metal complexes, in most cases, can be used in the distinction between *cis* and *trans* isomers. Thus, from the crystal field theory, it is found that the two absorption bands ($^1A_{1g} \rightarrow {}^1T_{1g}$, $^1A_{1g} \rightarrow {}^1T_{2g}$) are expected for octahedral symmetry (Fig. 14.6). Theory shows that if the octahedral symmetry is changed to tetragonal by replacement of two of the equal ligands by different ones, the resulting geometrical isomers can be *cis* or *trans*. In this case the $^1T_{1g}$ state is split into two states with the splitting for the *trans* isomer being approximately twice that of the cis isomer [24]. If the splitting is large, which depends on the position of the ligands in the spectrochemical series, three absorptions bands are possible, mainly for the *trans* isomers where the first two bands are result of the splitting of the $^1T_{1g}$ state. These three bands appear mostly when the original ligand in the complex is substituted by those that are separated from it in the spectrochemical series. Since the split of the $^1T_{1g}$ state is usually smaller for the *cis* isomers only broadening of the band occurs. While the $^1T_{2g}$ state is also split into two states, no resolution is observed even for *trans* isomers.

A classical example of *cis-trans* isomers that can be identified by electronic spectroscopy is the $[CoCl_2(en)_2]^+$ complex. In this case the *trans*-$[CoCl_2(en)_2]^+$ isomer presents bands at 16 425 cm^{-1}, 22 250 cm^{-1}, and 25 000 cm^{-1} while for the *cis*-$[CoCl_2(en)_2]^+$ the bands are at 19 850 cm^{-1} and 27 000 cm^{-1} [25]. A diagram of energy for *cis-trans* isomers of general formula $[MX_4Y_4]$ can be seen in Fig. 14.6. This diagram explains the transitions observed for the *cis-trans* cobalt isomers.

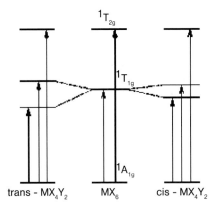

Fig. 14.6 Diagram of energy levels for the bands in $[MX_6]$ complexes and *cis/trans* $[MX_4Y_2]$ isomers. Adapted from ref. [25].
Reproduced by permission of the American Chemical Society.

Group theory combined with infrared spectroscopy is very useful in the determination of the stereochemistry of metal complexes. There is a general rule for active vibrations in the infrared spectra: the more symmetric the molecule, the fewer infrared active bands are to be expected. Thus for the *cis*-$[M(CO)_4(L)_2]$, that has a C_{2v} symmetry, four CO frequencies are observed, while only one CO frequency is active in the infrared spectrum for the *trans* isomer, which belongs to a D_{4h} symmetry [26–27].

The change in the pattern of spectra is also observed using ^{1}H NMR, as can be observed in the Fig. 14.7 for the *cis* and *trans* $[Os(O)_2(bipy)_2]ClO_4$ complexes. This change in the pattern of resonances is consistent with the change from C_2 to D_{2h} symmetry upon isomerization from the *cis* to *trans* isomers [28].

A

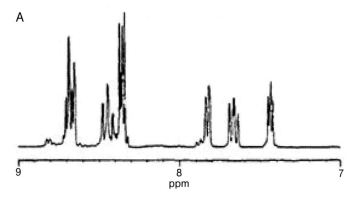

B

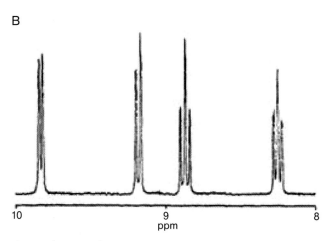

Fig. 14.7 ^{1}H NMR of $[Os(O)_2(bipy)_2](ClO_4)_2$: (A) *cis*-$(CD_3CN)$ and (B) *trans*-(Me_2SO-d_6).

14.2

The *Cis-Trans* Isomerization of Metal Complexes: Mechanisms and Effects

14.2.1

***Cis* or *Trans* Isomer?**

As shown above, several factors can determine the preference for the *cis* or *trans* isomer. Although kinetic and thermodynamic effects drive the reactions, the electronic and steric influences present determine the products to be formed. Thus, surprisingly, the reaction of [IrH(CO)(PPh$_3$)$_3$] with equivalent amount of the 1,3-bis[(diisopropylphosphino)methyl]benzene at 60 °C in THF or C$_6$H$_6$ gives the *trans*-[IrH$_2$(CO){C$_6$H$_3$(CH$_2$P(i-Pr$_2$)$_2$}] isomer [29] (Fig. 14.8).

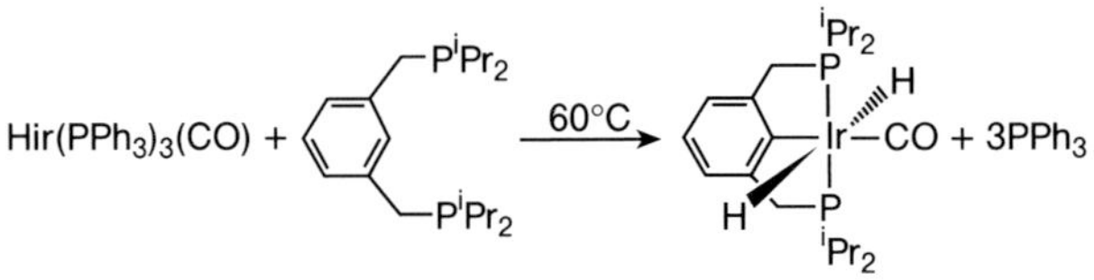

Fig. 14.8 Synthetic reaction of the *trans*-[IrH$_2$(CO){C$_6$H$_3$(CH$_2$P(i-Pr$_2$)$_2$}] isomer [29].

It is known that while *cis*-dihydride complexes are abundant, few stable *trans*-dihydride complexes have been reported [29]. In this case, in spite of the large *trans* influence of the hydride ligand, which should afford the more stable *cis*-dihydride configuration, the dominant steric factors of the CO with the bulk phosphine ligand disfavors it. For the same reason, when the *cis* isomer is heated at 90 °C under 35 psi of H$_2$ it is quantitatively converted to *trans*-[IrH$_2$(CO){C$_6$H$_3$(CH$_2$P(i-Pr$_2$)$_2$}], showing that in this special case the dihydride complex is thermodynamically more stable than the *cis* isomer (Fig. 14.9). The hydride ligands are forced to be *trans* to each other when the phosphines present in the complex are bulky, but when a less bulky phosphine, such as PMe$_3$, is used the electronically more favorable *cis* isomer is the major product in solution. This process was also observed for platinum complexes where steric factors play a major role in the stabilization of four-coordinate platinum *trans*-dihydrides [30–33]. In this case the higher thermodynamic stability of the *trans* isomer in comparison to the *cis* isomer does not follow the well-known very strong *trans* influence of the hydride ligand.

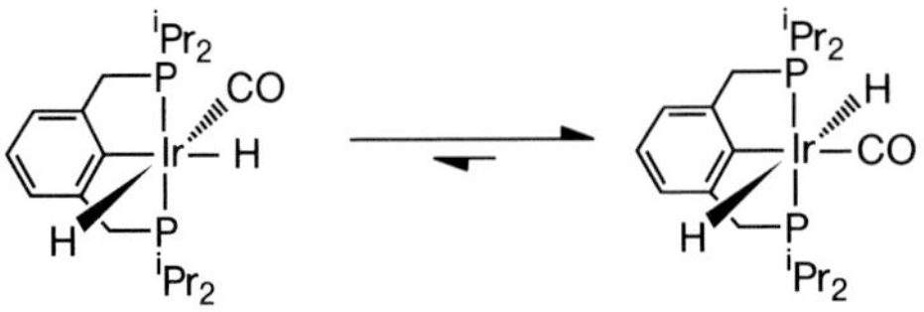

Fig. 14.9 Isomerization process of the [IrH$_2$(CO){C$_6$H$_3$(CH$_2$P(i-Pr$_2$)$_2$}] complex [29].

This isomerization process is thought to occur through the phosphine arm opening, but the Ar–H reductive–elimination–oxidative–addition pathway is not excluded [29].

There is great interest in understanding the preference of different transition metals for *cis* or *trans* isomers or the way in which these complexes undergo geometrical isomerization. Indeed, the course of many reactions of these species, such as nucleophilic substitution, electron transfer, oxidative addition, reductive elimination, thermal decomposition, interaction with molecules of biological interest, and so on, is dictated by the geometry of these compounds.

14.2.2
Isomerization Processes

The relative reactivity of *cis-trans* isomeric pairs is related to their physical and chemical properties. Thus, information concerning isomerization barriers is essential for rational synthetic design and for understanding the reactivity of the isomers.

Isomerization of transition metal complexes is frequently observed to occur for octahedral d^6 or square-planar d^8 systems. Although the interconversions between isomers are commonly achieved thermally, they are also induced photochemically or by oxidation/reduction. In the latter case, the thermodynamically favored isomeric distribution is dependent upon the oxidation state of the metal center of the complex.

Some complexes, such as the $[RuCl_2(dppm)_2]$ (dppm = bis(diphenylphosphine)-metane) can be interconverted by all three types of processes: photochemical, thermal, and redox reactions [34]:

1. Thermal interconversion (*trans/cis*) – *trans* to *cis* occurs quantitatively in 1,2-dichloroethane under reflux (bp 83.3 °C) (Fig. 14.10).

2. Photochemical interconversions (*trans/cis*) – The photolysis of *cis*-$[RuCl_2(dppm)_2]$ in CH_2Cl_2 solution with white light or with the 436 nm line of a high-pressure Xe arc lamp with a monochromator resulted in a quantitative conversion to *trans*-$[RuCl_2(dppm)_2]$. The photoisomerization in this case is induced by direct photolyses of a d-d excited state (Fig. 14.10).

3. Oxidative isomerization (*trans/cis*) – The one-electron oxidation process of *trans*-$[RuCl_2(dppm)_2]$ and subsequent reduction produces the *cis* isomer (Fig. 14.11).

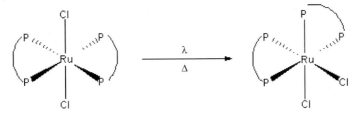

Fig. 14.10 General process of thermal and photochemical interconversions.

The isomerization processes of *trans–cis* in the [RuCl$_2$(dppm)$_2$] complex are summarized in Fig. 14.11.

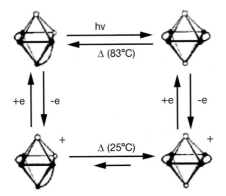

Fig. 14.11 *Cis-trans* isomerization processes in the [RuCl$_2$(dppm)$_2$] complex. From Ref. [34]. Reproduced by permission of the American Chemical Society.

The inorganic chemistry is so sensitive that while the ruthenium complex can be converted to the *cis* form from the *trans* form by all the three types of processes mentioned above, the *trans*-[OsCl$_2$(dppm)$_2$] complex undergoes the same conversion either with more difficulty or not at all. Thus, *trans*-[OsCl$_2$(dppm)$_2$] can also be thermally converted from *trans* to *cis* in the same way as shown for the ruthenium complex, but it requires more drastic conditions than its ruthenium analog, i.e., heating at reflux in 1,2-dichlorobenzene (bp 180.5 °C). Under the same conditions used for the photochemical or electrochemical isomerization of the ruthenium complex, isomerization was not observed for the osmium compound. Almost certainly these isomerization processes involve the dissociation of the chlorine in both complexes and in this case the *trans* effect plays a dramatic role. Indeed, the substitution reaction occurs much faster for the ruthenium complex than for the osmium (Fig. 14.12). Thus, the easy labilization of chloride ion in the *cis* Ru isomer provides an good example of a great kinetic *trans* effect in Ru^{II}-phosphine complexes [35]. Additionally, the contrast in substitutional labilities between *cis*-[RuCl$_2$(dppm)$_2$] and *cis*-[OsCl$_2$(dppm)$_2$] demonstrates the diminution in the *trans* effect going from the harder to softer metal, which is in accordance with the general increase in substitution inertia for heavier transition metal down a general series.

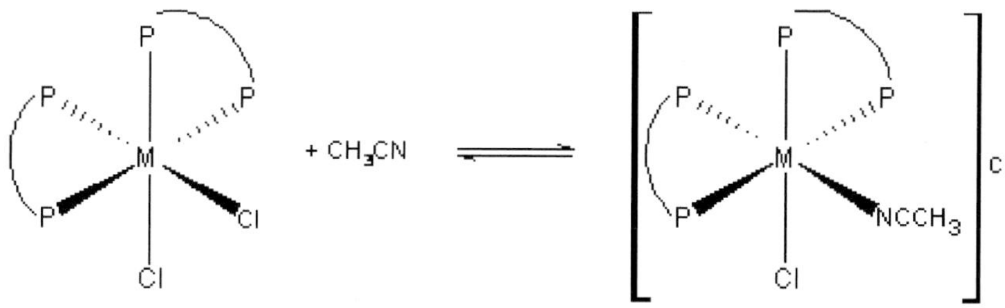

Fig. 14.12 Substitution *trans* effect.

Thermodynamically, [RuCl$_2$(dppm)$_2$] prefers the *cis* orientation. There are at least two reasons for this fact. First, since chlorine is a good σ/π donor and is not a π acceptor, when two of them are mutually *trans*, the axis containing them and the metal center becomes very rich in electrons, and as a consequence the involved Ru–Cl bonds are expected to be weaken, facilitating the isomerization process. On the other hand, the σ/π donation from the chlorine through the ruthenium to a *trans* located phosphorus, a good π acceptor atom, stabilizes the *cis* isomers in comparison to the *trans*. This could lead the reader to think that there is a contradiction in that the more stable isomer allows easier dissociation of the chlorine in substitution reactions. It is interesting to bear in mind that according to recent work of the authors *trans* to *cis* isomerization from the [RuCl$_2$(dppb){4–4′-(X)$_2$-2,2′-bipy}] (X = -H, -NO$_2$, -Me, -COOH, -SMe, -O=Me, -Cl, -OMe) complexes occur through a twist mechanism where there was no real dissociation of the chlorine, but just a relaxation of the Ru-Cl bonds [36]. The suggested mechanism was supported mainly by the thermodynamic data ($\Delta H^{\ddagger}$ and $\Delta S^{\ddagger}$) of the isomerization processes.

The authors have also recently used pulse differential voltammetry to follow the rate of substitution of one chlorine in the *cis*-[RuCl$_2$(bipy)(dppb)] complex by heterocyclic ligands, according to the reaction shown in Fig. 14.13 [37]. This reaction efficiently occurs only with the *cis* isomer. In all reactions, the dissociated chlorine was always that one *trans* to the phosphorus, showing the effective *trans* effect of this atom. The lack of the strong *trans* effect of the nitrogen atom is shown in the substitution reactions in the *cis*-[RuCl$_2$(bipy)$_2$] complex, where the chlorines are *trans* to nitrogen atoms. In this case, a drastic condition of reaction, such as temperature, is necessary for the substitution of a chlorine by a DMSO or pyridine ligand [38–40]. In both case the *cis* configuration of the precursor is retained.

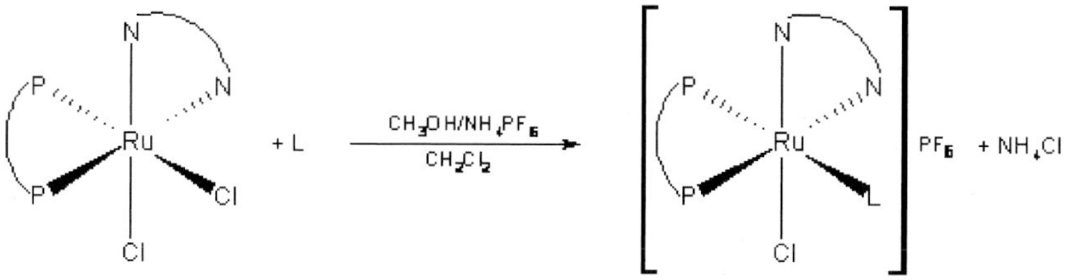

Fig. 14.13 Substitution reaction in the [RuCl$_2$(bipy)(dppb)] by *N*-heterocyclic ligand (L).

14.3

Cis-Trans Isomers of Metal Complexes as Potential Therapeutics

Recent advances in medicinal inorganic chemistry have demonstrated significant prospects for the utilization of metal complexes as drugs, presenting a prosperous field for inorganic chemistry. The compounds to be used as drugs have to present low toxicity and high specificity. Some geometrical isomers of different transition metals can present these characteristics and they have been extensively studied, mainly after the discovery by Rosenberg that the *cis*-diamminodichloroplatinum(II), *cis*-[PtCl$_2$(NH$_3$)$_2$], and *cis*-diamminotetrachloroplatinum(IV), *cis*-[PtCl$_4$(NH$_3$)$_2$], complexes may play role in cell division [41–43]. The reactivities and efficiencies of cis/transplatin complexes in cancer treatment are not the same. While cisplatin is effective in treating a variety of cancers, and is one of the most widely used anticancer drugs, especially for testicular cancer, for which it has a greater than 90% cure rate, the *trans* isomer does not have the same effectiveness. This may originate in part from its kinetic instability and consequent susceptibility to deactivation. Substitution of one or both ammine ligands in transplatin by more bulky ligands, which can slow down ligand substitution reactions of the two chloride ions, has produced compounds with better anticancer efficacy. Efforts have been directed toward a rational design of compounds with specific characteristics that could allow them to be administered orally or to avoid known mechanisms of platinum drug resistance.

cis-Amminedichloro(2-methylpyridine)platinum(II), ZD0473, is hydrolyzed more slowly than cisplatin. This is a consequence of an effect well described in the literature, which depends on the structure of the compound; the pyridine ring is tilted by 102.7° with respect to the PtCl$_2$N$_2$ square plane which leads to the slower rate of substitution reaction [44–49].

The majority of the complexes tested as drugs to date have been structural analogs of cisplatin, having a *cis* geometry, two or at least one amine donor group and two anionic leaving groups, as shown in Fig. 14.14.

There are a few platinum compounds in clinical trials or recently approved for clinical use (Fig. 14.15) but none of them have yet demonstrated significant advantages over cisplatin.

In this new class of platinum complexes some examples do not present the necessary *cis* configuration with respect to the chlorines. Oxaliplatin has been approved for clinical use in France and China for colorectal cancer. The interest in developing platinum complexes that bind to DNA in a fundamentally different manner to cisplatin is an attempt to overcome the resistance pathways that have evolved to eliminate the drug.

The cytotoxity of complexes depends not only on the isomer used but on the entire coordination sphere of the metal center. Thus, the sterically crowded platinum complex *cis*-[PtCl$_2$(bmic)] [bmic = bis-(*N*-methylimidazol-2-yl)carbinol)] was reported to have significant cytotoxicity in L1210 leukemia-bearing mice, while *cis*-[PtCl$_2$(bmi)] (bmi = N^1,$N^{1'}$-dimethyl-2,2'-biimidazole), which has less steric bulk around the metal, was found to be inactive. The greater steric hindrance around the platinum metal in *cis*-[PtCl$_2$(bmic)] caused it to be less reactive than cis-[PtCl$_2$(bmi)] and rendered it less susceptible to deactivation by cellular thiols [48].

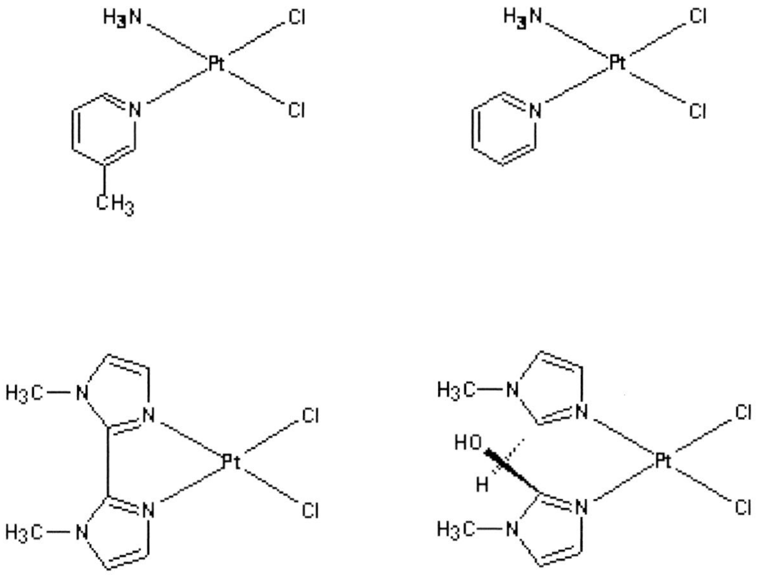

Fig. 14.14 Some geometrical cis-platinum drugs [48,49].

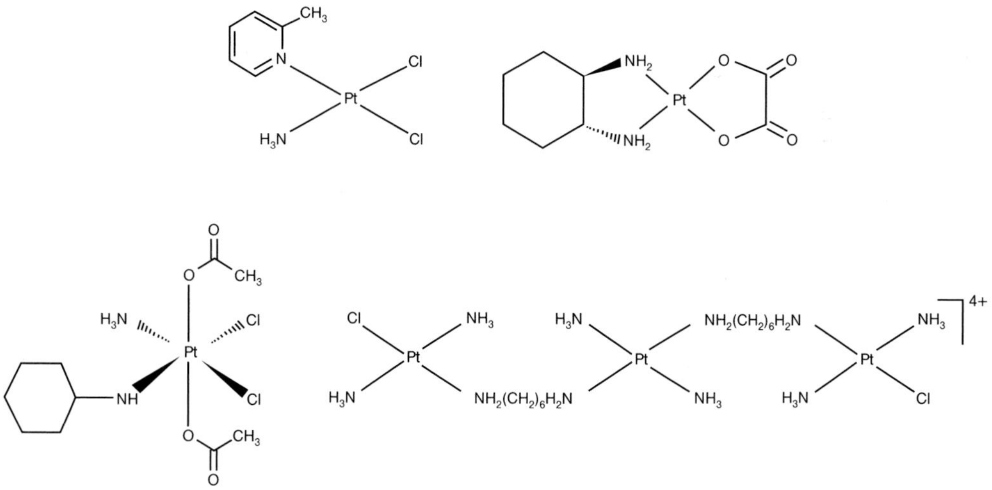

Fig. 14.15 Some platinum compounds in clinical trials [50].

Trans complexes with the general formulas $[PtCl_2(L)_2]$ (L = pyridine, N-methyl-imidazole and thiazole) or $[PtCl_2(L)(L')]$ (L = quinoline and L' = RR'SO where R = methyl and R' = methyl, phenyl benzyl or L = quinoline and L' = NH_3) (Fig. 14.16) showed comparable activity to cisplatin and greater activity than trans-platin in cisplatin-sensitive and -resistant cell lines [51–54].

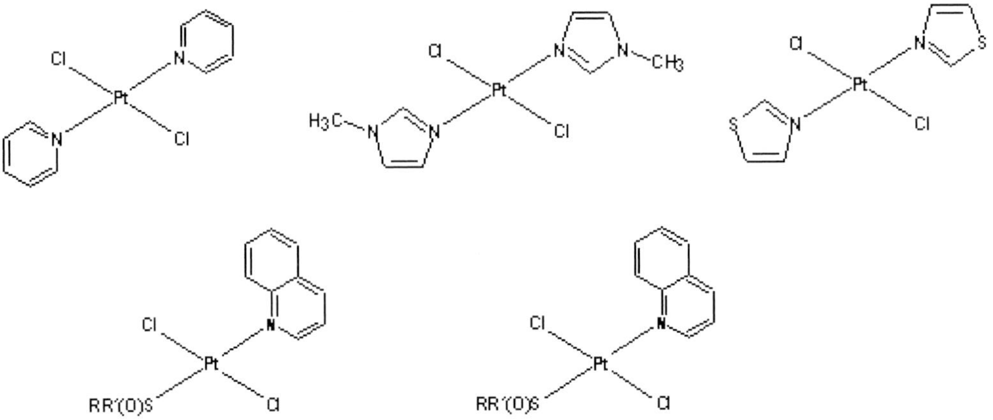

Fig. 14.16 *Trans*-$[PtCl_2(L)(L')]$ complexes (L = L' = pyridine, N-methylimidazole, thiazole, quinoline; L = quinoline and L' = NH_3 or RR'SO where R = methyl and R' = methyl, phenyl benzyl) [51–54].

In cisplatin-sensitive and -resistant L1210 cell lines, the *trans* isomers exhibited greater activities than their *cis* counterparts [51–52].

The cytotoxicity of cisplatin originates from the hydrolysis facility of the chlorine, its binding to DNA, and the formation of covalent cross-links. One disadvantage of cisplatin is its limited solubility in aqueous solutions and its intravenous administration. Platinum complexes with distinctively different DNA binding modes from those of cisplatin may provide higher antitumor activity against cisplatin-resistant cancer cells.

From the above it is clear that geometrical isomerism is far from being of only academic interest, since complexes can bind to the bases of DNA and can be effective as drugs in cancer treatment. Ruthenium complexes with oxidation states of +2 or +3, for example, exhibit antitumor activity, especially against metastatic cancers [55]. The Ru(III) complex Na [*trans*-$RuCl_4$(Im)(Me_2SO)] (NAMI) is currently in a second clinical phase. For Ru(III) compounds, in vivo reduction to Ru(II) may be required for activity. This requirement probably has to do with the fact that the cytotoxicity of ruthenium complexes correlates with their ability to bind DNA, but its analog (ImH)[*trans*-$RuCl_4$(Im)(Me_2SO)] (NAMI-A) does not appear to involve DNA binding. It has been suggested that the presence of less basic N-heterocyclic than imidazole in this complex reduces the loss of DMSO thereby increasing its antitumor action [56]. The *trans* effect is also present in the promising pharmacological properties of nitrosyl ruthenium complexes. Thus, *trans-*

$[Ru(NH_3)_4P(OEt)_3(NO)](PF_6)_3$ exhibits reduced toxicity and similar hypertensive activity compared with sodium nitroprusside, which is used to treat cardiovascular disorders in animal studies. This property is due to the strong *trans* effect of the $P(OEt)_3$ ligand, which is responsible for the release of the NO group [57]. With *trans*-$[RuCl(cyclam)(NO)](PF_6)_2$, prolonged blood pressure reduction was observed (cyclam = 1,4,8,11-tetraazacyclotetradecane) [58].

14.4
Applications of *Cis-Trans* Isomerization of Metal Complexes in Supramolecular Chemistry

Supramolecular chemistry is a subject of current research interest with regard to information storage, molecular electronic devices, as models of biological energy. The ability of a metal ion to organize a ligand around its coordination sphere has led to the design of several intramolecularly organized recognition sites. Exceptionally impressive self-assembled species have been generated from tetraphenyl-porphyrin derivatives, in which two adjacent or opposite phenyl groups are replaced by *para*-substituted pyridines. Thus starting from the *cis/trans*-$[PtCl_2(NCPh)_2]$ or *cis/trans*-$[PdCl_2(NCPh)_2]$ complexes the corresponding porphyrin derivatives can be obtained (Fig. 14.17) [59].

Four-coordinate complexes are particularly well suited to act as square tetramer supramolecular structures formed by the coordination of *cis* and *trans* square-planar Pt^{II} and Pd^{II} complexes by the 4′-pyridyl groups of the mixed phenyl/pyridyl porphyrins (Fig. 14.18).

Polymetallated porphyrins obtained by attaching $[Ru(bipy)_2Cl]^+$ complexes to the peripheral pyridyl residues are able to act as efficient catalysts in multi-electron transfer reactions [60–66] Thus, porphyrins bound to $[Ru(bipy)_2Cl]^+$ in *cis* or *trans* positions (Fig. 14.19) can also be used for the electrocatalysis of sulfite.

Ruthenium square species where the bridging ligands are in *cis* positions were also obtained by the authors, starting from the $[RuCl_2(dppb)PPh_3]$ complex with alterdentate ligands such as pyrazine, 4,4′-bipyridine and 1,2-bis-*trans*-(4-pyridyl)-ethylene (Fig. 14.20) [68].

The size of the cavity of these molecularly organized complexes can be controlled by the length of the 4,4′-bipyridine ligands used in the syntheses.

The controlled mechanism of substitution of the chlorine in the *cis*-$[RuCl_2-(bipy)(dppb)]$ allows the synthesis of polymethalated porphyrins such as $\{H_2TPyP-[RuCl(N-N)(dppb)]_4\}^{4+}$ and $\{M-TPyP[RuCl(N-N)(dppb)]_4\}^{4+}$ [M = Co^{2+}, Zn^{2+} and Mn^{2+}; N-N = 2,2′-bipyridine] with the structure shown in Fig. 14.21.

These species have been used to modify electrodes to be used as sensors for the detection of molecules of pharmacological interest, such as dopamine.

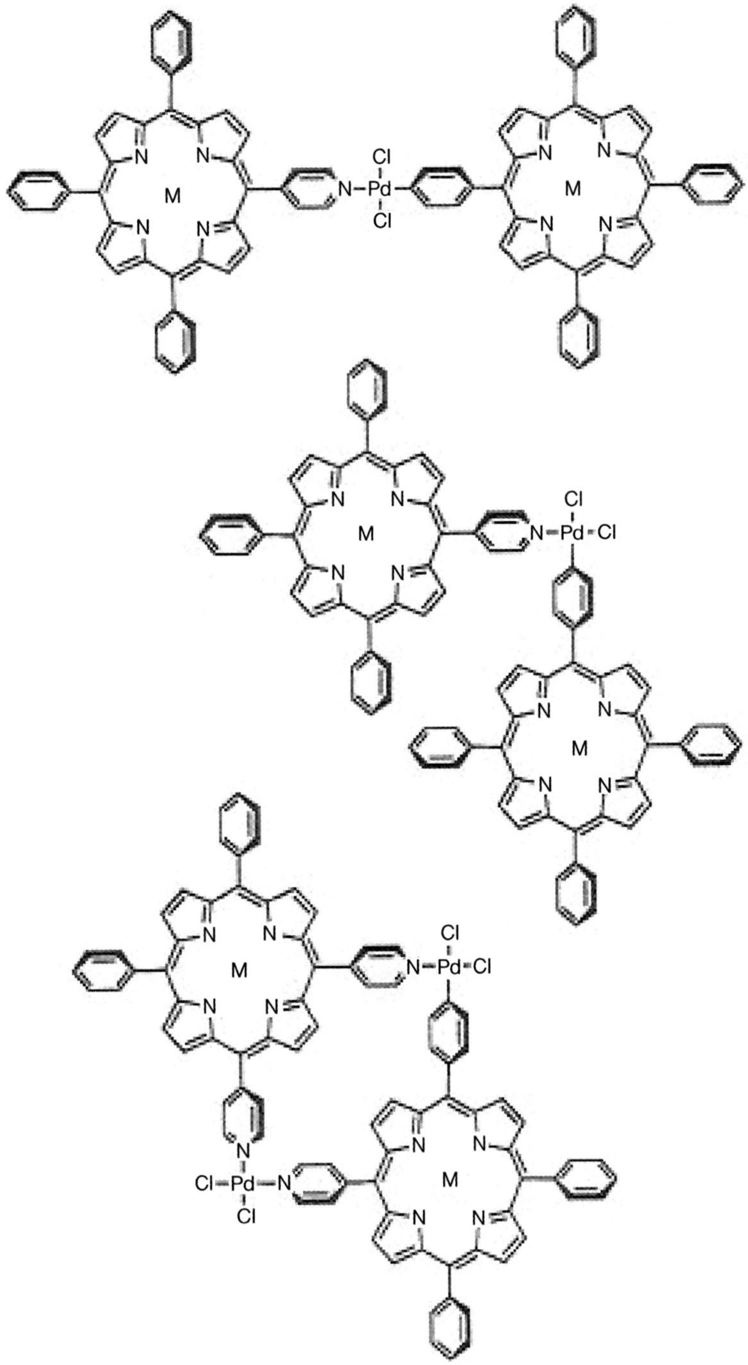

Fig. 14.17 *Cis/trans*-[PdCl$_2$(phenyl/pyridyl)] porphyrin dimers.
From Ref. [59]. Reproduced by permission of the Royal Society of Chemistry.

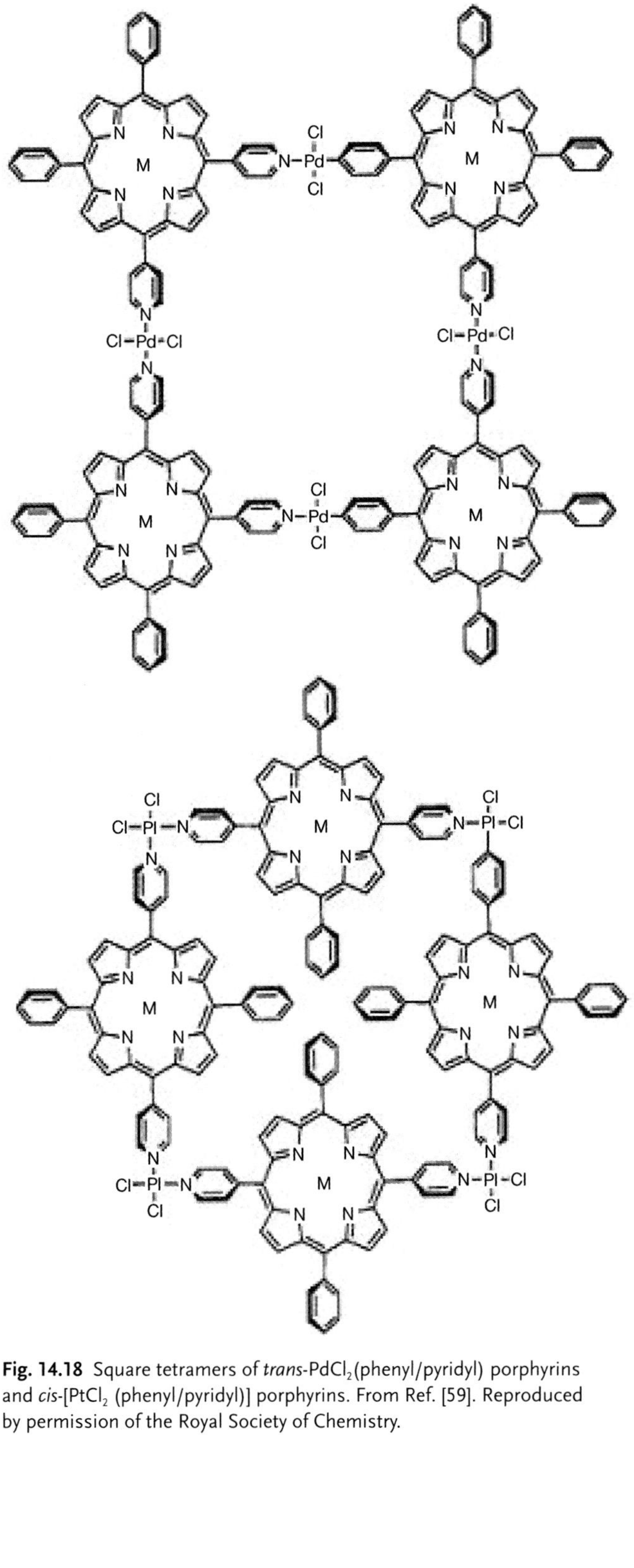

Fig. 14.18 Square tetramers of *trans*-PdCl$_2$(phenyl/pyridyl) porphyrins and *cis*-[PtCl$_2$ (phenyl/pyridyl)] porphyrins. From Ref. [59]. Reproduced by permission of the Royal Society of Chemistry.

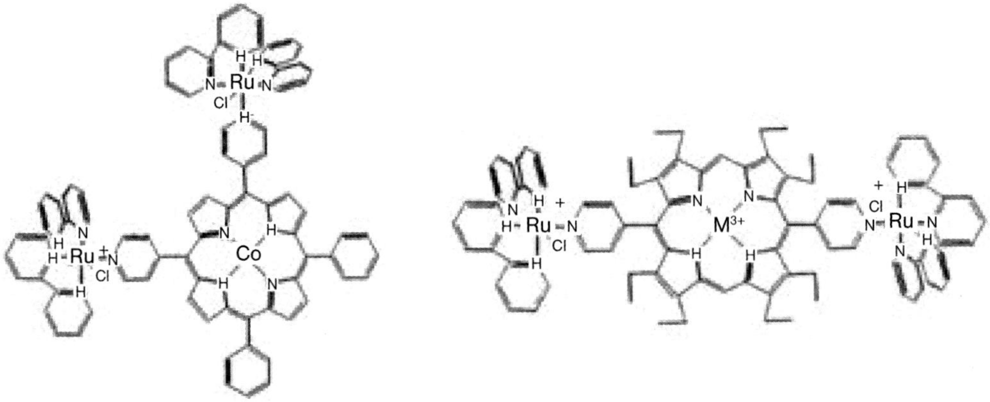

Fig. 14.19 Structures of *cis*-{Co[5,10-di(4-pyridil)-15,20-(phenyl)porphyrin}
[RuCl(bipy)$_2$]$_2$ and *trans*-{M^{2+}[5,15-di(4-pyridil)-octaethylporphyrin]}[RuCl(bipy)$_2$].
From Ref. [67]. Reproduced by permission of Elsevier Science.

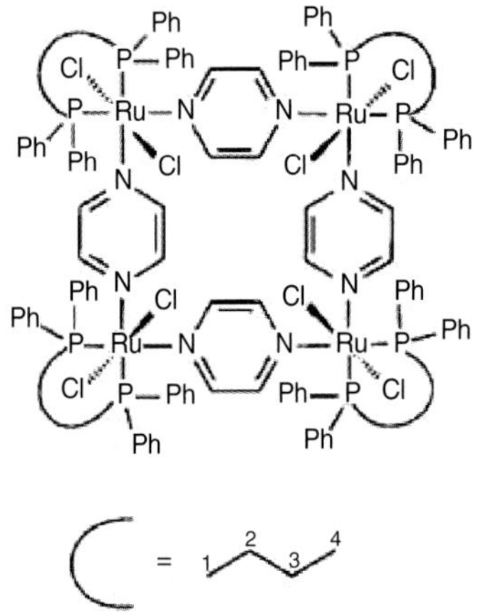

Fig. 14.20 Structure for the [RuCl$_2$(μ-pz)(dppb)]$_4$ complex [68].

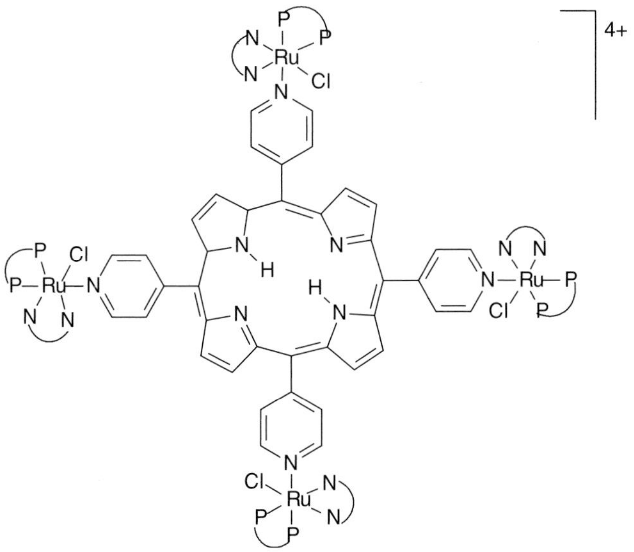

Fig. 14.21 Structure for the {TPyP[RuCl(N-N)(dppb)]₄}⁴⁺ complex [69].

14.5
Final Remarks

In this chapter the discussion on isomerization in transition metal complexes had to be limited to a choice of certain subjects. There is no difficulty in finding other interesting faces of isomerism but here we tried to give some information on the synthesis, purification, stability, and applications for isomers of transition metal chemistry. The goal in this chapter was to show the remarkable and widespread potential for *cis/trans* isomers, including supramolecular chemistry.

List of Abbreviations

Ar	aromatic
bipy	2,2'-bipyridine
bmi	$N^1,N^{1'}$-dimethyl-2,2'-biimidazole
bmic	bis-(N-methylimidazol-2-yl)carbinol)
bp	boiling point
CF	crystal field theory
CFSE	crystal field stabilization energy
cisplatin	*cis*-dichlorodiamminoplatinum(II)
CTI	*cis-trans* isomerization.
cyclam	1,4,8,11-tetraazacyclotetradecane

DMSO	dimethyl sulfoxide
dppb	1,4-bis(diphenylphosphine)butane
dppe	1,2-bis(diphenylphosphine)etane
dppm	bis(diphenylphosphine)metane
en	ethylenediamine
Im	imidazole
LVT	ligand valence theory
MO	molecular orbital theory
NAMI	Na [*trans*-RuCl$_4$(Im)(Me$_2$SO)]
NAMI-A	(ImH)[*trans*-RuCl$_4$(Im)(Me$_2$SO)]
pz	pyridine
THF	tetrahydrofurane

References

1 A.Werner, *Ber. Dtsch Chem. Ges.* **1911**, *44*, 1887–1898.

2 A. von Zelewsky, *Stereochemistry of Coordination Compounds.* John Wiley & Sons, New York.

3 S. H. C. Briggs, *J. Chem. Soc.* **1929**, 685–690.

4 W. Breuel, K. Gleu, *Z. Anorg. Allgem. Chem.* **1938**, *237*, 326–334.

5 W. Breuel, K. Gleu, *Z. Anorg. Allgem. Chem.* **1938**, *237*, 335–349.

6 S. Anderson, F. Basolo, *J. Am. Chem. Soc.* **1960**, *82*, 4423–4424.

7 S.L. Queiroz, A. A. Batista, K. S. MarcFarlane, G. Oliva, M. T. do P. Gambardella, A. H. R. Santos, S. J. Rettig, B. R. James, *Inorg. Chim. Acta* **1998**, *267*, 209–221.

8 A. A. Batista, S. L. Queiroz, M. P. de Araújo, K. S. MacFarlane, B. R. James, *J. Chem. Ed.* **2001**, *78*, 89–90.

9 A. A. Batista, S. L. Queiroz, M.P. de Araújo, K. S. MacFarlane, B. R. James, *J. Chem. Educ.* **2001**, *78*, 87–88.

10 F. Basolo, R. G. Pearson, *Prog. Inorg. Chem.* **1962**, *4*, 381–453.

11 A. A. Batista, E. E. Castellano, G. V. Poelhsitz, J. Ellena, R. C. L. Zampieri, *J. Inorg. Biochem.* **2002**, *1*, 82–88.

12 M. Linhard, M. Weigel, H. Flygare, *Z. Anorg. Allg. Chem.* **1950**, *263*, 233–244.

13 H. Seiler, C. Biebricher, H. Erlenmeyer, *Helv. Chim. Acta* **1963**, *46*, 2636–2638.

14 Y. Tsunoda, T. Takeuchi, Y. Yoshino, *Sci. Papers Coll. Gen. Educ.* Univ. Tokyo, **1964**, *14*, 55–62.

15 L. F. Druding, R. B. Hagel, *Anal. Chem.* **1966**, *38*, 478–480.

16 L. F. Druding, R. B. Hagel, *Separ. Sci.* **1969**, *4*, 89–94.

17 I. Y. Yamamoto, A. Nakahara, R. Tsuchida, *J. Chem. Soc. Jpn* **1954**, *75*, 232–234.

18 G. Stefanovic, T. Janjic, *Anal. Chim. Acta* **1954**, *11*, 550–553.

19 G. Stefanovic, T. Janjic, *Anal. Chim. Acta* **1958**, *19*, 488–492.

20 F. Jursík, *J. Chromatogr.* **1967**, *26*, 339–341.

21 D. T. Haworth, M. J. Zetmeisl, *Separ. Sci.* **1968**, *3*, 145–150.

22 J. Bielza, J. Casado, R. Ribas, *An. Quim.* **1969**, *55*, 217–220.

23 R. K. Ray, G. B. Kauffman, *Inorg. Chim. Acta* **1989**, *162*, 45–48.

24 B. N. Figgs, *Introduction to Ligands Fields.* Interscience, USA, **1966**.

25 D. T. Haworth, K. M. Elsen, *J. Chem. Educ.* **1973**, *50*, 301–302.

26 M. Y. Darensbourg, D. J. Darensbourg, *J. Chem. Educ.* **1970**, *47*, 33–34.

27 R. K. Pomeroy, K. S. Wijesekara, *Inorg. Chem.* **1980**, *19*, 3729–3735.

28 J. C. Dobson, K. J. Takeuchi, D. W. Pipes, D. A. Geselowitz, T. J. Meyer, *Inorg. Chem.* **1986**, *25*, 2357–2365.

29 B. Rybtchinski, Y. Ben-David, D. Milstein, *Organometallics* **1997**, *16*, 3786–3793.

30 A. Immirzi, A. Musco, G. Carturan, U. Belluco, *Inorg. Chim. Acta* **1975**, *12*, L23.

31 B. L. Shaw, M. F. J. Uttley, *Chem. Soc., Chem. Commun.* **1974**, 918–919.

32 T. Yoshida, S. Otsuka, *J. Am. Chem. Soc.* **1977**, *99*, 2134–2140.

33 R. S. Paonessa, W. C. Trogler, *J. Am. Chem. Soc.* **1982**, *104*, 1138–1140.

34 B. P. Sullivan, T. J. Meyer, *Inorg. Chem.* **1982**, *21*, 1037–1040.

35 B. P. Sullivan, J. M. Calvert, T. J. Meyer, *Inorg. Chem.* **1980**, *19*, 1404–1407.

36 M. O. Santiago, C. L. Donicci-Filho, I. de S. Moreira, R. M. Carlos, S. L. Queiroz, A. A. Batista, *Polyhedron* **2003**, *22*, 3205–3211.

37 E. M. A. Valle, Ms. thesis, Federal University of São Carlos, São Carlos (SP), Brazil, **2004**.

38 F. Maspero, G. Ortaggi, *Ann. Chim. (Rome)* **1974**, *64*, 115–117.

39 J. D. Birchall, T. D. O'Donoghue, J. R. Wood, *Inorg. Chim. Acta* **1979**, *37*, L461–L463.

40 O. V. Sizova, Yu. Ershov, N. V. Ivanova, A. D. Shashko, A. V. Kuteikina-Teplyakova, *Russian J. Coord. Chem.* **2003**, *29*, 494–500.

41 B.Rosenberg, L. Van Camp, T. Krigas, *Nature* **1965**, *205*, 698–699.

42 B. Rosenberg, L. Van Camp, E. B. Grimley, A. J. Thomson, *J. Biol. Chem.* **1967**, *242*, 1347–1352.

43 B. Rosenberg, *Plat. Met. Rev.* **1971**, *15*, 42–51.

44 F. Basolo, F. J. Chatt, H. B. Gray, R. G. Pearson, B. L. Shaw, *J. Chem. Soc.* **1961**, 2207–2215.

45 G. Wilkinson, R. D. Gillard, J. A. McCleverty, *Comprehensive Coordination Chemistry*. Pergamon Press, New York, **1987**.

46 J. F. Holford, F. I. Raynaud, B. A. Murrer, K. Grimaldi, J. A. Hartley, M. J. Abrams, L. R. Kelland, *Anti-Cancer Drug Des.* **1998**, *13*, 1–18.

47 E. Wong, C. M. Giandomenico, *Chem. Rev.* **1999**, *99*, 2451–2466.

48 M. J. Bloemink, H. Engelking, S. Karentzopoulos, B. Krebs, J. Reedijk, *Inorg. Chem.* **1996**, *35*, 619–627.

49 Y. Chen, Z. Guo, S. Parsons, P. J. Sadler, *Chem. Eur. J.* **1998**, *4*, 672–676.

50 C. X. Zhang, S. J. Lippard, *Curr. Opin. Chem. Biol.* **2003**, *7*, 1–9.

51 M. Van Beusichem, N. Farrell, *Inorg. Chem.* **1992**, *31*, 634–639.

52 N. Farrell, *Met. Ions Biol. Syst.* **1996**, *32*, 603–639.

53 N. Farrell, T. T. B. Ha, J. P. Souchard, F. L. Wimmer, S. Cros, N. P. Johnson, *J. Med. Chem.* **1989**, *32*, 2240–2241.

54 N. Farrell, In: Platinum and other Metal Coordination Compounds in Cancer Chemotherapy. S. B. Howell (ed), Plenum Press, New York, **1991**.

55 M. J. Clark, *Coord. Chem. Rev.* **2003**, *236*, 209–233.

56 A. Bergamo, B. Gava, E. Alessio, G. Mestroni, B. Serli, M. Cocchietto, S. Zorzet, G. Sava, *Int. J. Oncol.* **2002**, *21*, 1331–1338.

57 A. S. Torsoni, B. F. de Barros, J. C. J. Toledo, M. Haun, M. H. Krieger, E. Tfouni, D. W. Franco, *Nitric Oxide* **2002**, *6*, 247–254.

58 F. G. Marcondes, A. A. Ferro, A. S.Torsoni, M. Sumitani, M. J. Clarke, D. W. Franco, E. Tfouni, M. H. Krieger, *Life Sci.* **2002**, *70*, 2735–2752.

59 M. D. Charles, J.-M. Lehn, *J. Chem. Soc. Chem. Commun.* **1994**, 2313–2315.

60 K. Araki, H. Toma, *J. Coord. Chem.* **1993**, *30*, 9–17.

61 K. Araki, H. Toma, *J. Photochem. Photobiol.* **1994**, *83*, 245–250.

62 K Araki, H. Toma, *J. Chem. Res.* **1994**, *7*, 1501–1515.

63 K. Araki, L. Angnes, C. Azevedo, H. Toma, *J. Electroanal. Chem.* **1995**, *397*, 205–210.

64 K. Araki, L. Angnes, H. Toma, *Adv. Mater.* **1995**, *7*, 554–559.

65 C. Azevedo, K. Araki, L. Angnes, H. Toma, *Electroanalysis* **1998**, *7*, 467–471.

66 L. Angnes, C. Azevedo, K. Araki, H. Toma, *Anal. Chim. Acta* **1996**, *329*, 91–96.

67 N. Rea, B. Loock, D. Lexa, *Inorg. Chim. Acta* **2001**, *312*, 53–66.

68 S. L. Queiroz, E. Kikuti, A. G. Ferreira, M. O. Santiago, A. A. Batista, E. E. Castellano, J. Ellena, *Supramol. Chem.* **2004**, *16*, 255–262.

69 L. R. Dinelli, PhD thesis, Federal University of São Carlos, São Carlos (SP), Brazil, **2003**.

70 N. M. Barbosa Neto, L. Boni, C. R. Mendonça, L. Misogutti, S. L. Queiroz, L. R. Dinelli, A. A. Batista, S. C. Zílio, *J. Phys. Chem. B.* **2005**, *36*, 17340–17345.

Index

a

ab initio 86, 114, 143, 262
absorbance difference 156
absorption 53, 68, 81
absorption spectroscopy 47–51, 60, 62
acceptor 333
acetamide 145, 146, 152, 160, 162
acetate 144, 150, 169
Achantamoeba 209
AchR 310
acid-base catalysis 174, 176, 276
π-acid 325
acidity 144
acryloyl 304
activation energy 182
active site 201, 210, 216
acylnitroso 300
addition 331
adjacent 3
Aequorea 80
Aeromonas 209
aggregation 262, 267
agonist 246
AIMD 114
albonoursin 306, 307
alkenes 9
alkylproline 273
allergy 201, 203, 207
Alternaria 306
Alzheimer's disease 266, 267
amide 12, 143, 144–166, 167–169, 295
α-amylase 170
aminimide 298
o-aminobenzoate 157
α-aminobutyric acid 106
aminomethanoproline 235
aminoproline (Amp) 232, 233
amyloid 106, 180, 253, 311
amyotrophic lateral scerosis 203

androgen 269
angiotensin 160, 176, 188, 235
anion 84, 85
antamanide 279
anthracene 35
antibiotic 100
antitumor 315
AP-1 263
APIase 196, 214, 272, 288
apoptosis 204, 208, 264, 269, 270, 279
Arabidopsis 277
arachidonic acid 96, 102–108
archeal rhoddopsin 66–71
arylazo 309–315
ascomycin 261, 263, 281
asCP 88, 89
Aspergillus 306
assymetric movement 171
ATP 99, 210, 213
autoantibodies 201
autocatalytic 80
aza-amino acid 247
azabicycloalkane 239
azapeptide 148
azaproline 148
aza-proline 247
azeridine 273
azetidine 273
azetidine carboxylate (Aze) 232, 233, 235
azobenzene 9, 310–316
azosulfide 10

b

Bacillus 306
bacteria 47, 77, 96, 99, 100, 197, 206
bacterial reaction center 35, 39–47
bacteriochlorophylls 37
bacteriophage 184, 185
bacteriorhodopsin 301

cis-trans Isomerization in Biochemistry. Edited by Christophe Dugave
Copyright © 2006 WILEY-VCH Verlag GmbH & Co. KGaA, Weinheim
ISBN: 3-527-31304-4